43 Springer Series in Solid-State Sciences

Edited by Hans-Joachim Queisser

Springer Series in Solid-State Sciences

Editors: M. Cardona P. Fulde K. von Klitzing H.-J. Queisser

Managing Editor: H. K. V. Lotsch   Volumes 1–89 are listed at the end of the book

Johann-Martin Spaeth   Jürgen R. Niklas
Ralph H. Bartram

Structural Analysis of Point Defects in Solids

An Introduction to Multiple Magnetic Resonance Spectroscopy

With 165 Figures

Springer-Verlag
Berlin Heidelberg New York
London Paris Tokyo
Hong Kong Barcelona
Budapest

Professor Dr. Johann-Martin Spaeth
Priv.-Doz. Dr. Jürgen R. Niklas

Fachbereich Physik, Universität-Gesamthochschule
Warburger Strasse 100, W-4790 Paderborn, Fed. Rep. of Germany

Professor Ralph H. Bartram, Ph. D.

Department of Physics, University of Connecticut
Storrs, CT 06268, USA

Series Editors:

Professor Dr., Dres. h. c. Manuel Cardona
Professor Dr., Dr. h. c. Peter Fulde
Professor Dr., Dr. h. c. Klaus von Klitzing
Professor Dr., Dres. h. c. Hans-Joachim Queisser

Max-Planck-Institut für Festkörperforschung, Heisenbergstrasse 1
W-7000 Stuttgart 80, Fed. Rep. of Germany

Managing Editor:

Dr. Helmut K. V. Lotsch

Springer-Verlag, Tiergartenstrasse 17, W-6900 Heidelberg, Fed. Rep. of Germany

ISBN-13 : 978-3-642-84407-2   e-ISBN-13 : 978-3-642-84405-8
DOI : 10.1007 / 978-3-642-84405-8

Typesetting: Camera ready by authors
54/3140 – 5 4 3 2 1 0 – Printed on acid-free paper

Preface

The investigation of point defects in model systems, such as F centers in alkali halides, has a venerable history in condensed matter physics. However, in recent years, the study of defects has acquired a much greater practical importance for materials science, since defects even in low concentration have a controlling influence on the bulk properties of solids such as semiconductors and laser materials. It follows that the demand for reliable methods of defect structural analysis and of defect engineering has increased enormously.

The advent of multiple magnetic resonance techniques revolutionized the structural analysis of point defects in solids. The extension and elaboration of these techniques in recent years have greatly enhanced their discrimination, reliability and applicability. With the help of modern experimental methods such as computer controlled experiments and computer aided data acquisition and analysis, these techniques were shown to be applicable to practical problems in materials science, as well as to the study of model defects. A correlation between magnetic resonance spectra and bulk properties of materials is often achievable with these multiple magnetic resonance techniques.

Although many multiple resonance techniques have been developed over the course of a generation, the principles and procedures involved in their application to structural analysis of point defects have never been collected in a single volume. The objective of the present book is to make these principles and procedures, including recent developments, which are, of course, familiar to the dedicated practitioners of this seemingly arcane discipline, accessible to a much more comprehensive class of materials scientists. The intention is to provide the reader with a working knowledge which will enable him either to apply multiple magnetic resonance spectroscopy to the investigation of defects in a wide variety of host materials, or at least to appreciate the power of the method and the implications of its findings.

We should like to express our appreciation to Dr. F. Lohse for providing information on ODMR spectroscopy, to Mr. P. Alteheld and Mr. M. Rac for reading and correcting the manuscript, and to Mrs. E. Henrichs and Mrs. W. Kriete for their skillful and patient typing of the manuscript. One of the authors (J.–M. S.) gratefully acknowledges a one–semester stipend from the Volkswagen–Stiftung without which the completion of this book would not have been possible.

Paderborn   *J.–M. Spaeth   J.R. Niklas*
Storrs, March 1992   *R.H. Bartram*

Table of Contents

1. Introduction

In the first section of this chapter, the term "structure of point defects in solids" will be defined. The various aspects of the term structure of point defects as well as the range of defect systems which can be investigated by magnetic resonance spectroscopy are discussed. The kinds of questions that can be addressed and answered with the methods currently available are outlined. Subsequent sections contain general descriptions of the basic ideas concerning how electron paramagnetic resonance (EPR) and the various methods of multiple magnetic resonances, such as electron nuclear double resonance (ENDOR) and optically detected EPR and ENDOR, can be used to investigate and determine defect structures. The reader is given an overview of what this book is intended to deal with and what it is intended to emphasize. As mentioned in the preface, the major purpose of the book is to provide the reader with a working knowledge enabling him to apply multiple magnetic resonance spectroscopy to the investigation of defects in a large variety of hosts. It is, therefore, not our intention to carefully justify all concepts from first principles. For this the reader is referred to text books on electron paramagnetic resonance and nuclear magnetic resonance. It is not supposed that the reader is completely familiar with EPR and experienced with its use. On the other hand, it is not possible within the framework of this book to outline EPR in as much detail as is done in text books on EPR. Only those basic concepts which are necessary to understand the ENDOR method are briefly discussed. Therefore, it may be helpful in cases where the EPR spectra are difficult to understand to additionally consult EPR text books when applying ENDOR methods (Ref. [1.1–5]). A very brief outline is given of the theoretical interpretation of the experimental results. Ideally, the aim is a quantitative understanding of the electronic defect structure. However, there are important cases where even a less accurate theoretical account of the measured data is helpful in achieving an unambiguous analysis of the experimental spectra. In that sense, theory is instrumental in structure determination and is, therefore, properly included in this book. Two chapters are included which contain practical information on the design of ENDOR and ODMR (optically detected magnetic resonance) spectrometers. Finally, several appendices are included which may be helpful in the application of multiple magnetic resonance techniques.

1.1 Structure of Point Defects

Defects very often determine the bulk properties of solids such as the optical absorption or emission, the mechanical, or the electrical properties. Even defects with a concentration as low as 10^{16} cm^{-3} often predominantly determine the bulk properties of solids. For example, the F center in potassium chloride, which consists of an electron trapped in a chloride ion vacancy, renders the normally colorless potassium chloride crystal dark blue, even at a concentration of the order of 10^{16} cm^{-3} [1.6]. Another example is provided by the well–known donors and acceptors in semiconductors such as phosphorus and boron, respectively, doped in silicon, which determine the electrical conductivity, and are the basis of semiconductor technology [1.7]. Apart from analyzing materials, there is also a growing interest in engineering particular properties of solids by producing specific defects which give rise to the desired properties. The attempts to produce tunable solid state lasers are an example of this. In many areas of solid state physics there is, therefore, a great interest in methods which allow the determination of the structures of defects, and permit correlation of these structures with bulk properties of the solid.

An obvious meaning of the word structure is the atomic configuration of a point defect. Generally, a point defect is a defect with "zero dimension" on a macroscopic scale, in contrast to defects with "one dimension", such as dislocations, or "two dimensions", such as grain boundaries. An impurity atom occupying a regular lattice site of a single crystal, is for example, such a point defect. Our discussion is concerned with defects in single crystals, since the methods to be described are less suitable for polycrystalline or amorphous material. A typical question would be on which of the lattice sites the impurity is located. This is usually not clear from the start in a crystal with different types of atoms in the unit cell. As an additional possibility, the impurity could reside on one of the various interstitial sites. In many cases the chemical identity of the impurity is not clear either, particularly in solids with "unintentional" impurities which may be incorporated during crystal growth in an uncontrolled manner. Another class of point defects includes the intrinsic point defects such as vacancies, interstitials or antisite defects. The latter are crystal atoms of type A residing on crystalline sites B in a binary crystal AB. Antisite defects exist in as–grown crystals, while vacancies and interstitials often are produced by ionizing radiation.

A particular problem is usually the low concentration of defects, which requires very sensitive methods for their investigation.

In this book other defects, which are small aggregates of the point defects just discussed, are also dealt with under the concept of a "point defect".

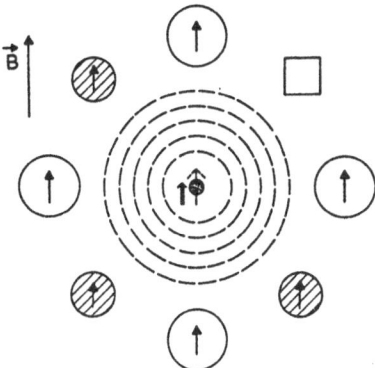

Fig. 1.1. Schematic representation of a point defect in a binary solid. The unpaired electron spin, the electron spin density distribution and the nuclear spins of the lattice nuclei are indicated. The spins which are aligned in an external static magnetic field B are indicated by *arrows*. The *open square* represents a vacancy. The *hatched* and *open circles* represent two kinds of atoms or ions

In Fig. 1.1, a small aggregate consisting of an impurity next to a vacancy is shown schematically. A few other recent examples are: F centers in alkali halides associated with molecules such as OH^- or CN^- [1.8], or antisite defects in III–V semiconductors associated with interstitials or impurities [1.9–10]. Small oxygen clusters in silicon [1.11] have recently been described, which influence the electrical properties of silicon grown from quartz crucibles after thermal treatments (thermal donors). There is growing interest in the study of small aggregate defects, which greatly influence important properties of the host crystals, particularly in the semiconductor field.

The determination of the atomic configuration of aggregates is not an easy task, and represents the state–of–the–art of the methods described here. Sometimes aggregates are too small to be seen by high resolution electron microscopy, and too large and complicated to be easily dealt with using magnetic resonance methods.

At a closer look, one is also interested in the determination of lattice relaxations around point defects, preferably not only of the nearest neighbors. Finally, knowledge of the electronic structure, i.e., a precise quantum mechanical description of the electronic states of the defect, will be the ultimate aim of a structure determination.

Many defects possess dynamical properties which are of interest as a function of temperature. For example, an impurity can be in an off–center position at low temperature; i.e., it does not occupy a regular substitutional lattice site, but it tunnels between various off–center sites and eventually transforms into a thermally activated local vibrational mode upon raising the temperature.

A number of dynamical effects can also be investigated with paramagnetic resonance methods.

There are many methods to *"characterize"* a solid with respect to its defects, a term often currently used in semiconductor research and technology. However, only very few methods can say something about the defect structures, let alone determine them accurately on the atomic scale. Local vibrational mode (LVM) spectroscopy can often determine which impurity is involved in a defect, locate the impurity site and state what the defect symmetry is, but no correlation to other properties is easily possible. The *Mössbauer* effect, perturbed angular correlation (PAC) and other methods of so–called nuclear solid state physics usually give information about the symmetry of a defect and about nuclear quadrupole interactions. However, they cannot say more, and they are restricted to a few suitable *Mössbauer* active or γ–emitting nuclei, which usually have to be incorporated deliberately into the solid for such an investigation. This restriction also limits their use. These methods have the advantage of a very high sensitivity and investigations can be performed at any temperature. There is no restriction as to the type of solid.

Electron paramagnetic resonance methods are the most powerful techniques for the determination of defect structures. Their restriction is, of course, that the defects must be paramagnetic. Fortunately, this is very often the case. In particular, the transition metal ions and rare earth ions are paramagnetic. In radiation damage, where electron and hole pairs are created and separated upon defect formation, one obtains paramagnetic defects. In semiconductors the position of the *Fermi* level can often be suitably shifted by co–doping or by applying ionizing radiation [1.7–12] to change the charge states of the defects. It is then often possible to make the defect of interest paramagnetic. Another constraint is that a static magnetic field and a microwave or a radio–frequency field can penetrate into the crystal in order to induce magnetic resonance transitions.

Therefore, only crystals which are non–metallic, and neither superconducting nor highly conducting, can be investigated with these methods. In spite of their power to analyze defect structures, the methods are limited by their sensitivity and the usual need to work at low temperature.

1.2 Basic Concepts of Defect Structure Determination by Electron Paramagnetic Resonance

One of the most important clues about the defect structure can be obtained from the EPR spectrum of a defect, namely, the symmetry of the defect. For example, if the EPR lines do not vary upon rotating the magnetic field within any of the crystal planes, then the defect is isotropic. An example of an isotropic defect is a hydrogen atom on a substitutional site in an alkali halide lattice. For lower symmetry, the EPR lines may vary upon rotation of the magnetic field with respect to a crystal axis, and they may split into several lines. The details of such splitting patterns depend on the spin of the defect and on the symmetry properties of the defect and the host system (Chap. 3). A low symmetry defect will usually be present in several orientations in the crystal. For example, a defect consisting of a pair of impurities in a cubic crystal with C_{4v} symmetry, i.e., having a fourfold symmetry axis, can be aligned parallel to any of the six cubic $<100>$ crystal axes. In this case, for an arbitrary orientation of the magnetic field with respect to the crystal, one measures the superposition of six EPR spectra according to the six center orientations. The angular variation of the EPR lines, when rotating the magnetic field within a particular crystal plane, can be used to determine the symmetry of the defect (Chap. 3). The symmetry of the defect reflects the symmetry of its wavefunction. It yields information concerning whether there is a mirror plane, a rotation axis or a combination of several of these symmetry elements. This knowledge greatly facilitates the analysis of multiple magnetic resonance spectra. However, the knowledge of the defect symmetry alone does not suffice to determine the chemical identity of the defect. The analysis of the EPR spectra yields information on the nature of the defect, but this information is usually not sufficient to identify an impurity or an intrinsic defect [1.1–5] (also Chap. 3).

The identification of the "chemical" identity of a defect is based on the measurement of the magnetic interaction between the magnetic moment of the unpaired electron(s) or hole(s) of the defect, and the magnetic moments of nuclei belonging to the "impurity" atom of the defect (if any), and of the atoms of the surrounding lattice. In the case of antisite defects, the "chemical" impurity is, of course, an intrinsic lattice atom on a site normally not occupied by this atom. This interaction is usually called hyperfine (hf) interaction if it occurs between the unpaired electron (hole) and the central nucleus of the impurity atom; it is called superhyperfine (shf) or ligand hyperfine (lhf) interaction if it occurs between the magnetic moment of the unpaired electron (hole) and the magnetic moments of the nuclei of the lattice surroundings. The arrows in Fig. 1 symbolize the electron (hole) spin and the nuclear spins.

The hf interaction gives rise to an energy level splitting and, hence, to a splitting of the EPR lines. This hf splitting is often resolved in an EPR spectrum. If the magnetic nucleus occurs in several isotopes, then a chemical identification of the impurity is easily made possible by taking advantage of the ratio of the observed hf splittings. A hf interaction is proportional to the size of the nuclear magnetic moment, which can be used to identify the corresponding isotope. The EPR intensities usually roughly reflect the natural abundance of the different isotopes. Unfortunately, there are important and common impurities, like oxygen, where the majority of the isotopes are nonmagnetic. In such a case, a specific doping with a magnetic isotope (e.g., ^{17}O) may be necessary. The shf interactions with nuclei of the surrounding lattice could give the information required to determine the site of the defect. The shf interactions are usually smaller than the hf interactions and give rise to an additional splitting of the hf EPR lines. The analysis of the number of shf lines and their intensity pattern, as well as their evolution upon crystal rotation, usually allows for the determination of the site of the impurity and other details of the structure. This will be shown in detail in Sect. 3.6. However, in the EPR spectra of solid state defects the resolution of the shf interactions is often not very good. In many cases only the shf interactions with nearest neighbors are resolved. Very often, however, perhaps in the majority of cases, there is no resolution of the shf interactions at all. This makes EPR spectroscopy in most solids very different from that of e.g. radicals in solution, a spectroscopy widely used in chemistry to identify molecular structures, see for example [1.3]. The reason for this is not the so-called homogeneous EPR line width, which in solids is also usually of the order of 10^{-5} T. The reason for the large EPR line width and the lack of resolution of the shf structure is the large number of superimposed homogeneous EPR lines due to the many nuclei of the surrounding lattice, which interact with the unpaired electron. Very often the size of the shf interactions falls off very slowly with distance from the center of the defect, while the number of participating nuclei increases very rapidly. To give a specific example: Suppose the neighbor nuclei have a nuclear spin of 1/2 like ^{29}Si or ^{19}F, and there are N nuclei having a shf interaction. Then there are 2^N homogeneous EPR lines contributing to the spectrum. For many defects in solids N may be of the order of 30 or even bigger. If there is a small number of nuclei which have a shf interaction much bigger than all the others, then they show up in a shf structure, all the other nuclei contributing only to the line width. If, however, the shf interactions are all very similar and change from neighbor shell to neighbor shell only gradually, then there is no shf resolution observable. In the EPR spectroscopy of chemical radicals in solution there are only a small number of nuclei interacting (often a few hydrogen atoms), and, therefore, the shf structure is mostly well resolved. Molecular crystals are, in a sense, intermediate cases. The shf interactions within the molecule, where the unpaired electron or hole is located, are much larger than those with the neighboring molecules. Therefore, the shf structure of the one molecule having the unpaired electron spin is usually resolved. For most defects in solids like ionic crystals or semiconductors, the unpaired electron is rather delocalized, with the result that many nuclei contribute to the shf structure and the EPR line width.

If it were possible to measure the nuclear magnetic resonances (NMR) of the nuclei of the surrounding lattice directly, the local magnetic field seen by the nuclei would just give the required information on the defect structure, since it is a superposition of the applied static magnetic field and the equivalent field of the shf interactions. If the local fields of all the nuclei and their symmetry relations to the defect site were known, the structure of the defect, including the presence of vacancies and lattice distortions could be derived. One of the essential features is that in NMR the number of shf lines is greatly reduced compared to EPR. In EPR N nuclei having the same shf interaction cause 2^N shf EPR lines, while all these nuclei cause only 2 NMR lines if the electron spin of the defect is 1/2 [in general $(2S+1)$ NMR lines]. Unfortunately, however, NMR is not sensitive enough. One needs at least 10^{19} nuclei like protons to measure NMR. Usually, in conventional NMR the nuclei of the host lattice are measured. Here, we consider only the nuclei in the surroundings of a defect. One typically deals with defect concentrations of 10^{16} cm^{-3} or less and sample volumes usually well below 1cm^3. In EPR about 10^{11} electron spins can be detected, if the EPR line width is 10^{-4} T.

In 1956 *Feher* [1.13,14] invented a way out of the difficulties described above for the measurement of the shf interactions of solid state defects, by using the electron nuclear double resonance (ENDOR) method for the first time. The idea of this method is that the NMR transitions of a neighbor nucleus coupled to an unpaired defect electron by a shf interaction, cause changes of the electron spin polarization under suitable experimental conditions (partly saturated EPR), which can be detected as changes of the EPR line intensity; these changes are the so-called "ENDOR" signals (Chap. 5). Thus, the NMR transitions are detected using a quantum transformation to higher quanta, the EPR microwave quanta, which results in considerably higher sensitivity than in conventional NMR. The ENDOR effect for solid state defects is, at most, a few percent of the EPR effect, usually only 0.1–1%. From experience, the sensitivity gain relative to NMR is approximately 10^3–10^4.

The idea of a quantum transformation to enhance sensitivity, and of the observation of multiple magnetic resonances can be carried further. EPR can be observed with optical detection by measuring EPR–induced changes of the optical absorption [1.15,16], or optical emission bands of a defect (ODEPR) [1.17–19] (Chap. 4).

ENDOR can also be observed with optical detection (ODENDOR) [1.20]. It was recently shown that the absorption detected ENDOR effect can be of the same order of magnitude as the optically detected EPR effect [1.21,22], which makes this triple resonance method particularly powerful for the investigation of defects present at low concentrations.

In practical materials science there are specific problems not encountered when investigating a defect specifically prepared in a nice single crystal. Usually, there are many defect species simultaneously present with overlapping EPR spectra of very different signal intensities. The measurement of ENDOR then yields very complicated many–line spectra, the analysis of which is difficult, if not impossible. Here, new developments in ENDOR spectroscopy can be advantageous. One is the measurement of a kind of excitation spectrum of ENDOR lines, the so–called ENDOR–induced EPR or "field swept" ENDOR (Chap. 5). With this method it is possible to separate the EPR spectra of various defects in analogy to an optical excitation spectrum of an optical emission. The application of electron nuclear triple resonance or double–ENDOR allows the separate measurement of the ENDOR spectra of each defect. This method is applicable in lattices where the nuclei have a high abundance of magnetic nuclei (Chap. 5). Another very convenient way to measure the EPR and ENDOR spectra of each defect separately is by optical detection, if this is applicable. The absorption or emission bands of different defect species seldom overlap completely. Therefore, an optical transition specific to one particular defect can be used for its exclusive measurement (Chap. 4).

In materials science, another problem is connected with the correlation between bulk properties and the defects causing them. Obviously, a correlation can be made with optical properties by means of optical detection. Another question is that of electrical properties determined by the energy levels of a particular defect in the energy gap of a semiconductor. As is well known, the occupancy of a level can be influenced by changing the *Fermi* energy level. This can be done, for example, by doping of either donors or acceptors, to either raise or lower the level in the gap. If the level of the paramagnetic defect is empty, it can be filled by light of sufficient energy to excite electrons from the valence band into this level. By illuminating the crystal with light of variable energy and monitoring the appearance of the EPR, ENDOR, ODEPR or ODENDOR signals one can determine the energy level position of the defect. These methods are known under the acronyms photo–EPR, photo–ENDOR, photo–ODEPR and photo–ODENDOR, respectively. The defect structure is usually studied first in a sample where the paramagnetic level is occupied and its level position is then determined in a crystal co–doped such that the level is empty (Sect. 5.7).

The defect structure determination is based on the investigation of the hf and shf structure, as discussed above. The major information comes from the anisotropy of these interactions which is determined by measuring the angular dependence of the shf structure of the ENDOR spectra. From this it is obvious that the most precise and reliable results for defect structures are obtained from single crystals. The discussion in this book will, therefore, be focused on single crystal work. However, information can also be obtained on defect structures in a polycrystalline sample by the various methods discussed. It is usually less precise and reliable. Comparatively little ENDOR work exists for polycrystalline samples. Amorphous samples are even less suitable for EPR and ENDOR methods.

1.3 Superhyperfine and Electronic Structures of Defects in Solids

The analysis of EPR and ENDOR spectra yields the symmetry of the defect and the shf interaction tensors of the neighboring lattice nuclei, as well as their orientation with respect to the defect center. The elements of the shf tensors contain information on the electronic structure of the defect. In a simple one–particle approximation of the wave function of the defect, the isotropic part, which is also called the *Fermi* contact term [1.1], is proportional to the probability of finding the unpaired electron (or hole) at the site of the nucleus whose isotropic shf interaction was measured. The anisotropic part is essentially given by the expectation value $\langle 1/r^3 \rangle$ calculated with the one–particle defect wave function, r, being measured from the nucleus of the neighbor atom. Thus, by determining the shf tensors, important elements of the defect wave functions are known. However, one cannot determine the distance of a particular nucleus from the defect center from the shf interactions without any prior knowledge of the defect wave function. The relative distance of the measured nuclei can be determined if it is known that the radial part of the wave function falls off monotonically with distance from the defect center (Chap. 7). This is usually the case for defects in wide gap materials such as ionic crystals, and for deep level defects in semiconductors. Then, a clear defect model can be derived. If nothing is known about the wave function, one cannot decide on the site from ENDOR alone. This situation was discussed recently for chalcogen defects in silicon [1.23,24].

There are, however, complicated wave functions, which make the assignment of nuclei to particular neighbor shells very difficult, if not impossible. For example, for shallow defects in semiconductors there are oscillations of the unpaired spin density with distance. In such a case, it is very difficult to decide on a defect model.

A calculation of the defect wave function made with sufficient precision to allow a satisfactory interpretation of the measured shf structures is a formidable task in view of the many–particle system which one deals with in a solid state defect. This calculation is possible for "simple" defects such as hydrogen atoms in alkali halides [1.25] or color centers in alkali halides. In covalent crystals like silicon, or nearly covalent crystals like the III–V semiconductors such as gallium arsenide, the situation is much more difficult [1.26]. Nevertheless, it can be helpful to make a simple theoretical estimate of the shf constants for the assignment of shf interactions to particular nuclei of the surrounding lattice, and for defect structure determination, particularly in largely ionic crystals. For example, if it can be expected from general arguments that the amplitude of the wave function will decrease with distance from the defect core, then the so–called anisotropic shf constants b (Chap. 7) of more distant nuclei are explained rather well by the simple classical point dipole–dipole approximation, in which both the electron magnetic moment (at the defect core) and the nuclear magnetic moments are replaced by point dipoles. Then, the point dipole analysis of the b values can be used to assign the shf constants to particular neighbor nuclei. Therefore, simple approximations for the theoretical interpretation of the shf structure are discussed in Chap. 7, which in many cases can be very helpful and "instrumental" for structure determination and for description of lattice distortions around impurities. In Chap. 7

an account of a rigorous approach to the theoretical interpretation of the shf data is also presented.

The quadrupole interactions can also be determined from the ENDOR analysis if the lattice contains nuclei with $I > 1/2$ which have electrical nuclear quadrupole moments. A simple analysis of these quadrupole interactions can be very helpful in determining the charge state of defects, which is of particular importance in semiconductor physics. This analysis is based on an account of the electrical field gradient at the sites of the surrounding lattice nuclei. Various simple approximations are also discussed in Chap. 7.

2. Fundamentals of
Electron Paramagnetic Resonance

In this chapter the basic features of electron paramagnetic resonance (EPR) are briefly discussed. This chapter cannot replace a textbook on EPR (for example [2.1–5]). The discussion is restricted to the fundamentals needed to understand the phenomena of electron paramagnetic resonance and nuclear magnetic resonance, and the application of EPR spectroscopy to defects in solids. Typical EPR spectra and the information gained from them for defect structures are discussed in Chap. 3. Experimental aspects of EPR will be discussed in Chap. 8, together with those of ENDOR. The methods of optical detection of EPR are described in Chap. 4, while their experimental aspects are described in Chap. 9. SI units are used throughout this book, except in Chap. 7, where the cgs system is employed.

2.1 Magnetic Properties of Electrons and Nuclei

Each electron in an atomic orbit carries an orbital angular momentum due to its motion about the nucleus. Associated with the angular momentum is a magnetic dipole moment $\boldsymbol{\mu}_L$ which is proportional to the orbital angular momentum, which, in units of $\hbar = h/2\pi$, is denoted by $\mathbf{L}$.

$$\boldsymbol{\mu}_L = -\mu_\mathrm{B}\mathbf{L}, \tag{2.1.1}$$

where μ_B is the *Bohr* magneton, which is given by

$$\mu_\mathrm{B} = e\hbar/2m_\mathrm{e} = 9.274015 \times 10^{-24}\ \mathrm{Am^2}. \tag{2.1.2}$$

It also possesses an angular momentum from the electron spin with the resulting magnetic dipole moment $\boldsymbol{\mu}_S$ given by

$$\boldsymbol{\mu}_S = -g_\mathrm{e}\mu_\mathrm{B}\mathbf{S}, \tag{2.1.3}$$

where $g_\mathrm{e} = 2.002319$ is the electronic g factor and $\mathbf{S}$ is the angular momentum of the electron spin.

The quantities **S** and **L** are operators in quantum mechanics. In terms of **L** and **S** the magnetic dipole moment operator $\boldsymbol{\mu}$ is

$$\boldsymbol{\mu} = -\mu_\mathrm{B}(\mathbf{L} + g_\mathrm{e}\mathbf{S}). \tag{2.1.4}$$

The minus sign arises from the electronic charge and indicates that the magnetic dipole moment and the angular moment of the electron are antiparallel. Equation (2.1.4) is rigorously valid only for spherical symmetry, for which the angular momentum is well defined. Magnetic nuclei also have an angular momentum **I**, due to the nuclear spin, and a magnetic dipole moment $\boldsymbol{\mu}_I$ related to each other by a scalar gyromagnetic ratio. It is convenient to write this relation in the form

$$\boldsymbol{\mu}_I = g_\mathrm{n}\mu_\mathrm{n}\mathbf{I}, \tag{2.1.5}$$

where g_n is the nuclear g factor, which can be positive or negative, and μ_n is the nuclear magneton,

$$\mu_\mathrm{n} = \frac{e\hbar}{2m_\mathrm{p}} = \frac{m_\mathrm{e}}{m_\mathrm{p}}\mu_\mathrm{B} \approx \frac{\mu_\mathrm{B}}{1836}, \tag{2.1.6}$$

m_e is the mass of the electron ($m_\mathrm{e} = 9.109390 \times 10^{-28}\mathrm{g}$) and m_p is the mass of the proton. The nuclear magneton is roughly 2000 times smaller than the electronic *Bohr* magneton.

A table of g_n values is given in appendix A. To give a few examples: $g_\mathrm{n}(^1\mathrm{H}) = +5.58569$, $g_\mathrm{n}(^{19}\mathrm{F}) = +5.257732$, $g_\mathrm{n}(^{35}\mathrm{Cl}) = +0.5479157$, $g_\mathrm{n}(^{37}\mathrm{Cl}) = +0.4560820$, $g_\mathrm{n}(^{29}\mathrm{Si}) = -1.1106$.

Many nuclei with a nuclear spin $I > 1/2$ (see below) possess an electrical nuclear quadrupole moment Q, which characterizes the electrical charge distribution in the nucleus. A spherical charge distribution has $Q = 0$, one of an ellipsoidal shape with the long axis parallel to the angular momentum axis has $Q > 0$ (prolate), and one with an ellipsoidal shape with the short axis parallel to the angular momentum axis (oblate) has $Q < 0$.

2.2 Electrons and Nuclei in an External Magnetic Field

In atomic systems angular momenta are quantized and, therefore, also the respective magnetic dipole moments. The eigenvalues are:

$$\begin{aligned}
\mathbf{L}^2 \ &: L(L+1)\hbar^2, \quad L_z : m_L\hbar, m_L = -L, (-L+1)\ldots, +L; \\
\mathbf{S}^2 \ &: S(S+1)\hbar^2, \quad S_z : m_S\hbar, m_S = -S, (-S+1)\ldots, +S; \quad (2.2.1) \\
\mathbf{I}^2 \ &: I(I+1)\hbar^2, \quad I_z : m_I\hbar, m_I = -I, (-I+1)\ldots, +I.
\end{aligned}$$

The z–direction is chosen as the quantization axis by convention, and m_L, m_S, m_I are commonly called orientation quantum numbers or magnetic quantum numbers.

A free electron has $S = 1/2$ and $m_S = \pm 1/2$. The eigenfunctions for S_z are denoted by $|+1/2\rangle$ or $|+\rangle$ and $|-1/2\rangle$ or $|-\rangle$. In general, there are $(2S+1)$ states. The energy E of a magnetic dipole moment in a static magnetic field $\mathbf{B}$ is

$$E = -\boldsymbol{\mu} \cdot \mathbf{B}. \tag{2.2.2}$$

By convention the static magnetic field $\mathbf{B_0}$ is oriented along the z–axis, $\mathbf{B_0} = (0, 0, B_0)$. Therefore,

$$E = -\mu_z B_0. \tag{2.2.3}$$

For a free electron one obtains

$$E_S = +g_e \mu_B B_0 m_S, \tag{2.2.4}$$

and for a magnetic nucleus

$$E_I = -g_n \mu_n B_0 m_I. \tag{2.2.5}$$

In Fig. 2.1 the variation of energy according to (2.2.4,5) is shown for $S = 1/2$ and a nucleus with $I = 3/2$, assuming $g_n > 0$. The zero of energy refers to zero magnetic energy. The energy differences for a particular value of the static magnetic field B_0 between subsequent orientation quantum numbers are indicated by arrows. For the electron they are:

$$\Delta E_S = g_e \mu_B B_0, \tag{2.2.6}$$

and for the nucleus,

$$\Delta E_I = g_n \mu_n B_0. \tag{2.2.7}$$

An electromagnetic wave inducing a resonant transition would therefore require a quantum of energy satisfying the following resonance condition:

$$h\nu = \frac{hc}{\lambda} = \Delta E, \tag{2.2.8}$$

where ν is the frequency of the radiation and λ the corresponding wavelength. For $B_0 = 0.35$ T, the wavelength for an electron would be $\lambda_S = 3.06$ cm, that is a radiation in the microwave range (X–band, $\nu = 9.8$ GHz), while for a proton the wavelength would be $\lambda_p = 20.4$ m, a wavelength in the radio frequency range (rf) ($\nu_p = 14.7$ MHz). Note that the energies involved in electron spin transitions and nuclear spin transitions are in the ratio of approximately $10^3 : 1$. Both energies are very small compared to the energies involved in optical transitions, which are typically of the order of 1 to several eV, i.e., of the order of 10^4 cm^{-1}. Thus, the quantum transformation for an NMR transition detected by ODENDOR (Chap. 5) would be a transformation

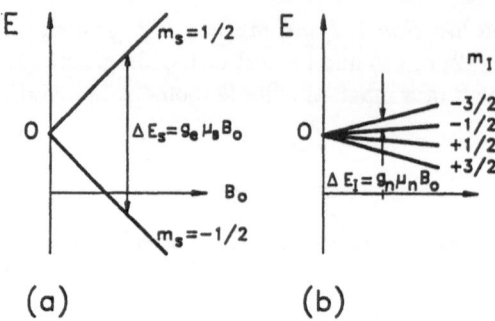

Fig. 2.1. a Electron *Zeeman* levels for $S = 1/2$ as a function of the static magnetic field B_0. The magnetic dipole transition of the basic EPR experiment is indicated by an *arrow* at $B_0 = \Delta E_S/g_e\mu_B$, $\Delta E = h\nu_{\mathrm{EPR}}$. **b** Nuclear *Zeeman* levels for $I = 3/2$ $(g_n > 0)$ as a function of the static magnetic field B_0. The energy difference for a NMR transition $B_0 = \Delta E_I/g_n\mu_n$ is indicated, $\Delta E = h\nu_{\mathrm{NMR}}$

from 5×10^{-4} cm^{-1} (NMR) through 3×10^{-1} cm^{-1} (EPR) to 10^4 cm^{-1}, or in the ratio of roughly $1:10^3:10^8-10^9$. It is also interesting to note for the following discussion that the thermal energy kT at room temperature is $1/40$ eV, or approximately 200 cm^{-1}. At room temperature the thermal energy far exceeds all magnetic energy differences ΔE of (2.2.6,7) for both electrons and nuclei. At the temperature of liquid helium (4.2 K) it is still about an order of magnitude larger than ΔE_S! This is very different from considerations in optical spectroscopy where ΔE is usually very large compared to kT. As will be discussed later, the inequality $kT \gg (\Delta E_S, \Delta E_I)$ results in a very small occupation difference between the various m_S, m_I states.

2.3 Some Useful Relations for Angular Momentum Operators

Let **J** stand for any of the angular momentum operators **L**, **S**, and **I**. The operator **J** is decomposed into

$$\mathbf{J} = \begin{pmatrix} J_x \\ J_y \\ J_z \end{pmatrix} = J_x\hat{\mathbf{x}} + J_y\hat{\mathbf{y}} + J_z\hat{\mathbf{z}}. \tag{2.3.1}$$

$\hat{\mathbf{x}}$, $\hat{\mathbf{y}}$ and $\hat{\mathbf{z}}$ are unit vectors in the x, y and z directions of the Cartesian coordinate system, respectively.

Two basic eigenvalue equations follow from the quantum mechanics of angular momenta (see textbooks on quantum mechanics)

$$\mathbf{J}^2|J, m_J\rangle = \hbar^2 J(J+1)|J, m_J\rangle, \tag{2.3.2}$$

Equations (2.4.12,13) describe the motion of the magnetization in a static magnetic field and are identical with the classical equations of motion for the *Larmor* precession. The precession frequency is given by:

$$\omega_L = \frac{|M|}{|J|}|\mathbf{B}_0|. \tag{2.4.14}$$

Thus, for the electrons it has the value

$$\omega_S = \frac{g_e \mu_B B_0}{\hbar}, \tag{2.4.15}$$

and for the nuclei,

$$\omega_I = \frac{g_n \mu_n B_0}{\hbar}. \tag{2.4.16}$$

Note, that the classical *Larmor* frequency is identical with the frequency which follows from (2.2.6–8) (by replacing $h\nu$ by $\hbar\omega$).

2.5 Basic Magnetic Resonance Experiment

In the basic magnetic resonance experiment, the magnetic dipole transition is induced between energy levels of electron spins or nuclear spins in a static magnetic field, which are characterized by the magnetic quantum numbers m_S and m_I, respectively. In order to induce such a magnetic dipole transition, a microwave field (for EPR) or radio frequency (rf) field (for NMR) of frequency ω must be applied, such that the oscillating magnetic field amplitude B_1 is perpendicular to the static magnetic field B_0 (which is parallel to the z–axis). Figure 2.2a shows this field arrangement schematically for B_1 parallel to the x–axis. Experimental details will be discussed in Chap. 8. In Fig. 2.2b the energy levels for a free electron in a certain static magnetic field B_0, and the magnetic dipole transitions for absorption and emission are indicated. In Fig. 2.2c the occupation of the two levels is assumed to be different (Sect. 6). The total magnetic field at the sample is

$$\mathbf{B} = \begin{pmatrix} 2B_1 \cos(\omega t) \\ 0 \\ B_0 \end{pmatrix}. \tag{2.5.1}$$

Therefore, the time dependent spin *Hamilton* operator is:

$$\mathcal{H}(t) = g_e \mu_B B_0 S_z + 2 g_e \mu_B B_1 S_x \cos(\omega t)$$

$$= \mathcal{H}_0 + \mathcal{H}_W \cos(\omega t). \tag{2.5.2}$$

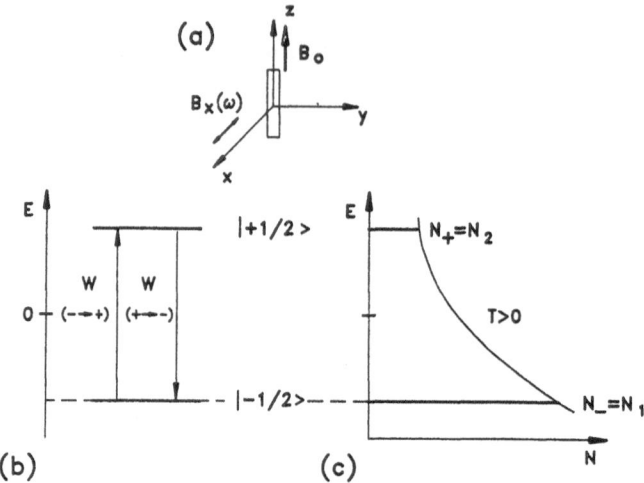

Fig. 2.2a–c. Basic EPR experiment: **a** Geometry of sample, static magnetic field B_0 and oscillating microwave field amplitude $B_x(\omega)$ $(= B_1)$. **b** Electron *Zeeman* levels for a certain field B_0 and microwave–induced magnetic dipole transitions. **c** Occupation of the two *Zeeman* levels of the spin system in thermal equilibrium with the lattice for $T > 0$

The transition probability from $|-1/2\rangle$ to $|+1/2\rangle$ $W(- \Rightarrow +)$ (absorption), which is equal to that of $W(+ \Rightarrow -)$ (induced emission), can be calculated by time dependent perturbation theory, as long as $B_1 \ll B_0$, which is usually the case for EPR and NMR spectroscopy. The general result is given by the "Golden Rule" (see textbooks on quantum mechanics) to be

$$W(- \Rightarrow +) = W(+ \Rightarrow -)$$

$$= \frac{1}{4}\hbar^2|\langle -1/2|\mathcal{H}_W|+1/2\rangle|^2 g(\nu), \qquad (2.5.3)$$

where $g(\nu)$ is a form function (see below).

By using the lowering and raising operators [(2.3.4–10)] it can easily be seen that for electrons one obtains the EPR transition probability, W_{EPR}

$$W_{\text{EPR}} = \frac{1}{4}\gamma_e^2 B_1^2 g(\nu), \qquad (2.5.4)$$

where $\gamma_e = -g_e\mu_B/\hbar$, the gyromagnetic ratio of the electrons. Similarly, for NMR transitions one obtains

$$W_{\text{NMR}} = \frac{1}{4}\gamma_I^2 B_1^2 g(\nu), \qquad (2.5.5)$$

with $\gamma_I = g_n\mu_n/\hbar$, the gyromagnetic ratio of nuclei. $g(\nu)$ is a form function of the transition, which is normalized according to

$$\int_0^\infty g(\nu)d\nu = \int_0^\infty g(\omega)d\omega = 1. \qquad (2.5.6)$$

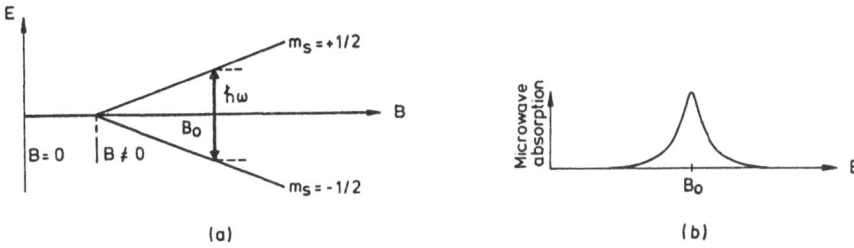

Fig. 2.3. a Electron *Zeeman* levels for $S = 1/2$, for $B = 0$ and $B \neq 0$. The magnetic dipole transition of the basic EPR experiment is indicated and occurs at at $B = B_0$ for the microwave energy $\hbar\omega$. **b** Microwave absorption at $B = B_0$: the EPR line

The introduction of the form function $g(\nu)$ acknowledges the fact that the energy levels are not infinitesimally sharp. If they were, then W_{NMR} and W_{EPR} would increase in proportion to t^2 and eventually diverge (for example, [2.2] and [2.5]). All levels have a finite lifetime because of spin–lattice relaxation (Sect. 2.6) and are thus broadened. This is formally described by the form function $g(\nu)$.

The matrix element $\langle -1/2|S_z| + 1/2\rangle$ vanishes. Therefore, no magnetic dipole transitions are induced for a rf or microwave field with B_1 parallel to B_0. The polarization B_1 of the rf field must always be perpendicular to B_0. It can, of course, also be parallel to the y–direction, since the same result is obtained as for the x–direction.

If both levels, $m_S = +1/2$ and $m_S = -1/2$, had the same occupation $N_+ = N_-$ (Fig. 2.2c), by inducing magnetic dipole transitions no microwave energy could then be transferred from the microwave field to the spin system, since the probabilities for absorption and induced emission are equal. The probability for spontaneous emission is very small and can be neglected. Therefore, an energy transfer is only possible if the occupation numbers satisfy $N_+ < N_-$. This is indicated in Fig. 2.2c assuming a *Boltzmann* distribution for the occupation (Sect. 2.7). In this case, a microwave absorption can be observed, if the microwave frequency ω fulfills the resonance conditions $\hbar\omega = g_e\mu_B B_0$ of (2.2.6). The microwave frequency is kept constant (Chap. 8); therefore, the resonance is measured by varying the magnetic field B_0. Figure 2.3 shows this for a paramagnetic electron. A microwave absorption (Fig. 2.3b) can be obtained only as long as an occupation difference $N_+ - N_-$ is maintained. If the magnetic dipole transitions are the only transitions present, then the EPR transition is quickly "saturated" to zero as soon as $N_+ = N_-$ is reached. However, the impurity atoms and the nuclei are embedded in a crystalline environment which gives rise to "spin–lattice relaxation", and provides the basis for the possibility of stationary observation of EPR and NMR.

The magnetic field $\mathbf{B}$ is measured in *Tesla* (T) throughout this book. In the older literature the cgs–unit *Gauss* (G) is often used (10^4 G=1 T). Sometimes the magnetic energy is calculated by using $\mathbf{H}$ (in *Oersteds*) instead of $\mathbf{B}$ according to

$$E = -\boldsymbol{\mu} \cdot \mathbf{H}. \tag{2.5.7}$$

In this case, the definition of the *Bohr* magneton contains μ_0 and becomes

$$\mu_B = \frac{\mu_0 e \hbar}{2m_e}. \tag{2.5.8}$$

2.6 Spin–Lattice Relaxation

In a crystalline environment an occupation difference $(N_+ - N_-)$ between the electronic *Zeeman* levels is maintained by an electron–phonon interaction, and is therefore dependent on the temperature. The equilibrium occupation of both states is determined by a *Boltzmann* distribution (Sect. 2.7).

One basic mechanism for the spin–lattice relaxation is simply the fact that, as a consequence of vibrations of magnetic neighbor atoms, the local magnetic field at the site of the unpaired electron due to the neighbor atoms contains an oscillatory component with the resonance frequency, which can induce magnetic dipole transitions. The relaxation due to dipole–dipole interactions, which are a function of $1/r^3$, is called the *Waller* mechanism. The spin–orbit coupling is also a function of the vibrational coordinates via the orbital functions, and their interactions with moving neighbor atoms also gives rise to relaxation (the *Van Vleck* mechanism). The transition probability for such a relaxation transition depends on the density of the phonon states $g(\nu_{12})$ at transition frequencies near ν_{12}, denoting state $|-1/2\rangle$ by 1, state $|+1/2\rangle$ by 2,

$$W_{12} = W_{21} = \frac{1}{\hbar^2}|\mathcal{H}'_{12}|^2 g(\nu_{12}). \tag{2.6.1}$$

To discuss the relaxation mechanism in detail would be beyond the scope of this book (e.g., [2.1,2,5]). For the practical application of EPR and ENDOR it is, however, important to consider the temperature dependence of the spin–lattice relaxation probability, which comes through the temperature dependence of the phonon density of states $g(\nu_{12})$. A few of the prominent processes are summarized below.

Usually one discusses the spin–lattice relaxation in terms of the spin–lattice relaxation time (T_{1e} for electrons and T_{1n} for nuclei, respectively). This is the characteristic time which is needed for a spin system to return to the thermal equilibrium occupation of the *Zeeman* levels after application of a perturbation to the system, for instance, a microwave or rf pulse.

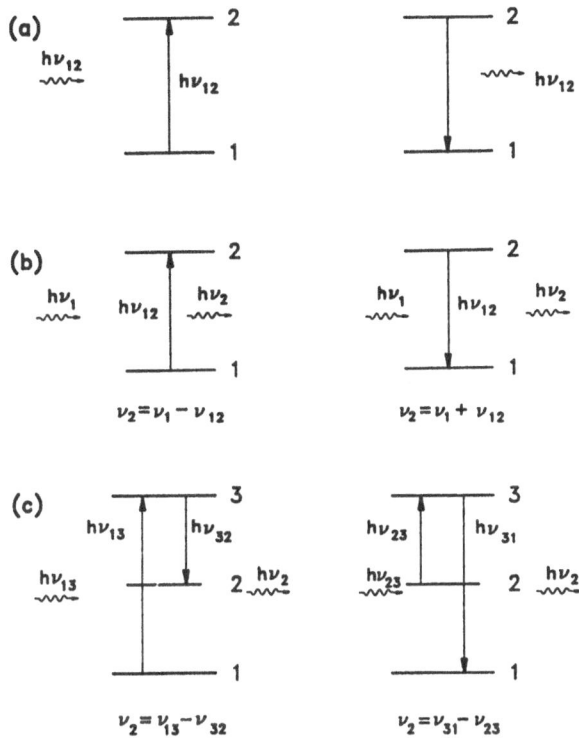

Fig. 2.4a–c. Schematic representation of different spin–lattice relaxation mechanisms **a** direct process, **b** *Raman* process, **c** *Orbach* process

The *direct process* involves a resonant absorption or emission of a phonon with frequency ν_{12}, in which $T_1 \propto 1/T$ (Fig. 2.4a). In the *Raman* process a phonon is scattered at the spin system. The phonon can gain or lose energy (Fig. 2.4b). The process is comparable to the optical *Raman* process, in which *Stokes* and Anti-*Stokes* lines occur. The effect is a second order effect involving a virtual intermediate state. $T_1 \propto (T^{-5}, T^{-7}, T^{-9})$ depending on the system and temperature range. In the *Orbach* process a phonon is absorbed by the spin system into a real excited spin state. The spin loses memory of where it came from, and is relaxed into another spin state under phonon emission. $T_1 \propto \exp(h\nu_{13}/kT)$ (Fig. 2.4c). The phonon frequency ν_{13} is smaller than the *Debye* frequency. In solids, nuclei with quadrupole moments are more efficiently relaxed than those without, since fluctuating electrical fields contribute to the nuclear spin–lattice relaxation via the electrical quadrupole interaction experienced by the nucleus in an electrical field gradient (Chap. 5) [2.6].

T_1 increases with decreasing temperature and, in practice, several of the processes discussed can occur for one defect, depending on the temperature range. At very low temperature the direct process dominates because only few phonon states are occupied. The spin–lattice relaxation time T_1 also depends on the magnetic field value, the power of the B–dependence being different for the various processes. For details the reader is referred to the literature, for example [2.6-8]. As will be seen in Chap. 5, T_1 of both the electron spin system and the nuclear spin system are important for ENDOR measurements. Since a prediction of T_1 is not possible for the complicated solid state systems at a given temperature, it is best to be able to vary the sample temperature in a wide range (1.5 K–300 K).

2.7 Rate Equations for a Two–Level System

The EPR signal depends on the occupation difference of the two electronic *Zeeman* levels and, therefore, one must discuss rate equations, which determine the occupation of the two levels. As shown schematically for a two–level system in Fig. 2.5 (which would also apply for a nuclear spin with $I = 1/2$), one must consider the two magnetic dipole transitions W_{12} and W_{21}, induced by the microwave field and the two relaxation transitions R_{12} and R_{21}. The spin system has a total number of N spins with N_2 spins in the state 2 for $m_S = +1/2$, while N_1 spins are in the state 1 with $m_S = -1/2$. One has

$$N = N_1 + N_2, \tag{2.7.1}$$

$$\frac{dN_1}{dt} = -(W_{12} + R_{12})N_1 + (W_{21} + R_{21})N_2, \tag{2.7.2}$$

$$\frac{dN_2}{dt} = +(W_{12} + R_{12})N_1 - (W_{21} + R_{21})N_2. \tag{2.7.3}$$

Equations (2.7.2,3) shall be discussed first for the case in which there are microwave–induced dipole transitions but no spin–lattice relaxation ($R_{12} = R_{21} = 0$). The time evolution of the occupation difference $\Delta N = N_1 - N_2$ (for $N_1 > N_2$) is then

$$\frac{d(\Delta N)}{dt} = -2 W_{EPR} \Delta N, \tag{2.7.4}$$

where $W_{EPR} = W_{12} = W_{21}$. The net transition rate is proportional to the occupation difference and proportional to the transition probability. Since the latter depends on B_1^2 (2.5.4), an occupation difference ΔN is decreased efficiently for a large microwave power incident on the sample,

$$\Delta N \propto \exp(-\frac{t}{\tau}) \tag{2.7.5}$$

with

$$\tau = 2 W_{EPR}. \tag{2.7.6}$$

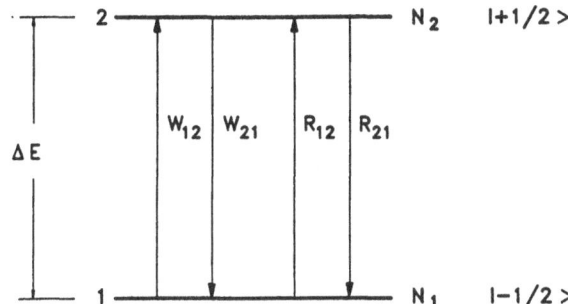

Fig. 2.5. Two–level system ($S = 1/2$) with the microwave transitions W_{12} (absorption) and W_{21} (emission) and the spin–lattice relaxation transitions R_{12} and R_{21}

For $t \to \infty$, (2.7.5) predicts $\Delta N \to 0$, and there is no net microwave absorption any more. The EPR is then said to be *"saturated"*. Without spin–lattice relaxation, there would be no possibility of observing a stationary EPR signal, as was already mentioned.

Let us now consider thermal equilibrium and no microwave–induced transitions, that is $W_{12} = W_{21} = 0$. In thermal equilibrium the occupations N_1 and N_2 are stationary, that is $dN_1/dt = 0$ and $dN_2/dt = 0$. Therefore, it follows from (2.7.2,3) that

$$R_{12}N_1 = R_{21}N_2, \tag{2.7.7}$$

$$\frac{R_{12}}{R_{21}} = \frac{N_2}{N_1} = \exp(-\Delta E/kT) \equiv \exp(x). \tag{2.7.8}$$

Since the occupation ratio is given by the *Boltzmann* statistics, ΔE being the energy difference between levels 2 and 1, the relaxation probability R_{12} is smaller by $\exp(x)$ than R_{21}. We can now calculate the occupation difference $\Delta N_0 = N_1 - N_2$ in thermal equilibrium:

$$\Delta N_0 = N_1 - N_2 = N_1 - N_1 \exp(x) = N_1[1 - \exp(x)], \tag{2.7.9}$$

$$N = N_1 + N_2 = N_1[1 + \exp(x)]. \tag{2.7.10}$$

A combination of (2.7.9,10) yields

$$\Delta N_0 = N \left(\frac{1 - e^x}{1 + e^x}\right) = N \tanh \left(\frac{\Delta E}{2kT}\right). \tag{2.7.11}$$

As mentioned earlier, one usually deals with $h\nu = \Delta E \ll kT$, the so–called high temperature approximation. Therefore, ΔN_0 is given approximately by

$$\Delta N_0 = N \frac{\Delta E}{2kT}. \tag{2.7.12}$$

To give an example for the occupation difference ΔN_0, let us consider: $T = 300$ K, $B_0 = 0.35$ T. For electrons we obtain $\Delta N_0 \approx 2 \times 10^{-3}$, for protons $\Delta N_0(p) \approx 3 \times 10^{-6}$. These are very small numbers, which make the observation of microwave and rf transitions difficult, and require high spectrometer sensitivity. The paramagnetic susceptibility, χ is calculated as

$$\chi = \frac{M}{B_0} = \frac{\Delta N_0 \mu_B}{B_0} = \frac{N g_e \mu_B^2 B_0}{2kT B_0} = \frac{N \mu_B^2}{kT}. \tag{2.7.13}$$

Equation (2.7.13) is the *Curie* law of paramagnetism (for $g_e \approx 2$).

At very low temperature, at which the ODMR experiments are usually carried out (typically at 1.5 K), (2.7.11) must be applied (the *Langevin* function) or, for a system with angular moment **J** the more general *Brillouin* function B_J (for example [2.9] or [2.10]). For $S = 1/2, L = 0$ we obtain the *Langevin* function

$$M_S = \Delta N_0 \, \mu_B = N \mu_B \tanh(\Delta E/2kT). \tag{2.7.14}$$

For an angular momentum **J**, (2.7.14) must be replaced by

$$M_J = N J g_J \mu_B B_J(J g_J \mu_B B_0/kT) = N J g_J \mu_B B_J(x), \tag{2.7.15}$$

$$B_J(x) = \frac{(2J+1)}{2J} \coth \left[\frac{(2J+1)x}{2J} \right] - \frac{1}{2J} \coth \left(\frac{x}{2J} \right). \tag{2.7.16}$$

For $S = 1/2$, and $L = 0$, (2.7.15) is identical with (2.7.14).

We now consider the stationary occupation resulting from simultaneous microwave–induced and relaxation transitions between levels 1 and 2.

How the spin system returns to thermal equilibrium from a non–equilibrium state $(W_{12} = W_{21} = 0)$ can be calculated from (2.7.8) and (2.7.2,3). One obtains

$$\frac{d\Delta N}{dt} = R[1 - \exp(x)]N - R[1 + \exp(x)]\Delta N, \tag{2.7.17}$$

where $R = R_{21}$. The solution of the differential equation is an exponential

$$\Delta N = A \exp(-t/T_1) + \Delta N_0 = A \exp(-t/T_1) + N \tanh(\Delta E/2kT), \tag{2.7.18}$$

where

$$T_1 = \frac{1}{[1 + \exp(x)]} = \frac{1}{(R_{12} + R_{21})}. \tag{2.7.19}$$

For $\Delta E \ll kT$, this can be approximated by

$$T_1 = \frac{1}{2R}. \tag{2.7.20}$$

From (2.7.18) it is seen that if the spin system were pushed off thermal equilibrium, for example, by an rf or microwave pulse, it would return to the equilibrium occupation exponentially with characteristic time T_1. Often T_1 is called the *longitudinal* spin–lattice relaxation time, since it refers to the magnetization along the magnetic field direction (z). The rate of return to equilibrium is proportional to the deviation of the occupation difference ΔN, from that at thermal equilibrium, ΔN_0.

Finally, we can solve the rate equations (2.7.2,3) for the stationary state in the presence of both microwave–induced transitions and spin–lattice relaxation. The condition for a stationary state is,

$$\frac{dN_1}{dt} = \frac{dN_2}{dt} = 0. \tag{2.7.21}$$

It follows from (2.7.2) or (2.7.3) that

$$-(W_{12} + R_{12})N_1 + (W_{21} + R_{21})N_2 = 0. \tag{2.7.22}$$

By means of (2.7.1), (2.7.4), (2.7.8), (2.7.11), (2.7.19) and (2.7.20), this can be written as

$$2W_{\mathrm{EPR}}\Delta N + 2R(\Delta N - \Delta N_0) = 0, \tag{2.7.23}$$

$$\Delta N = \frac{\Delta N_0 R}{(W_{\mathrm{EPR}} + R)} = \frac{\Delta N_0}{1 + (W_{\mathrm{EPR}}/R)}. \tag{2.7.24}$$

Inserting (2.5.4) and (2.7.20) one obtains:

$$\Delta N = \frac{\Delta N_0}{1 + \frac{1}{2}\gamma_e^2 B_1^2 T_1 g(\nu)},$$

$$\Delta N = \frac{\Delta N_0}{1 + s}. \tag{2.7.25}$$

The last term in (2.7.25) is called the saturation factor s. If s satisfies $s \gg 1$, then the EPR signal decreases, in the extreme case almost to zero. s depends on both B_1^2 and T_1. If T_1 is long (e.g., at low temperature) saturation can be reached with moderate microwave power levels of the order of 1 mW or less, while at room temperature T_1 is usually so short, that saturation is very difficult. A partially saturated EPR signal is the condition for the detection of ENDOR (Chap. 5).

The microwave power P, which can be absorbed by the paramagnetic sample is given by:

$$P_{MW} = W_{\mathrm{EPR}}\Delta N\, h\nu = \frac{1}{4}\gamma_e^2 B_1^2 g(\nu)\frac{\Delta N_0}{1 + s} h\nu$$

$$= \frac{N(h\nu)^2\, \gamma_e^2\, B_1^2\, g(\nu)}{8kT\,[1 + \frac{1}{2}\gamma_e^2\, B_1^2\, T_1\, g(\nu)]}. \tag{2.7.26}$$

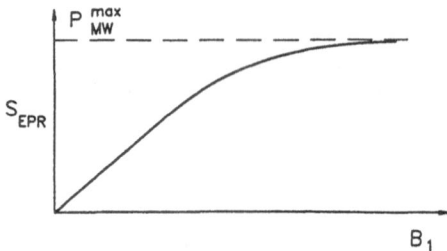

Fig. 2.6. Saturation of the EPR signal as a function of B_1 (i.e., square root of microwave power) for a two–level system assuming a diode as microwave detector

As long as s satisfies $s \ll 1$, P increases in proportion to B_1^2. The maximum of the EPR signal, which is proportional to P_{MW}, is limited by the spin–lattice relaxation time T_1. For $s \gg 1$, (2.7.26) can be approximated by

$$P_{MW}(max) = \frac{N(h\nu)^2}{2kT \cdot 2T_1}.\tag{2.7.27}$$

Figure 2.6 shows the schematic increase of the EPR signal $S_{\mathrm{EPR}} \propto \sqrt{P_{MW}}$ as a function of B_1 for a given temperature, that is, for a given T_1 (Sect. 8.3.2). From (2.7.26,27) it is also seen that P_{MW} increases with increasing microwave frequency. It is, therefore, advantageous to use high frequencies such as 35 GHz, or higher. However, technical problems arising at high frequencies do limit the gain (Chap. 8).

2.8 Bloch Equations

In Sect. 2.4 the equation of motion for a classical paramagnetic magnetization M_S was found to be (the index S is omitted)

$$\frac{d\mathbf{M}}{dt} = \gamma(\mathbf{M} \times \mathbf{B}).\tag{2.8.1}$$

Equation (2.8.1) does not yet reflect the fact that the paramagnetic defect is within a crystalline lattice, and therefore experiences spin–lattice relaxation. It is necessary to distinguish between two spin-lattice relaxation times. T_1 is associated with the component of the magnetization which is parallel to the static magnetic field B_0, while T_2 is associated with the transverse components of the magnetization. T_2 is the analogue of T_1 for a magnetization moving in the x–y plane ($B_0 \parallel z$), which can be achieved by a suitable microwave pulse (a so–called 90° pulse). The transverse components of the magnetization do not influence the energy; they can change without coupling to the lattice. The mechanism responsible for T_2 is often a magnetic

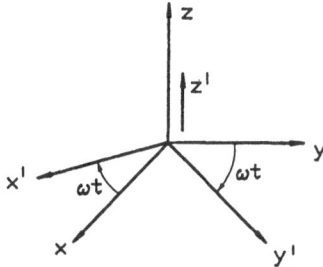

Fig. 2.7. Rotating frame x', y', z' and laboratory frame x, y, z

dipole–dipole interaction. In solids one usually finds $T_2 \ll T_1$. (For details see textbooks on EPR such as [2.2] and [2.5]).

The relaxation loss for the component of the transverse magnetization is also assumed to occur exponentially, which leads to:

$$\frac{dM_x}{dt} = -\frac{M_x}{T_2}, \tag{2.8.2}$$

$$\frac{dM_y}{dt} = -\frac{M_y}{T_2}. \tag{2.8.3}$$

The solutions of (2.8.1) supplemented by (2.8.2,3), are readily obtained in a reference frame which rotates with the microwave frequency about the z–axis. In this rotating frame x', y', z' (Fig. 2.7) (2.8.1) becomes

$$\frac{d\mathbf{M}'}{dt} = \gamma \mathbf{M}' \times \mathbf{B}_{\text{eff}}, \tag{2.8.4}$$

$$\mathbf{B}_{\text{eff}} = \begin{pmatrix} B_1 \\ 0 \\ B_0 + (\omega/\gamma) \end{pmatrix}, \tag{2.8.5}$$

where ω is the frequency of the microwave field. $\mathbf{B}_{\text{eff}}$ is called an effective field. The linearly polarized microwave field $B_1(t) = 2B_1 \cos(\omega t)$ driving the magnetic dipole transitions can be decomposed into right and left circularly polarized microwave fields about the z–axis,

$$\mathbf{B}(t) = \begin{pmatrix} B_1 \cos(\omega t) \\ B_1 \sin(\omega t) \\ 0 \end{pmatrix} + \begin{pmatrix} B_1 \cos(\omega t) \\ -B_1 \sin(\omega t) \\ 0 \end{pmatrix}. \tag{2.8.6}$$

In the rotating frame, the second component in (2.8.6) has the same sense of rotation as the *Larmor* precession $\omega_0 = -\gamma B_0$, and the first component moves in the opposite direction. This component has, however, no noticable effect on the spin system.

With the abbreviation $\omega_0 = -\gamma B_0$, ω_0 being the resonance frequency of the spin system, the *Bloch* equations with relaxation in the rotating frame x', y' and z' are:

$$\frac{dM'_x}{dt} = (\omega - \omega_0)M'_y - \frac{M'_x}{T_2}, \qquad (2.8.7)$$

$$\frac{dM'_y}{dt} = -(\omega - \omega_0)M'_x - \frac{M'_y}{T_2} + \gamma B_1 M'_z, \qquad (2.8.8)$$

$$\frac{dM'_z}{dt} = -\gamma B_1 M'_y - \frac{M'_z - M_0}{T_1}. \qquad (2.8.9)$$

M_0 is the magnetization in thermal equilibrium in the absence of microwave transitions.

The solutions of (2.8.7–9) for the stationary state, where $dM_{x'}/dt = dM_{y'}/dt = dM_{z'}/dt = 0$ are found in a straightforward way to be:

$$M'_x = \frac{\gamma B_1(\omega_0 - \omega)T_2^2}{1 + (\omega_0 - \omega)^2 T_2^2 + \gamma^2 B_1^2 T_1 T_2} M_0, \qquad (2.8.10)$$

$$M'_y = \frac{\gamma B_1 T_2}{1 + (\omega_0 - \omega)^2 T_2^2 + \gamma^2 B_1^2 T_1 T_2} M_0, \qquad (2.8.11)$$

$$M'_z = \frac{1 + (\omega_0 - \omega)^2 T_2^2}{1 + (\omega_0 - \omega)^2 T_2^2 + \gamma^2 B_1^2 T_1 T_2} M_0. \qquad (2.8.12)$$

The transformation back into the laboratory frame x, y, z gives

$$\mathbf{M} = \begin{pmatrix} M_x \\ M_y \\ M_z \end{pmatrix} = \begin{pmatrix} M'_x \cos(\omega t) + M'_y \sin(\omega t) \\ -M'_x \sin(\omega t) + M'_y \cos(\omega t) \\ M'_z \end{pmatrix}. \qquad (2.8.13)$$

The transverse magnetization has one component, $M_{x'}$, which rotates synchronously with B_1 about the z-axis, while $M_{y'}$ is shifted in its phase by 90° (x' is parallel to B'_1). Therefore, $M_{x'}$ will constitute the dispersive part χ' of the complex magnetic susceptibility $\chi = \chi' - i\chi''$, while $M_{y'}$ determines the absorptive part χ''

$$\chi' = \frac{M'_x}{2B_1}, \qquad (2.8.14)$$

$$\chi'' = \frac{M'_y}{2B_1}. \qquad (2.8.15)$$

Without saturation, that is, for $\gamma^2 B_1^2 T_1 T_2 \ll 1$, we obtain:

$$\chi'' = \frac{\gamma T_2 M_0}{2[(1 + (\omega - \omega_0)^2 T_2^2]}, \qquad (2.8.16)$$

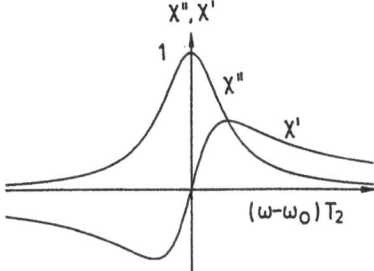

Fig. 2.8. Line shape for the real and imaginary part of the complex susceptibility χ' and χ'', respectively. χ'' has a *Lorentzian* line shape, which is the line shape measured for a homogeneous EPR line

$$\chi' = \frac{\gamma(\omega - \omega_0)T_2^2 M_0}{2[(1 + (\omega - \omega_0)^2 T_2^2]}, \tag{2.8.17}$$

$$M_z = M_0. \tag{2.8.18}$$

The absorption χ'' and dispersion χ' are plotted versus $(\omega - \omega_0)T_2$ in Fig. 2.8. χ'' follows a *Lorentzian* line shape of the form $f(x) = 1/(1 + x^2)$. The *Lorentzian* shape is the line shape of a homogeneous EPR line. We shall discuss in a later chapter that the line shape of EPR lines in the presence of hyperfine interactions is generally *Gaussian* (Sect. 3.7). The half width of the homogeneous line is

$$\Delta\omega_{1/2} = \frac{2}{T_2}, \tag{2.8.19}$$

while the maximum value for χ'' is for $\omega = \omega_0$

$$\chi''_{max} = \frac{1}{2}\gamma T_2 M_0 = \frac{1}{2}\chi_0\omega_0 T_2 = \frac{\chi_0\omega_0}{\Delta\omega_{1/2}}, \tag{2.8.20}$$

where

$$\chi_0 = M_0/B_0. \tag{2.8.21}$$

It can be seen from (2.8.19), that the transverse spin–spin relaxation time T_2 determines the line width of the homogeneous line. The maximum value of χ'' at $\omega = \omega_0$ can be interpreted as the susceptibility of the system in thermal equilibrium multiplied by the quality factor $\omega_0/\Delta\omega_{1/2}$ of the resonance system. Note that χ'' satisfies the relation

$$\int_{-\infty}^{\infty} \chi'' d\omega = \frac{1}{4}\omega_0\chi_0. \tag{2.8.22}$$

Consequently, the area under the EPR line is independent of the relaxation times T_1 and T_2, and since χ_0 is proportional to the total number of spins, the area is a measure of the total number of spins, or paramagnetic defects. By calibration of the area (2.8.22) one can determine the concentration, or total number of paramagnetic defects, provided the measurement is done without saturating the EPR signal, as assumed in (2.8.22). The average power absorbed by the spin system is:

$$P = \frac{1}{T}\int_0^T \mathbf{H} \cdot d\mathbf{B} = \mu_0 \int_0^T H_z \dot{H}_z dt$$

$$+ \frac{\mu_0}{T}\left[\int_0^T H_x \dot{M}_x dt + \int_0^T H_y \dot{M}_y dt\right]. \qquad (2.8.23)$$

One obtains

$$\int_0^T H_x \dot{H}_x dt = \int_0^T \cos(\omega t)\sin(\omega t)dt = 0, \qquad (2.8.24)$$

and by use of (2.8.13) one finds

$$P = \omega B_1 M_y' = 2\omega B_1^2 \chi''. \qquad (2.8.25)$$

The absorbed power will first increase by increasing B_1. However, under saturating conditions comparison with (2.8.11) shows that P will reach a constant value, the saturation value. This was already seen in the discussion of the rate equations of a two–level system (Sect. 2.7).

Under saturating conditions, that is, large B_1 amplitudes, (2.8.14,15), together with (2.8.10,11) can be brought into the forms

$$\chi' = \frac{1}{2}\left(\frac{\gamma M_0 T_2}{\sqrt{1+s^2}}\right)\left(\frac{T_2(\omega_0 - \omega)/\sqrt{1+s^2}}{1 + [T_2(\omega_0 - \omega)/\sqrt{1+s}\ ]^2}\right), \qquad (2.8.26)$$

$$\chi'' = \frac{1}{2}\left(\frac{\gamma M_0 T_2}{1+s^2}\right)\left(\frac{1}{1 + [T_2(\omega_0 - \omega)/\sqrt{1+s^2}]^2}\right), \qquad (2.8.27)$$

where s^2 is the saturation factor, $s^2 = \gamma^2 B_1^2 T_1 T_2$. The line still has a *Lorentzian* shape. It is, however, broadened by a factor $\sqrt{1+s^2}$ and the amplitudes are diminished, χ' by $\sqrt{1+s^2}$ and χ'' by $(1+s^2)$. It follows that

$$\Delta\omega_{1/2} = \frac{2}{T_2}\sqrt{1+s^2}. \qquad (2.8.28)$$

Finally, by comparison of the results of the microscopic theory of the two–level system with those obtained here, we get an interpretation of the form function $g(\nu)$ introduced in (2.5.3) in Sect. 2.5. An expression for the absorbed power follows from (2.8.25) and (2.8.27),

$$P = B_0 M_0 \frac{\gamma B_1^2 T_2}{1 + (\omega - \omega_0)^2 T_2^2 + \gamma^2 B_1^2 T_1 T_2}. \qquad (2.8.29)$$

In Sect. 2.7 we obtained (2.7.26) for the absorbed power at resonance, which is equivalent to

$$P = \frac{1}{2} \frac{B_0 M_0 \gamma^2 B_1^2 g(\nu)}{1 + \frac{1}{2}\gamma^2 B_1^2 g(\nu) T_1}. \tag{2.8.30}$$

Comparison of (2.8.29,30) shows that

$$g(\nu) = \frac{2T_2}{1 + 4\pi^2(\nu - \nu_0)^2 T_2^2}. \tag{2.8.31}$$

This is the normalized form function of the unsaturated resonance line, i.e., a *Lorentzian* line. T_2 processes limit the life time of the quantum mechanical energy levels. Such a broadening was assumed when introducing $g(\nu)$ in Sect. 2.5.

2.9 Conventional Detection of Electron Paramagnetic Resonance and Its Sensitivity

The detection of paramagnetic resonance transitions is achieved by measuring the microwave power absorbed by the sample under resonance conditions. In principle, this could be done in a quasi–optical way by irradiating the sample with plane, polarized microwaves of a certain power, and detecting the loss of power behind the sample. Such an experiment is not sensitive enough. As discussed in Sect. 2.7 the occupation difference ΔN_0 is very small, and hence the paramagnetic susceptibility. Furthermore, the concentration of paramagnetic defects is often less than ppm, with the result that one has to resort to spectrometer designs, which have the highest possible sensitivity. One uses a "bridge" method to measure the change of χ'' induced by the resonance, and places the sample into a microwave cavity in order to increase the amplitude B_1 of the microwave field, making use of a high quality factor of the cavity of the order of 10000. (For details see Chap. 8.)

Since the microwave sources cannot be varied much in their frequency, one keeps the microwave frequency constant and varies the magnetic field, as indicated in Fig. 2.2. However, one must not scan the magnetic field too rapidly, if one wants to avoid the so–called "passage" effects. The solutions (2.8.10–12) of the *Bloch* equations are only valid in the stationary state, that is, in thermal equilibrium. Therefore, in going through the resonance by varying the static magnetic field $B \Rightarrow B_0 = \omega_0/\gamma$, one must always try to maintain thermal equilibrium. But, of course, this cannot actually be achieved. What can be maintained is the adiabatic condition. The external field is varied slowly so as to allow the continuous adaptation of the energy level system, i.e., the system always behaves in a quasi–stationary fashion. It can be shown that this condition is fulfilled for

$$\frac{dB_0}{dt} \ll \gamma B_1^2; \tag{2.9.1}$$

see *Abragam* [2.2] for further details.

For very narrow lines and long spin–lattice relaxation times it can be difficult to fulfill (2.9.1). For sensitivity reasons, it is also quite common in practice not to measure the signal $\chi''(B_0)$ by slowly varying B_0, but to measure its derivative $d\chi''(B_0)/dB_0$ by superimposing on B_0 a small oscillating field along the z–axis. The spectra shown in the following chapters measured by conventional EPR are, therefore, all "derivatives" of EPR lines, and denoted simply as EPR lines. The frequency of this field modulation must not be too high, especially at low temperature, in order not to violate the slow adiabatic passage condition.

The sensitivity of EPR detection is an important feature for the application to materials science, since very often only very low defect concentrations are present. The sample size depends on the microwave band used; in X–band (10 GHz) it is typically $5 \times 5 \times 10$ mm^3 for a cylindrical cavity, and in K–band (24 GHz), $2 \times 2 \times 5$ mm^3. The sensitivity is limited by the noise amplitude U_R of the microwave detector,

$$U_R = \sqrt{4RFkT\Delta\nu}, \tag{2.9.2}$$

which is the *Nyquist* noise including a noise factor F from the diode, which depends on the diode, microwave frequency and detection frequency (field modulation frequency, Chap. 8). R is the ohmic resistance and $\Delta\nu$ is the band width of the detection system.

The signal amplitude U_S is given by the filling factor η and the quality factor Q of the microwave cavity. The filling factor is roughly the sample volume divided by the effective cavity volume,

$$U_S = \eta Q \chi'' \sqrt{RP_{MW}}, \tag{2.9.3}$$

where P_{MW} is the microwave power. P_{MW} is limited by the saturation of the spin system and therefore cannot be enhanced arbitrarily. Equation (2.9.3) is valid without saturation. The sensitivity is limited by the condition

$$U_S = U_R, \tag{2.9.4}$$

which would yield a signal–to–noise ratio of 1. One obtains

$$\chi''_{min} = \frac{2}{Q\eta} \sqrt{\frac{FkT\Delta\nu}{P_{MW}}}. \tag{2.9.5}$$

From (2.8.20) and (2.7.13) we have

$$\chi''_{max} = \frac{\omega_o}{\Delta\omega_{1/2}} \chi_0 = \frac{\omega_o}{\Delta\omega_{1/2}} \frac{NS(S+1)g_e^2\mu_B^2}{3kT}. \tag{2.9.6}$$

Equations (2.9.5,6) can be solved for N,

$$N = 6\sqrt{\frac{F k T \, \Delta\nu}{P_{MW}}} \cdot \frac{k T \, \Delta\omega_{1/2}}{\omega_0 S(S+1) g_e^2 \mu_B^2} \cdot \frac{1}{\eta \, Q}, \qquad (2.9.7)$$

where N is the minimum number of spins to be detected with a signal–to–noise ratio of 1. The first term in (2.9.7) is determined by the microwave power and the detection system. The second term depends on the sample containing defects, while the last term is given by the quality factor, size of the cavity and the sample size. The sensitivity is proportional to the EPR line width $\Delta\omega_{1/2}$. Most EPR lines of solid state defects are rather broad. Often $\Delta\omega_{1/2}$ is of the order of 5 – 50mT. This limits the sensitivity more than anything else, since most spectrometers reach the theoretical sensitivity quite well. The sensitivity is mostly given in terms of $N_{min}/\Delta\omega_{1/2}$. A typical sensitivity for electron spin resonance is $N_{min} = 5 \times 10^{11}$–$1 \times 10^{12}$ spins/mT for $T = 300$ K, $S = 1/2$, $\nu_0 = 10$ GHz, $\Delta\nu = 1$ Hz. In NMR the sensitivity is much smaller. For protons one obtains about 10^{18}–10^{19} spins/mT for $\omega_0 = 40$ MHz, $T = 300$ K, $I = 1/2$, and $\Delta\nu = 1$ Hz.

Therefore, to obtain reasonable EPR spectra, considering line splittings due to, for example, hyperfine interactions, a minimum spin concentration is of the order of 10^{14}–10^{16} spins/cm^3. This number depends, of course, on the details of the spin system.

Theoretically, the sensitivity should increase with increasing resonance frequency ω_0, which enters directly into the second term of (2.9.7) and indirectly in an enhanced filling factor in a smaller cavity. Experience teaches, however, that the theoretical gain, estimated to be $N_{min} \propto \omega_0^{-7/2}$, is seldom really achieved in practice. With higher microwave frequencies one loses part of the theoretical gain through other difficulties (Chap. 8).

3. Electron Paramagnetic Resonance Spectra

Electron Paramagnetic Resonance (EPR) spectra can be measured in unperturbed crystals which contain paramagnetic ions or molecules as constituents, such as $CuSO_4$ with paramagnetic Cu^{++} ions, and in crystals containing paramagnetic defects. In this chapter, typical EPR spectra of defects will be discussed in view of the application of multiple magnetic resonance methods to be discussed later. It is the purpose of this chapter mainly to discuss which kind of structural information can be obtained from the EPR spectra. Particular emphasis is given to a discussion of the EPR line width and its origin. It will be seen later, that the structural information from the EPR spectra is needed for an unambiguous analysis of the ENDOR spectra. We must restrict the discussion to "typical" spectra, in the sense that only those interactions of the unpaired electrons or holes which often occur are discussed, and the discussion is limited to simple quantum mechanical solutions of the *Schroedinger* equation. There is such a variety of possible spectra that a more comprehensive discussion would be beyond the scope and intention of this book (for more information see textbooks on EPR quoted in Chap. 1).

3.1 Spin Hamiltonian

The solution of the *Schroedinger* equation of an impurity atom or intrinsic defect in a crystalline matrix, considering all the interactions between the impurity and the lattice environment, as well as the "internal" interactions, such as the spin–orbit interaction, is too complicated to be applicable to the interpretation of EPR spectra. Therefore, a number of simplifications are introduced, which lead to the concept of an effective spin and the so–called spin *Hamiltonian*, see Sect. 7.3 for a further discussion.

The energies available for inducing magnetic dipole interactions are, at most, of the order of 10 cm^{-1}, and are usually less then 1 cm^{-1}. The energy difference between the ground state of the defect system and most of the excited states is usually several orders of magnitude higher. The energy separation is typically of the order of 10^4 cm^{-1}, which also greatly exceeds the thermal energy. Therefore, in conventional EPR spectroscopy we deal only

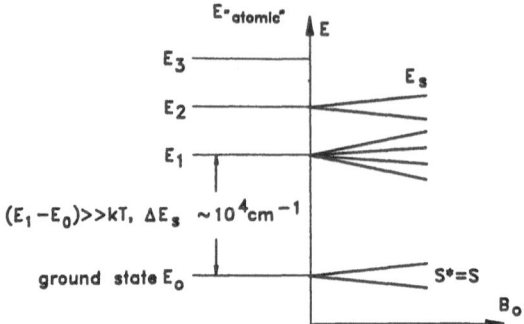

Fig. 3.1. Illustration of the energy levels of a paramagnetic defect in a solid to introduce the concept of the effective spin

with the electronic ground state of the defect system. This is not so when applying the optical detection of EPR (Chap. 4).

The situation is shown schematically in Fig. 3.1. For an uneven number of electrons the ground state is at least doubly degenerate. This degeneracy can only be lifted by a magnetic field (*Kramers* doublet). Some of the states of the ground state manifold will be orbital singlets. These are *Kramers* doublets and the total wave function will almost have the transformation properties of the spin angular momentum, at least for practical purposes. Then the ground state manifold is said to have an *effective* spin, which is usually simply called spin; in this particular case it equals the true spin. The concept is generalized. In order to avoid the detailed discussion of the complicated coupling scheme between spin, orbital momenta and B_0, one ascribes an effective spin S^* to the ground state manifold and a $(2S^* + 1)$ degeneracy, except for cases with non–*Kramers* doublets, which will not be discussed here [3.1,2]. The splitting in the magnetic field is

$$\Delta E_{S^*} = g_{\text{eff}} \mu_B B_0. \qquad (3.1.1)$$

In the following we will omit the distinction between S^* and S. Unfortunately, the word "effective spin" is used for different approximations. We will have to discuss another effective spin later in the analysis of ENDOR spectra, which should not be mixed up with the one discussed here (Chap. 6).

The *Hamiltonian* operator of the effective spin is called the spin *Hamiltonian*. It is introduced to describe the behavior of the ground state manifold as measured in the paramagnetic resonance spectra. Since the unpaired defect electron is not in free space but embedded in a crystalline environment, it will not have the simple isotropic behavior of a free spin, which was discussed in the previous chapter. At least this was always implicitly assumed. The energy levels of the ground state manifold will depend on the angle between the crystalline axis and the magnetic field. In fact, as mentioned in Chap. 1, this is the basis of the structure determination using magnetic resonance.

The spin *Hamiltonian* must contain the operators of the electronic spins and of the nuclear spins, which may interact and contribute to the energy levels. It also contains the components of the magnetic field and, if applicable, of external stress and external electrical fields. Each of the different terms in the spin *Hamiltonian* must be invariant under the point symmetry operations of the paramagnetic defect. The spin *Hamiltonian* must be invariant under symmetry operations of the defect system.

In principle, the terms of the spin *Hamiltonian* have the form $\mathbf{B}^i\mathbf{S}^j\mathbf{I}^k$, where i, j, k are the powers to which each of the operators is raised. It can be shown, that i, j, k must obey certain restrictions due to the symmetry requirements and the multiplicity of the various angular momentum states (e.g., see [3.1,2], and further literature therein).

For most cases, it is sufficient to consider the bilinear and quadratic form of $\mathbf{B}^i\mathbf{S}^j\mathbf{I}^k$. The most important of such terms constitute the following spin *Hamiltonian*:

$$\mathcal{H} = \mathcal{H}_{EZ} + \mathcal{H}_{FS} + \mathcal{H}_{HF} + \mathcal{H}_{NZ} + \mathcal{H}_Q, \tag{3.1.2}$$

$$\mathcal{H}_{EZ} = \mu_B \mathbf{S} \cdot \tilde{g} \cdot \mathbf{B}_0 \quad \text{electron } \textit{Zeeman} \text{ interaction,} \tag{3.1.3}$$

$$\mathcal{H}_{FS} = \mathbf{S} \cdot \tilde{D} \cdot \mathbf{S} \quad \text{fine–structure interaction,} \tag{3.1.4}$$

$$\mathcal{H}_{HF} = \mathbf{I} \cdot \tilde{A} \cdot \mathbf{S} \quad \text{hyperfine interaction,} \tag{3.1.5}$$

$$\mathcal{H}_{NZ} = g_n \mu_n \mathbf{I} \cdot \mathbf{B}_0 \quad \text{nuclear } \textit{Zeeman} \text{ interaction,} \tag{3.1.6}$$

$$\mathcal{H}_Q = \mathbf{I} \cdot \tilde{Q} \cdot \mathbf{I} \quad \text{nuclear quadrupole interaction,} \tag{3.1.7}$$

where $\tilde{g}$, $\tilde{D}$ and $\tilde{A}$ are symmetrical tensors in three–dimensional space. They can be transformed into their principal axes systems, and are then diagonal. The quantum mechanical description of the effective spin $\mathbf{S}^*$ is identical to that of the real spin, i.e., one uses the same *Pauli* matrices and other representations as for the real spin. As mentioned above, the effective spin can sometimes be identified with the real spin, but can also mean something very different. This is the case when significant orbital contributions to the total angular momentum have to be considered (then the diagonal elements of $\tilde{g}$ differ significantly from 2.0023).

The gain in introducing the concept of the spin *Hamiltonian* is the ability to analyze the EPR spectra without a detailed knowledge of the true wave function of the defect system. All the difficulties associated with this ignorance are, in a sense, "hidden" in the interaction tensors $\tilde{g}, \tilde{A}, \tilde{D}$. Their elements do contain the information on the true wave functions of the system. This will become apparent when the theoretical interpretation of the "interaction parameters" or "interaction constants" are discussed in Chap. 7. It will be seen there that an understanding of the tensor elements requires the knowledge of the true wave function of the ground state of the system.

In the following sections the influence of the various terms in (3.1.2) on the EPR spectra will be discussed and illustrated with a few typical examples. An attempt has been made to illustrate the influence of each term in (3.1.2) as clearly as possible, and, therefore, the examples are chosen in such a way, that one of the terms in (3.1.2) always has the major influence on the EPR spectrum.

3.2 Electron Zeeman Interaction

A free electron has an isotropic g factor of $g_e = 2.0023$ and the electron *Zeeman* interaction is adequately described by

$$\mathcal{H}_{EZ} = g_e \mu_B \mathbf{B_0} \cdot \mathbf{S}. \tag{3.2.1}$$

As already mentioned, (3.2.1) is usually not sufficient to describe the electron *Zeeman* interaction for a paramagnetic impurity. The reason is that the spin–orbit interaction $\mathcal{H}_{LS}$ has to be taken into account. Its influence on the g factor is already seen in a free atom, when spin and orbital angular momentum are coupled. Let us consider the simple case of an atomic ns^2 np electronic configuration. This configuration describes, for example, a Tl^0 atom. Tl^0 is also contained in a paramagnetic defect in alkali halides, which will be discussed later in more detail (Sect. 3.5). Figure 3.2 shows the energy levels. The g factors for the $^2P_{1/2}$ and $^2P_{3/2}$ states are described by the *Landé* factor g_j, given by

$$g_j = 1 + \frac{J(J+1) + S(S+1) - L(L+1)}{2J(J+1)}. \tag{3.2.2}$$

For $L = 1, S = 1/2$, one obtains $g_{1/2} = 2/3$ and $g_{3/2} = 4/3$.

The magnetic splitting is described by m_j. The $^2P_{3/2}$ state is a quartet, $^2P_{1/2}$ a doublet. It is apparent here that the effective spin S^* is the total angular momentum $S^* = J$. It is not necessary to construct the concept of the effective spin in this simple case, as it is for the solid state defect. It is discussed here only to illustrate the ideas.

Usually, the deviations from $g_e = 2.0023$ observed for paramagnetic defects are much smaller than those discussed above. Typical values of $\Delta g = (g - g_e)$ seldom exceed ca. 0.5 (see end of this section). The reason for this observation is that a partial quenching of the orbital angular momentum contributions to the g factor occurs under the influence of the electrical crystal field of the surrounding lattice. For total quenching of the orbital momentum, the paramagnetic electron, although bound to an impurity atom, behaves like a free electron. Also, all s–state paramagnetic electrons have $g \approx g_e$ due to the lack of spin–orbit interaction. However, small spin–orbit effects may come in by overlap effects between the impurity s–electron and the electron shells of the neighbor atoms [3.1,2].

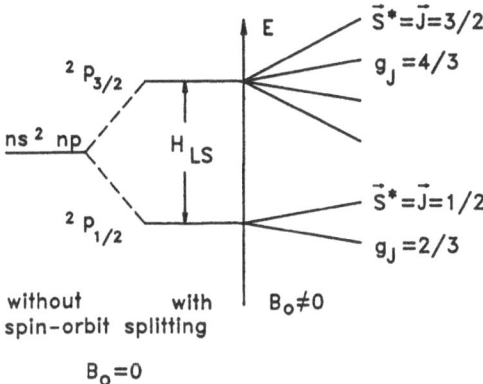

Fig. 3.2. Energy levels, g factors and effective spin of an ns^2 np electronic configuration with spin–orbit interaction with and without magnetic field

In order to illustrate the effect of orbital quenching, let us consider an atom with an unpaired outer p–electron in an orthorhombic crystal field produced by ionic charges (Fig. 3.3a). The crystal field operator $\mathcal{H}_{cryst}$ causes a splitting of the degenerate atomic states $|p_x\rangle, |p_y\rangle$ and $|p_z\rangle$ (Fig. 3.3b) given by

$$\langle p_x|\mathcal{H}_{cryst}|p_x\rangle = -\Delta,$$
$$\langle p_y|\mathcal{H}_{cryst}|p_y\rangle = 0, \qquad (3.2.3)$$
$$\langle p_z|\mathcal{H}_{cryst}|p_z\rangle = +\Delta.$$

The negative charges on the z–axis are repulsive and increase the energy of the p_z–orbital, while the positive charges on the x–axis lower the energy of the p_x–orbital. In the absence of charges on the y–axis, the energy of the p_y–orbital is not changed. The orbital *Zeeman* energy in the ground state $|p_x\rangle$ is proportional to the expectation value of L_z. Using (2.3.3–7)

$$\langle p_x|L_z|p_x\rangle = \frac{1}{2}\langle (p_+ + p_-)|L_z|(p_+ + p_-)\rangle$$

$$= \frac{1}{2}\{\underbrace{\langle p_+|L_z|p_+\rangle}_{1} + \underbrace{\langle p_-|L_z|p_-\rangle}_{-1} + \underbrace{\langle p_+|L_z|p_-\rangle}_{0} + \underbrace{\langle p_-|L_z|p_+\rangle}_{0}\}$$

$$= 0. \qquad (3.2.4)$$

Therefore, including the two spin functions $|+\rangle$ and $|-\rangle$, we obtain

$$\langle p_x|\langle +|\mu_B(L_z + g_e S_z)B_0|+\rangle|p_x\rangle = \frac{1}{2}g_e\mu_B B_0 \qquad (3.2.5)$$

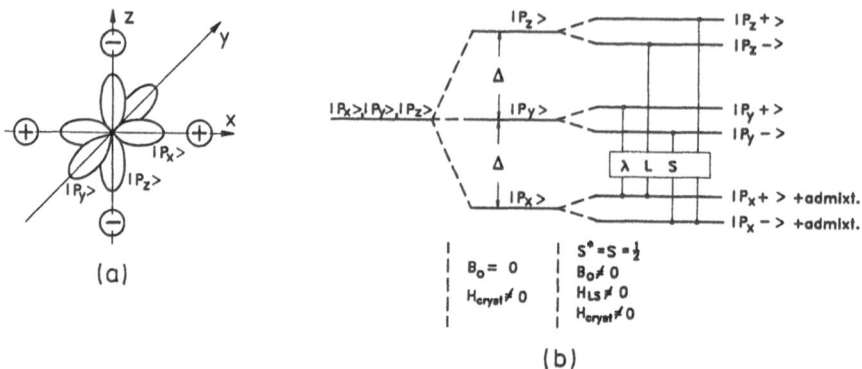

Fig. 3.3. a p–orbital in an orthorhombic crystal field. **b** Energy levels of p–electrons in the crystal field of **a** with and without a magnetic field. The orbital mixing because of spin–orbit interaction is indicated

The p–electron has a *Zeeman* energy like a free electron, since the orbital contribution (3.2.4) is totally quenched. This result is typical for an electronic configuration in which the p–orbitals are fixed by crystal fields, and not allowed to perform a free orbital rotation in one or several planes (as, for instance, $|p_x\rangle \pm i|p_y\rangle$). It is this rotation which has the angular momentum contribution to the *Zeeman* energy. Normally on observes a large degree of orbital quenching for solid state defects, the unpaired electrons or holes being in p, d or f orbital states. This fact facilitates both measurement and analysis of ENDOR spectra. In general, one observes a g shift: $g = g_e + \delta g$, where the g shift $\delta g \ll g_e$. δg is caused by the spin–orbit interaction $\mathcal{H}_{LS}$, which was neglected in the discussion above.

The discussion is limited here to $\mathcal{H}_{LS} \ll \mathcal{H}_{cryst}$. (For further details see Ref. [3.1,2]). The spin–orbit interaction operator $\mathcal{H}_{LS} = \lambda\mathbf{L}\cdot\mathbf{S}$ (λ being the spin–orbit constant) mixes the excited states $|p_y\rangle$ and $|p_z\rangle$ into the ground state *Kramers* doublet. As an example, these admixtures into $|p_x\rangle|+\rangle$ are calculated by perturbation theory of first order. One obtains:

$$(|p_x\rangle|+\rangle)^{(1)} = |p_x\rangle|+\rangle - \sum_n \frac{\langle n|\lambda\mathbf{L}\cdot\mathbf{S}|p_x\rangle|+\rangle}{(E_n - E_0)}|n\rangle$$

$$= |p_x\rangle|+\rangle - i(\lambda/2\Delta)|p_y\rangle|+\rangle - (\lambda/4\Delta)|p_z\rangle|-\rangle. \quad (3.2.6)$$

The ground state multiplet contains admixtures of the excited states with a weight factor of the order of λ/Δ. Therefore, if $\lambda \ll \Delta$, the g shift is small. By inserting $|p_+\rangle, |p_-\rangle$ and $|p_0\rangle = |p_z\rangle$ in (3.2.6) one obtains

$$(|p_x\rangle|+\rangle)^{(1)} = \sqrt{\frac{1}{2}}\left[(1 - \frac{\lambda}{2\Delta})|p_+\rangle|+\rangle + (1 + \frac{\lambda}{2\Delta})|p_-\rangle|+\rangle\right]$$

$$-\frac{\lambda}{4\Delta}|\mathrm{p_0}\rangle\,|-\rangle. \tag{3.2.7}$$

In (3.2.7) it becomes apparent that the orbital contributions of $m_l = +1$ and $m_l = -1$ do not cancel any more, as in (3.2.4),where the spin–orbit interaction was neglected. From (3.2.7) we obtain

$$(\langle\mathrm{p}_x|\,\langle+|\mu_\mathrm{B}(L_z + g_e S_z)B_0|+\rangle\,|\mathrm{p}_x\rangle)^{(1)} \approx \left(\frac{1}{2}g_e - \frac{\lambda}{\Delta}\right)\mu_\mathrm{B}\,B_0, \tag{3.2.8}$$

where terms in $(\lambda/\Delta)^2$ were neglected. For the effective spin S^* we obtain

$$\langle+^*|g_{zz}\mu_\mathrm{B}S_z^* B_0|+^*\rangle = \frac{1}{2}g_{zz}\mu_\mathrm{B}\,B_0. \tag{3.2.9}$$

Comparison with (3.2.8) yields

$$g_{zz} = g_e - \frac{2\lambda}{\Delta} \tag{3.2.10}$$

Similarly, for the other field orientations $\mathbf{B}_0 \parallel x$ and $\mathbf{B}_0 \parallel y$ one obtains the other two principal values of the g tensor

$$g_{xx} = g_e \tag{3.2.11}$$

$$g_{yy} = g_e - \frac{\lambda}{\Delta} \tag{3.2.12}$$

The results (3.2.10–12) obtained for the simple situation illustrated in Fig. 3.3 can be generalized as follows: The spin–orbit interaction brings in an orbital contribution to the angular momentum, while the crystal field has the tendency to quench the orbital angular momentum. Therefore, the g shift depends on the ratio of λ/Δ, i.e., the spin–orbit energy is compared to the crystal-field energy. In general for the principal values of the g tensor [3.1,2] one finds:

$$g_{ij} = g_e - \lambda \sum_{n\neq 0} \frac{\langle 0|\mathbf{L}_i|n\rangle\langle n|\mathbf{L}_j|0\rangle + \text{compl. conj.}}{(E_n - E_0)} \tag{3.2.13}$$

where $\langle 0|$ is the orbital wave function of the ground state, $\langle n|$ are the orbital wave functions for the excited states. It can be seen from (3.2.13) that a negative g shift is obtained, which is typical for "electron" centers, where $\lambda > 0$ (less than half filled shell of valence electrons), while a positive g shift is characteristic of "hole" centers, for which $\lambda < 0$,see also [3.1]. Typical g shifts δg are as follows:

(i) $\delta g \approx 10^{-5} \ldots 10^{-2}$

These values are typical of defects with a weak spin–orbit coupling $\lambda \approx 1 \ldots 10^2$ cm^{-1} in a strong crystal field. Light impurity atoms in ionic crystals and defects with an s–ground state typically have small δg values.

(ii) $\delta g \approx 10^{-2} \ldots$ several $\times \, 10^{-1}$

This is typical of transition metal ions with a 3d^n configuration ($\lambda \approx 10^2$ $\ldots 10^3$ cm^{-1} in a strong crystal field (for example in ionic crystals).

(iii) "δg" ≈ 1

Here, the concept of a g shift is no longer adequate. For the case $\lambda > \Delta$, one has to classify the states according to $|J, m_j\rangle$, and not with the effective spin S^*. For example the unpaired electrons of rare earth ions with a 4f^n orbital configuration ($\lambda \approx 10^3 \ldots 10^4$ cm^{-1}) do not "see" much of a crystal field, since the 4f^n orbitals are largely screened by the occupied outer orbitals (5s^{2}5p^6 for the threefold positive rare earth ions).

3.3 g–Factor Splitting of EPR Spectra

It is convenient to write the spin *Hamiltonian* in the principal axis system of the g tensor. It is clear from the preceding section that the g tensor reflects the local point symmetry of the defect. The principal axis system is denoted by x, y, z, while the laboratory axis system is X, Y, Z. In the principal axis system the *Zeeman* interaction becomes:

$$\mathcal{H}_{EZ} = \mu_B (g_{xx} S_x B_x + g_{yy} S_y B_y + g_{zz} S_z B_z). \tag{3.3.1}$$

In order to calculate the *Zeeman* energy, it is convenient to transform the magnetic field orientation $B_0 \parallel Z$ into the principal axis system of the g tensor. The *Zeeman* energy is:

$$E_{EZ} \;=\; g\mu_B B_0 m_S, \tag{3.3.2-a}$$

$$g \;=\; \sqrt{g_{xx}^2 l^2 + g_{yy}^2 m^2 + g_{zz}^2 n^2}, \tag{3.3.2-b}$$

where l, m, n are the direction cosines of the magnetic field in the principal axis system. For axial symmetry, g is given by

$$g = \sqrt{g_{\parallel}^2 \cos^2(\theta) + g_{\perp}^2 \sin^2(\theta)}, \tag{3.3.3}$$

where θ is the angle between (Z, z) and the common abbreviations $g_{xx} = g_{yy} = g_{\perp}$ and $g_{zz} = g_{\parallel}$ are adopted.

Figure 3.4 shows, as an example, the EPR spectrum of O_2^- centers in KCl. O_2^- centers are O_2^- molecules which occupy a vacant Cl$^-$ site. The defect is formed when KCl crystals are grown in air, or in an oxygen atmosphere. The O_2^- center has almost axial symmetry about a [110] axis, which is the direction of the line connecting the two oxygen nuclei of the O_2^- molecule (Fig. 3.5). For $\mathbf{B}_0 \parallel [100]$ (Fig. 3.4a) there are two EPR lines in the intensity ratio of

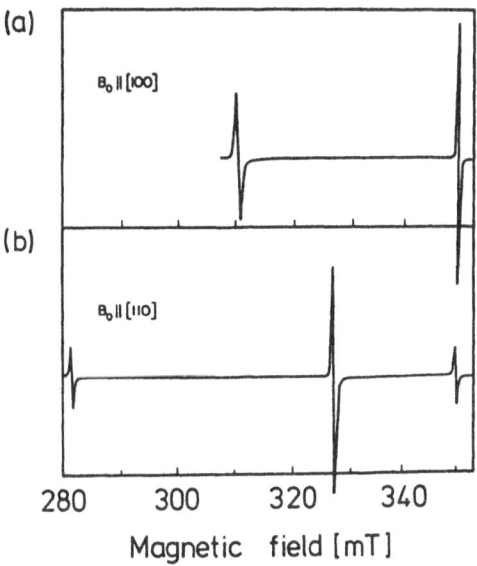

Fig. 3.4a,b. EPR spectrum of O_2^- centers in KCl. (After [3.3]) **a** $B_0 \parallel [100]$ **b** $B_0 \parallel [110]$

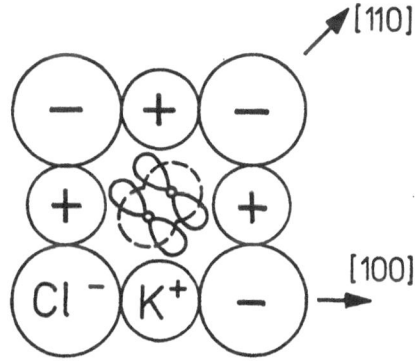

Fig. 3.5. Model of the O_2^- centers in alkali halides. (After [3.3])

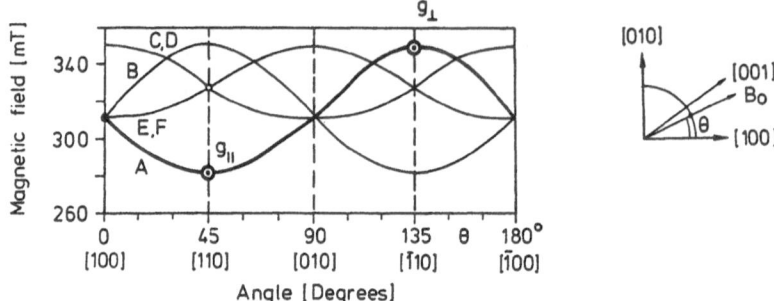

Fig. 3.6. Angular dependence of the EPR lines of O_2^- centers in KCl for rotation of the magnetic field in a {100} plane. (After [3.3])

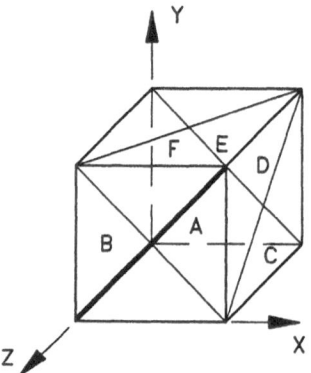

Fig. 3.7. O_2^- center orientations in alkali halides

1 : 2; for $\mathbf{B}_0 \parallel [110]$ (Fig. 3.4b) there are three lines in the intensity ratio of 1:4:1 (from low to high field). For an arbitrary field orientation there are six EPR lines of almost equal intensity [3.3,4].

Figure 3.6 shows the angular dependence of the EPR lines when rotating the magnetic field in a {100} plane of the crystal. The spectrum and the angular dependence are explained as follows, see also Fig. 3.7. Suppose that O_2^- was in only one orientation in the crystal, for example, with the axis parallel to the [110] orientation (A in Fig. 3.7). Then there would be only one EPR line, which would follow the angular variation marked "A" in Fig. 3.6. The line is at the lowest field position for the highest g value according to (3.3.3), $g = g_\parallel = 2.44$ and, correspondingly, at the highest field position for $g = g_\perp = 1.95$. The other observed lines occur because of the fact that the crystal contains six $\langle 110 \rangle$ orientations, which the O_2^- centers can be oriented along (EPR cannot distinguish between center orientations related by 180° to each other). For $\mathbf{B}_0 \parallel [100] \parallel X$, the axes of centers A, B, E and F all have the same 45° angle with $\mathbf{B}_0$; therefore, all their spectra coincide, while centers C, D have an angle of 90°$(g = g_\perp)$, and their spectra also coincide. This explains the intensity ratio of 2:4 in Fig. 3.4a. For $\mathbf{B}_0$ parallel to the A orientation (lowest field), center B is perpendicular (highest field), while the other four center orientations, C–F, all have the same intermediate angle (Fig. 3.6). EPR measurements with higher resolution reveal a superhyperfine structure (Sect. 3.6), and investigations of ^{17}O enriched samples (I(^{17}O)= 5/2)) show that the model of Fig. 3.5 is correct. The precise g values are: $g_{zz} = 2.436, g_{xx} = 1.951$, and $g_{yy} = 1.955$. The center is not precisely axially symmetric [3.5].

Whether or not the center symmetry can be analysed easily from the EPR spectrum largely depends on the line width of the EPR lines in comparison to the g values. In the case of the O_2^- center in KCl, the field splitting due to the g shift was large compared to the line width. If the g shift is only small, the analysis can become difficult, and the symmetry is not so apparent from the EPR spectra for different field orientations. Figure 3.8 shows, as an example, the EPR spectrum and its angular dependence of the so–called thermal donors in silicon for rotation of the magnetic field in a {110} plane [3.6]. These defects are formed in *Czochralski* grown silicon, which contains approximately 10^{18} cm^{-3} oxygen interstitials upon annealing at 450°C. The defects formed by annealing for times of the order of 30 min to several hours, are shallow double donors. To be able to measure their EPR spectra, one has to codope with acceptors like B, in order to singly ionize the thermal donors, and to bring them into a paramagnetic state. From Fig. 3.8a it is seen, that although the line width is rather small for a solid state defect $(\Delta B_{1/2} \approx 0.3$ mT), the g anisotropy is too small to separate the spectra well. The thermal donors, which are called "NL8" thermal donors, have C_{2v} symmetry, that is, a twofold symmetry axis along a $<100>$ axis and two {110} mirror planes (Fig. 3.8a), and $g_{[001]} = 1.99991, g_{[110]} = 2.00091$ and $g_{[1\overline{1}0]} = 1.99323$. Therefore, more lines are expected for the six possible center orientations. The center orientations are indicated by numbers in Fig. 3.8b. One observes just two EPR lines for $\mathbf{B}_0 \parallel$ [111], since the spectra of three center orientations coincide. The center symmetry is deduced from its EPR angular dependence. This symmetry information proves to be very vital for the analysis of the ENDOR spectra. It can be shown that the defect is an oxygen cluster with a core involving four oxygen atoms [3.7,8].

3.4 Fine–Structure Splitting of EPR Spectra

For paramagnetic defects with $S > 1/2$ an additional interaction can occur, which is commonly called the fine–structure (FS) interaction. As already mentioned in Sect. 3.1, its physical origin is the influence of the electrical crystal field felt by the spins through the spin–orbit interaction, as well as

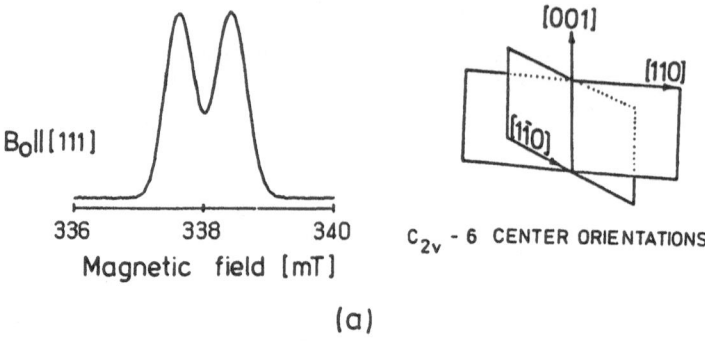

(a)

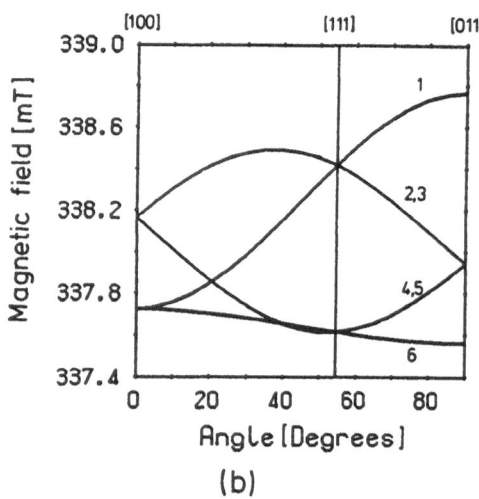

(b)

Fig. 3.8. a EPR spectrum of thermal donors in silicon (so–called NL8 spectrum) for $B_0 \parallel$ [111] (X–band) and representation of the C_{2v} center symmetry. b Angular dependence of the EPR spectum of thermal donors in silicon for rotation in a {110} plane between [100] and [011]. The different lines are due to the g factor anisotropy and different center orientations in the crystal. (After [3.6]) the magnetic dipole–dipole interaction between the unpaired electrons. This fine–structure leads to a splitting of the EPR lines in zero magnetic field ("zero–field splitting"). In EPR spectroscopy, only those fine–structure energies which are mostly of the order of the microwave energy (≈ 1 cm^{-1}) can be discussed. A proper discussion of the fine–structure effects on the EPR spectra is rather complicated. The complications depend on the symmetry of the defect, on the spin and on the size of the fine–structure interaction, with respect to the electron *Zeeman* energy. We will restrict our discussion to a few simple and typical cases. For more details the reader is referred to textbooks on EPR such as [3.1,2], as well as to a recent article by *Rudowicz* [3.9], in which the current situation in this area is reviewed and discussed.

The spin *Hamiltonian*, including the lowest order fine–structure term, becomes

$$\mathcal{H} = \mu_B \mathbf{S} \tilde{g} \mathbf{B}_0 + \mathbf{S} \tilde{D} \mathbf{S} \tag{3.4.1}$$

where $\tilde{D}$ is the fine–structure tensor. The trace of the fine–structure tensor is arbitrarily set to zero [3.2], since it only shifts the total ground state multiplet energy. Such a shift is not seen in the EPR transitions.

In its principal axis system x, y, z of the fine–structure term $\mathcal{H}_{FS}$ becomes

$$\mathcal{H}_{FS} = D_{xx} S_x^2 + D_{yy} S_y^2 + D_{zz} S_z^2. \tag{3.4.2}$$

Since the trace of $\tilde{D}$ is zero, there are only two independent diagonal elements:

$$\begin{pmatrix} D_{xx} & & \\ & D_{yy} & \\ & & D_{zz} \end{pmatrix} = \begin{pmatrix} -\frac{1}{3}D + E & & \\ & -\frac{1}{3}D - E & \\ & & +\frac{2}{3}D \end{pmatrix} \tag{3.4.3}$$

$$= \begin{pmatrix} E & & \\ & -E & \\ & & D \end{pmatrix} - \frac{1}{3}D \begin{pmatrix} 1 & & \\ & 1 & \\ & & 1 \end{pmatrix}. \tag{3.4.4}$$

This leads to

$$\mathcal{H}_{FS} = D[S_z^2 - \frac{1}{3}S(S+1)] + E[S_x^2 - S_y^2], \tag{3.4.5}$$

where D is the axially symmetric part, and E is the asymmetry parameter of the fine–structure interaction. If $\mathcal{H}_{EZ} \gg \mathcal{H}_{FS}$, the energy of (3.4.1) can be calculated in perturbation theory of first order. For simplicity we consider only axial symmetry and assume that g is isotropic:

$$E = g_e \mu_B B_0 m_S + \frac{D}{6}[3\cos^2(\theta) - 1][3m_S^2 - S(S+1)], \tag{3.4.6}$$

where θ is the angle between the z–axis of the fine–structure tensor and the orientation of $\mathbf{B}_0$. Figure 3.9a shows the energy according to (3.4.6) for $S = 1$,

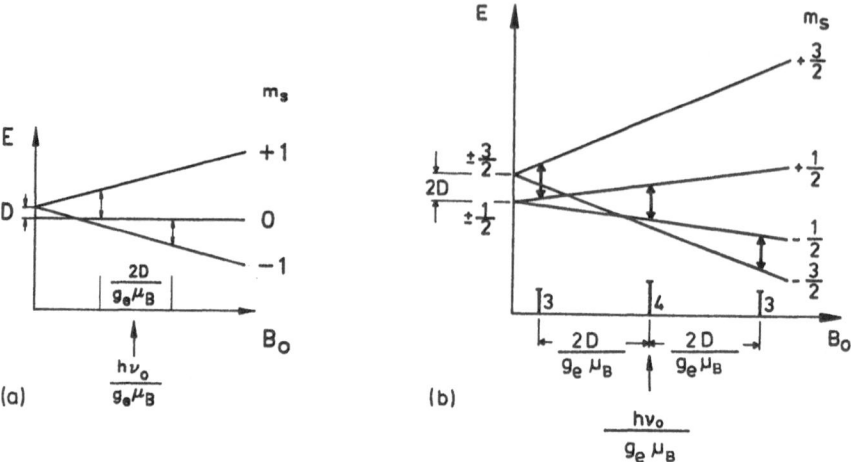

Fig. 3.9. a Energy level scheme for a triplet system, $S = 1$, in a static magnetic field B_0. The two fine–structure EPR transitions are indicated by *arrows*. Their field splitting is given in first order by twice the fine–structure splitting constant D. ν_0 is the microwave frequency. **b** Energy level scheme for a quartet system, $S = 3/2$, in a static magnetic field B_0. The three fine–structure EPR transitions are indicated by *arrows*. ν_0 is the microwave frequency, D the fine–structure splitting constant

and Fig. 3.9b shows the energy for $S = 3/2$, both figures are for $\mathbf{B}_0 \parallel z$, the principal axis of the FS tensor. For this orientation (3.4.6) becomes

$$E = g_e \mu_B B_0 m_S + \frac{D}{3}[3m_S^2 - S(S+1)]. \qquad (3.4.7)$$

The measurement is done for constant microwave frequency ν_0. Therefore, the resonance fields depend on the fine–structure interaction. For the example of $S = 1$ (a triplet system), one obtains two allowed EPR transitions for $\Delta m_S = \pm 1$:

$$h\nu_0 = E(m_S = 1) - E(m_S = 0)$$

$$= g_e \mu_B B_{01} + \frac{1}{3}D + \frac{2}{3}D = g_e \mu_B B_{01} + D, \qquad (3.4.8)$$

$$h\nu_0 = E(m_S = 0) - E(m_S = -1)$$

$$= \frac{2}{3}D + g_e \mu_B B_{02} - \frac{1}{3}D = g_e \mu_B B_{02} - D. \qquad (3.4.9)$$

The field difference is $(B_{02} - B_{01}) = 2D/g_e \mu_B$. This is a characteristic fine–structure splitting for axial symmetry with $\mathbf{B}_0$ parallel to the fine–structure z–axis. For the quartet system in Fig. 3.9b there are three lines. The central one is not shifted by the fine–structure interaction in first order.

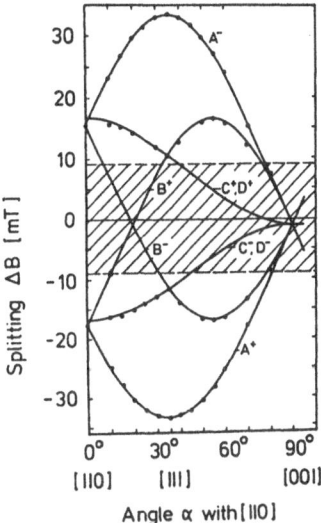

Fig. 3.10. Angular dependence of the EPR lines of F_3 centers in the quartet state for rotation of $\mathbf{B}_0$ in a $\{110\}$ plane. The *solid lines* are calculated with the appropriate spin *Hamiltonian* (see text). A–C denote the four possible center orientations along $<111>$. *Plus* and *minus signs* correspond to the transitions $m_s = +3/2 \Rightarrow +1/2$ and $m_s = -3/2 \Rightarrow -1/2$. The transition $m_s = +1/2 \Rightarrow -1/2$ coincides with the EPR line of F centers. In the *hatched field region* the lines are not resolved any more. (After [3.4])

The fine–structure splitting vanishes for $\theta = 54.7°$, for which $\cos^2(\theta) = 1/3$. As already mentioned, for $S = 1/2$ there is no fine–structure splitting. For $S=N/2$ one expects N EPR transitions for each defect orientation.

Examples for triplet EPR spectra will be discussed in Chap. 4 on the optical detection of EPR spectra. Triplet EPR spectra are typically found for excited states of two electrons, or an excited electron–hole defect system such as a bound exciton.

Figure 3.10 shows the angular dependence of the EPR spectra of so–called F_3 (or R) centers in KCl. This defect has $S = 3/2$; it is a spin quartet system. For $S = 3/2$, one expects three fine–structure transitions (Fig. 3.9b). The F_3 center is an aggregate of three F centers next to each other in a $\{111\}$ plane, each F center being an electron trapped at a Cl$^-$ vacancy. The F_3 center has a threefold symmetry axis, this axis being the z–axis of the fine–structure tensor (Fig. 3.11). There are four $<111>$ crystal directions and, therefore, the F_3 centers are distributed equally about all four orientations, which are labeled A–D in Fig. 3.10. The magnetic field is rotated in a $\{110\}$ plane such that $\mathbf{B}_0$ is parallel to the center orientation A for $\alpha = 30°$. Two of the three $m_s = \pm1$ fine–structure transitions are seen clearly: A$^+$ for $m_S = 3/2 \Rightarrow m_S = 1/2$ and A$^-$ for $m_S = -1/2 \Rightarrow m_S = -3/2$. The third

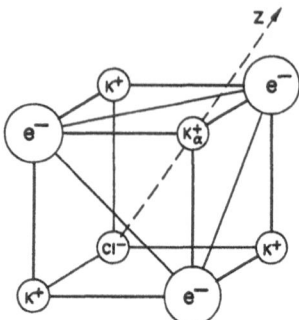

Fig. 3.11. Model of F_3 centers in KCl single crystals

transition for $m_S = 1/2 \Rightarrow m_S = -1/2$ is superimposed on the spectrum of F centers, which are present simultaneously in the middle part of Fig. 3.10. F_3 centers are produced by an aggregation of F centers whose optical absorption bands have been optically bleached at room temperature. No fine–structure transitions are separately resolved in the hatched part of the spectrum. The solid lines in Fig. 3.10 are calculated for the four center orientations according to (3.4.6) with $D/g_e\mu_B = 16.85$ mT, $E = 0$, $g = 1.996$ and $S = 3/2$.

For $B_0 \parallel [001]$ the angle to the <111> axes is $\theta = 54.7°$. Correspondingly, all fine–structure splittings vanish. From this observation the center symmetry becomes very clear.

An interesting feature of the fine–structure interaction of this particular defect is that its physical origin can largely be explained as being due to the magnetic dipole–dipole interaction between the three unpaired F center electrons forming the F_3 aggregate center [3.4].

Commonly, one observes fine–structure splittings for transition metal ions with the $3d^n$ configurations such as Mn^{2+} or Fe^{3+}. Very often these spectra also show a resolved hyperfine or superhyperfine structure, which then makes the spectrum rather complicated (Sect. 3.5).

The fine–structure angular dependence can greatly change its pattern if the solution of first order perturbation theory (3.4.6) breaks down. This is illustrated for Fe^{3+} impurities ($3d^5$ configuration) in a tetragonal symmetry, having an (almost) isotropic g factor and increasingly strong D values. Figure 3.12a shows the calculated EPR angular dependence for rotation in a {100} plane for $D/g_e\mu_B = 0.35$ mT and $\nu_{EPR} = 9.2$ GHz ($g = 2.0023$). B_0 is parallel to the fine–structure z–axis for $\theta = 0°$. For $\theta = 54.7°$ the fine–structure splitting vanishes. The same pattern is also obtained for axial, hexagonal or trigonal symmetry, if $\theta = 0°$ corresponds to the fine–structure z–axis. There is one c–axis in the crystal, which all centers are oriented along. Figure 3.12b shows the calculated EPR angular dependence for $D/g_e\mu_B = 35$ mT, that is the fine–structure energy is approximately 10% of the electron *Zeeman* energy. The single crossing point for $\theta = 54.7°$

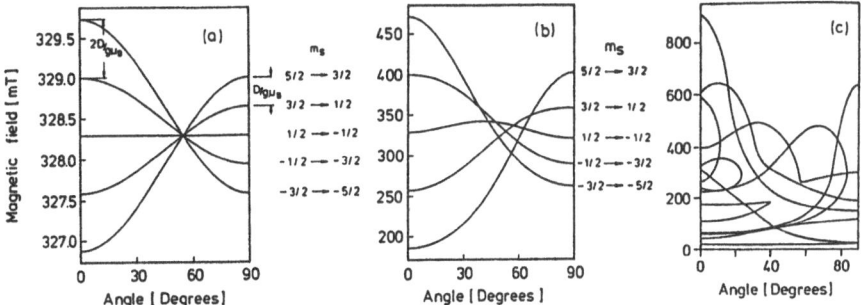

Fig. 3.12a–c. Angular dependence of the EPR spectrum of a $S = 5/2$ system (e.g., Fe^{3+}) in a tetragonal field. The angle θ is between $\mathbf{B}_0$ and the tetragonal axis. $B_0 = 328$ mT. g is assumed to be isotropic. **a** $D/g_e\mu_B \ll B_0$ $(D/g_e\mu_B = 0.35$ mT). First order perturbation theory suffices. **b** $D/g_e\mu_B \ll B_0$ $(D/g_e\mu_B = 35$ mT). First order theory does not suffice. Note, that there is no single crossing point any more for $\theta = 54.7°$. **c** $D/g_e\mu_B \approx B_0$ $(D/g_e\mu_B \approx 143$ mT). The electron quantization is largely influenced by the fine–structure interactions. See text

of all lines has vanished, and the transition $m_S = 1/2 \Rightarrow m_S = -1/2$ is no longer independent of angle. Still, the typical pattern of five fine structure transitions is recognized. In Fig. 3.12c, the transitions are calculated for $D/g_e\mu_B = 143$ mT, a fine–structure energy of about 40% of the electron Zeeman energy. The pattern becomes very complicated through the occurrence of "forbidden" transitions $(|\Delta m_S| > 1)$. The quantization of the electron spin is largely influenced by the fine–structure interaction; the m_S values $5/2, 3/2$, etc., are not good quantum numbers any more. The angular dependences of Figs. 3.12b,c were calculated by numerical diagonalization of the spin *Hamiltonian* (3.4.1). This fine–structure quantization scheme for the unpaired electron spin will have an influence on the analysis of the ENDOR spectra, and will be discussed there. EPR results on Fe^{3+} defects in $LiNbO_3$ are shown in Fig. 3.13, where, indeed, a case similar to the calculations of Fig. 3.12c was found experimentally. The analysis of the spectrum confirmed $S = 5/2$ of Fe^{3+}, $g \approx 2$, and an almost perfect axial symmetry of the fine–structure with $D/g_e\mu_B = 59.6$ mT. A similar result with an even larger D value was recently found for Fe^{3+} in $LiTaO_3$. The site of Fe^{3+} could only be determined by ENDOR [3.10–12].

If $D/g_e\mu_B$ is about as big as the *Zeeman* energy, then the spectra again become a little less complicated, since several forbidden transitions disappear. What is not seen in Figs. 3.12a–c is that the line intensities vary greatly. They are dependent on the fine–structure interactions.

Thus, the observation of a fine–structure splitting in the EPR spectra yields information on the spin multiplicity and, thus, often the charge state of a defect, as well as on the symmetry of the wave function of the defect.

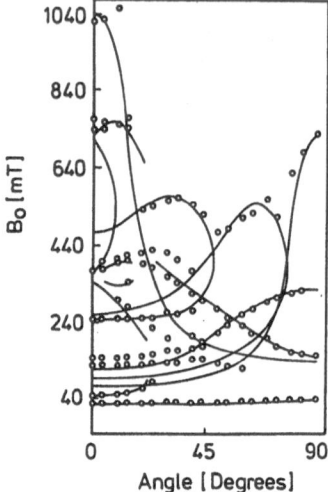

Fig. 3.13. Angular dependence of Fe^{3+} in $LiNbO_3$ single crystals. The *open circles* are the experimental data. The *solid lines* are calculated with an exact diagonalization of the spin *Hamiltonian*. $D/g_e\mu_B = 59.6$ mT, $\nu_0 = 8.9$ GHz, $g = 1.995$. (After [3.10,11])

Sometimes, as in the simple case of the F_3 centers in alkali halides, the theoretical interpretation of D also yields structure information: in that case, the distances between the three F centers can be inferred from the dipole–dipole interaction. In general, however, the interpretation of D is rather difficult.

Although there are general arguments for the occurrence or non–occurrence of fine–structure splittings (in particular [3.9]), like those based on symmetry, such arguments do not predict the size of the fine–structure splitting. Therefore, one often cannot resolve the fine–structure splitting if the EPR line width is too large compared to $(D/g_e\mu_B)$. Therefore, one cannot always easily determine the spin state of a defect, and thus its charge state, from the EPR spectra. Figure 3.14 shows, as an example, the EPR spectrum of Ni^{3+} in GaP [3.13]. Ni^{3+} occupies a Ga site and has the configuration $3d^7$. There is no resolved fine–structure splitting, and the spectrum looks as if it arises from an $S = 1/2$ defect. The structure indicated in Fig. 3.14a is a partially resolved superhyperfine structure due to an interaction with the four nearest ^{31}P neighbors. In fact, it is shown by ENDOR, that indeed, S has the value $3/2$ as expected. The fine–structure splitting is too small to be resolved [3.14].

The observation of the fine–structure splitting as a function of temperature can give very interesting information on the changes of the crystalline environment of an impurity with temperature. An example of this is the investigation of Mn^{2+} defects in $RbCdF_3$, a fluoride perovskite lattice,

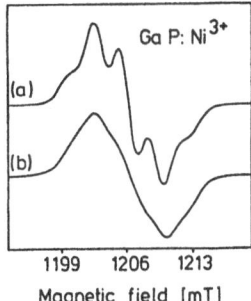

Fig. 3.14a,b. EPR spectrum of Ni^{3+} ($3d^8$, $S = 3/2$) on Ga sites in GaP. a $B_0 \parallel [100]$. b $B_0 \parallel [110]$. There is no fine–structure splitting observable. The partly resolved structure for $B_0 \parallel <100>$ is due to superhyperfine interactions with four equivalent nearest ^{31}P neighbors. (After [3.13])

which undergoes a structural phase transition from cubic to tetragonal symmetry at $T_c = 124$ K. In the cubic phase above 124 K, the Mn^{2+} spectrum (Fig. 3.15a) shows a pronounced sextet splitting due to a ^{55}Mn hyperfine structure and a ^{19}F superhyperfine structure due to the six nearest ^{19}F neighbors, see Sects. 3.5,6 for hyperfine and superhyperfine structure. There is only a weak crystal field term which is proportional to the fourth power of the spin operator, characteristic of a cubic (O_h^1) symmetry (for details see [3.2]). The different crystal field transitions are nearly superimposed. The spectrum was first analysed by *Rousseau* [3.15]. For temperatures below 124 K, the spectrum shows a dramatic change (Fig. 3.15b). There is now an axial fine–structure splitting, which leads to new transitions clearly seen at the low and high field sides of the spectrum. The small lines at around 290 mT and 372 mT are due to the fine–structure transitions $m_S = \pm 5/2 \Rightarrow \pm 3/2$, and the strong lines around 300 and 360 mT are those for $m_S = \pm 3/2 \Rightarrow \pm 1/2$. B_0 is parallel to the tetragonal axis of the crystal (which is the z–axis of the fine–structure tensor D). D increases from 0 for $T > T_c$ to 103 MHz for $T = 35$ K (for details see [3.16]).

Thus, the occurrence or changes of the fine–structure splittings yield information on the symmetry changes of the host crystal upon temperature variation. There are many examples in the literature where paramagnetic centers were used as probes to study structural phase transitions [3.17].

3.5 Hyperfine Splitting of EPR Spectra

The hyperfine (hf) interaction is the magnetic interaction between the magnetic moment of the unpaired electron or hole, and the magnetic moments of the nuclei. This interaction is described by the hf tensor $\tilde{A}$. If there is

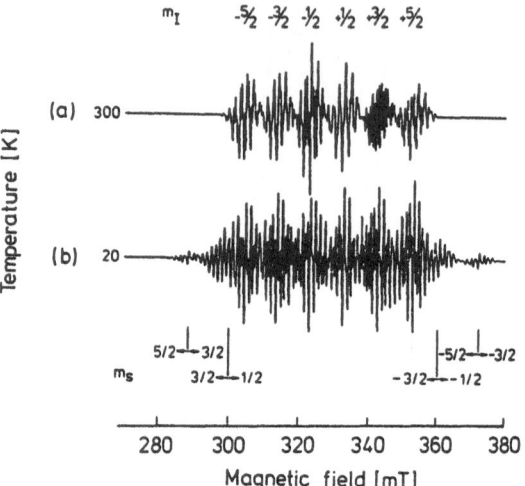

Fig. 3.15a,b. EPR spectrum of Mn^{2+} probes $(3d^5, S = 5/2)$ in $RbCdF_3$. $\mathbf{B}_0 \parallel [100]$. $\nu_0 = 9.26$ GHz. **a** $T > T_c$: cubic phase $(T = 300$ K) **b** $T < T_c$: tetragonal phase $(T = 20$ K) $T_c = 124$ K; $m_I(^{55}Mn) = 5/2$. (After [3.16])

hf interaction with the central nucleus of an impurity atom, then according to (3.1.2–7) the spin *Hamiltonian* becomes, for $S = 1/2$ (that is, no fine–structure interaction is included) and an isotropic g factor,

$$\mathcal{H} = \mu_B g \mathbf{B}_0 \cdot \mathbf{S} + \mathbf{I}\tilde{A}\mathbf{S} - g_n \mu_n \mathbf{B}_0 \cdot \mathbf{I}. \tag{3.5.1}$$

The second term in (3.5.1) is the hf interaction. The last term describes the nuclear *Zeeman* interaction. It is convenient to decompose the hf tensor $\tilde{A}$ into an isotropic and anisotropic part [3.18,2],

$$\tilde{A} = (a\tilde{\mathbf{1}} + \tilde{B}). \tag{3.5.2}$$

The scalar term a is the *Fermi* contact term. The anisotropic hf tensor $\tilde{B}$ is traceless and is often described in its principal axis system by two anisotropic hf interaction constants b and b'.

$$\tilde{B} = \begin{pmatrix} -b + b' & & \\ & -b - b' & \\ & & 2b \end{pmatrix}. \tag{3.5.3}$$

It follows that

$$b = \frac{1}{2}B_{zz} \tag{3.5.4}$$

$$b' = \frac{1}{2}(B_{xx} - B_{yy}). \tag{3.5.5}$$

b' describes the deviation from axial symmetry.

For the following discussion, it is useful to mention the connection between the hf interaction constants and the electronic structure of the paramagnetic electron. In the very simple approximation that the wave function of the unpaired electron or hole can be described by a one–particle wave function, such as a single orbital or an expansion of orbitals all belonging to the same spin (LCAO–approach), then there is a rather simple interpretation of the hf interaction constant. The *Fermi* contact term or isotropic hf interaction at a nucleus occupying a site r_l is:

$$a_l = \frac{2}{3}\mu_0 g_e \mu_B g_n \mu_n |\psi(r_l)|^2, \tag{3.5.6}$$

where $\psi(r)$ is the wave function of the defect in the one–particle approximation. Thus, the *Fermi* contact term describes the unpaired spin density at the nuclear site. It is easily seen that for a defect in which the unpaired electron is in an s–orbital, a_l is non–vanishing, while for all other orbitals, which have a node at the origin, a_l vanishes. The anisotropic tensor elements are given by:

$$B_{ij} = \frac{\mu_0}{4\pi} g_e \mu_B g_n \mu_n \int \left(\frac{3x_i x_j}{r^5} - \frac{\delta_{ij}}{r^3} \right) |\psi(r)|^2 dV \tag{3.5.7}$$

Thus, the anisotropic hf constants b and b' according to (3.5.4,5) essentially describe the average over $(1/r^3)$ calculated with the wave function of the defect, and reflect how the wave function falls off radially.

For the analysis of hyperfine structure and defect structure determination, it is important to realize that all hf constants are proportional to g_n. Therefore, the hf constants of different isotopes of the same element should be in the ratio of their respective g_n values. This is an important tool for chemical identification of impurities from the EPR spectra. If the hf tensor $\tilde{B}$ is in the principal axis system, it is easily seen from (3.5.3) that $\sum_i B_{ii} = 0$. Equations (3.5.6,7) can strictly be derived only from quantum mechanics (for details see Chap. 7). It is, however, often said that the hf interaction is a magnetic dipole–dipole interaction between electron and nucleus. It is a point dipole–dipole interaction only if the unpaired electron and the nucleus with which the hf interaction exists, are far away from each other and, therefore, found only for superhyperfine interactions.

Nevertheless, it is instructive to see how one can establish a connection between the classical dipole–dipole interaction and the hf interaction. The classical dipole–dipole interaction is:

$$W_{DD} = -3\frac{(\boldsymbol{\mu}_1 \cdot \mathbf{r}_{12})(\boldsymbol{\mu}_2 \cdot \mathbf{r}_{12})}{r_{12}^5} + \frac{\boldsymbol{\mu}_1 \cdot \boldsymbol{\mu}_2}{r_{12}^3}, \tag{3.5.8}$$

r_{12} is the distance between the two magnetic moments $\boldsymbol{\mu}_1$ and $\boldsymbol{\mu}_2$. Since $\boldsymbol{\mu}_1 = \boldsymbol{\mu}_I = g_n \mu_n \mathbf{I}$ and $\boldsymbol{\mu}_2 = \boldsymbol{\mu}_s = -g_e \mu_B \mathbf{S}$ one obtains:

$$\mathcal{H}_{DD} = \frac{g_e \mu_B g_n \mu_n}{r^3} \left[\frac{3(\mathbf{I} \cdot \mathbf{r})(\mathbf{S} \cdot \mathbf{r})}{r^2} - \mathbf{I} \cdot \mathbf{S} \right]. \tag{3.5.9}$$

r is the connection vector between electron and nucleus. If one now realizes that the unpaired electron does not have a point dipole moment, but rather can be extended in space, one must integrate (3.5.9) over the electron distribution:

$$\mathcal{H}_{HF} = \langle \psi(r) | \mathcal{H}_{DD} | \psi(r) \rangle$$

$$= g_e \mu_B g_n \mu_n \int \frac{1}{r^3} \left[\frac{3(\mathbf{I} \cdot \mathbf{r})(\mathbf{S} \cdot \mathbf{r})}{r^2} - \mathbf{I} \cdot \mathbf{S} \right] |\psi(r)|^2 dV \quad (3.5.10)$$

$$= g_e \mu_B g_n \mu_n \Sigma_{i,j} \mathbf{I}_i \mathbf{S}_j \int \frac{1}{r^3} \left(\frac{3x_i x_j}{r^2} - \delta_{ij} \right) |\psi(r)|^2 dV,$$

$$\mathcal{H}_{HF} = \Sigma_{i,j} B_{ij} \mathbf{I}_i \mathbf{S}_j + \tilde{A}(r=0) = \mathbf{I}\tilde{B}\mathbf{S} + \tilde{A}(r=0). \quad (3.5.11)$$

Comparison of (3.5.11) with (3.5.8) shows that the anisotropic hf interaction can be understood as a magnetic dipole–dipole interaction between electron and nucleus. If $\psi(r)$ becomes a δ–function then we have the classical point dipole–dipole interaction. The term $\tilde{A}(r=0)$ in (3.5.11) represents the *Fermi* contact term, which can only be understood with relativistic quantum mechanics. For a rigorous derivation of the hf interaction constants from the *Dirac* equation and further interpretation, see Chap. 7.

Often hf spectra are analysed in terms of the principal values of the hf tensor $\tilde{A}$. For axial symmetry the *Hamiltonian* (3.5.10) becomes

$$\mathcal{H}_{HF} = A_{\parallel} I_z S_z + A_{\perp} (I_x S_x + I_y S_y)$$

$$= a\mathbf{I} \cdot \mathbf{S} + b(3 I_z S_z - \mathbf{I} \cdot \mathbf{S}). \quad (3.5.12)$$

The different hf interaction parameters are related to each other:

$$A_{\parallel} = a + B_{\parallel} = a + 2b, \quad (3.5.13)$$

$$A_{\perp} = a + B_{\perp} = a - b. \quad (3.5.14)$$

Supposing only an isotropic hf interaction, an isotropic g factor and a hf interaction which is small compared to the electron *Zeeman* interaction, then (3.5.1) can be solved in perturbation theory of first order and for the energies one obtains:

$$E = g_e \mu_B B_0 m_S + a m_I m_S - g_n \mu_n B_0 m_I. \quad (3.5.15)$$

The wave functions are $\psi = |m_S\rangle |m_I\rangle$. For $S = 1/2, I = 1/2$ there are now four energy levels instead of two without the hf interaction because of the $m_S = \pm 1/2$ and $m_I = \pm 1/2$ quantum numbers (Fig. 3.16a).

Application of an oscillating magnetic field perpendicular to B_0 (along x) results in the time dependent perturbation operator

$$\mathcal{H}_{rf} = (g_e \mu_B S_x + g_n \mu_n I_x) 2 B_1 \cos(2\pi \nu t). \quad (3.5.16)$$

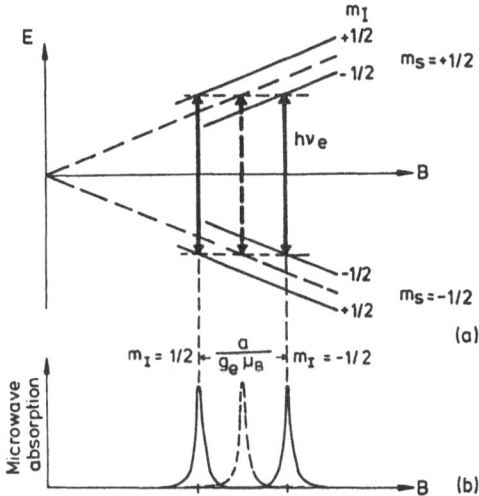

Fig. 3.16. **a** Electron *Zeeman* levels for an electron spin $S = 1/2$ with hyperfine coupling to a nuclear spin $I = 1/2$ and the resulting EPR transitions. The dashed arrow indicates the EPR transition without hyperfine interaction. **b** Hyperfine split EPR transition as a function of magnetic field

According to the discussion of magnetic dipole transitions in Chap. 2, either an EPR transition or a NMR transition will be induced, provided one or the other resonance condition $h\nu_e = g_e\mu_B B_0$ or $h\nu_n = g_n\mu_n B_0$ is fulfilled.

Application of a microwave field induces EPR transitions with the selection rule

$$\Delta m_S = \pm 1, \Delta m_I = 0. \qquad (3.5.17)$$

Since the microwave frequency is kept constant, the two transitions can be detected by variation of the static magnetic field B_0 (Fig. 3.16a,b). Instead of one transition observed without hf interaction, there are now two lines, the resonance fields of which follow easily from (3.5.15):

$$B_0(m_I = -1/2) = (h\nu_e + a/2)/g_e\mu_B,$$

$$B_0(m_I = +1/2) = (h\nu_e - a/2)/g_e\mu_B, \qquad (3.5.18)$$

The field separation of the two hf split lines $\Delta B = a/g_e\mu_B$. The assignment of the two hf lines to the m_I quantum numbers (Fig. 3.16) depends on the sign of a. If $a < 0$, it would be just the reverse. From the EPR spectra, one cannot decide on the sign of the isotropic hf constant from such a splitting (see further discussions in Chaps. 5,6).

As an example, for an hf–split EPR spectrum having the features just described, Fig. 3.17 shows that of Te$^+$ impurities in silicon [3.19], 92% of the

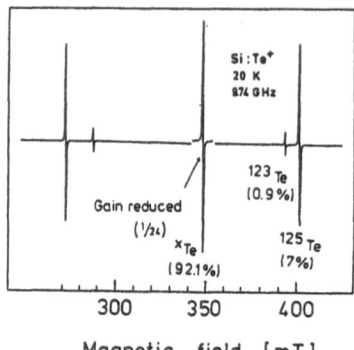

$$\text{Magnetic field [mT]}$$

Fig. 3.17. EPR spectrum of Te$^+$ impurities in silicon. The g factor is isotropic. The hyperfine split doublets are due to the two Te isotopes ^{125}Te and ^{123}Te with $I = 1/2$. The strong central line is due to the nonmagnetic Te isotopes. The hyperfine splitting is isotropic (X–band). (After [3.19])

Te isotopes are nonmagnetic and give rise to the central line near 350 mT (measured in X–band) corresponding to $g = 2$. The two magnetic isotopes ^{125}Te (7% abundant) and ^{123}Te (0.9% abundant) both have $I = 1/2$, and, therefore, should exhibit a doublet splitting. This is observed. The ratio of the line intensities approximately follows the isotope abundances. The ratio of the two hf interaction constants $a(^{125}\text{Te})/a(^{123}\text{Te}) = 1.2056$, $a(^{125}\text{Te}) = 1164.7 \times 10^{-4}$ cm^{-1}, $a(^{123}\text{Te}) = 966.1 \times 10^{-4}$ cm^{-1} is exactly that of the respective g_n values ($g_n(^{125}\text{Te}) = -1.7766$; $g_n(^{123}\text{Te}) = -1.4736$). Therefore, the hf splitting is used to identify the chemical nature of the impurity as being one Te atom. However, since further interactions are not resolved, it cannot be decided from the EPR spectrum, which site is occupied by the Te$^+$. To decide this, one has to perform ENDOR experiments. The positive charge follows from the simple argument that of the six outer electrons, Te needs four electrons for the bonding. It is, therefore, paramagnetic in its singly ionized state. By deep level transient spectroscopy (DLTS) and optical spectroscopy it is shown to be a double donor. The charge states Te^0 and Te^{++} are diamagnetic [3.19].

In favorable cases the hf interaction can reveal the presence of an aggregate center. For example, in Si one can form (S–S)$^+$ and (Se–Se)$^+$ aggregate centers.

Figure 3.18 shows the EPR spectrum for $\mathbf{B}_0 \parallel [100]$ for Se$^+$ and Se$_2^+$ defects in Si. ^{77}Se has $I = 1/2$ and is 7.6% abundant. All other isotopes are nonmagnetic. For the Se$^+$ defects the characteristic doublet hf splitting of ^{77}Se is seen (outermost lines). For Se$_2^+$ the probability that the unpaired electron will interact with the magnetic ^{77}Se is twice as high as for the isolated Se$^+$, which is observed as the intensity ratio between the hf split lines. Also the

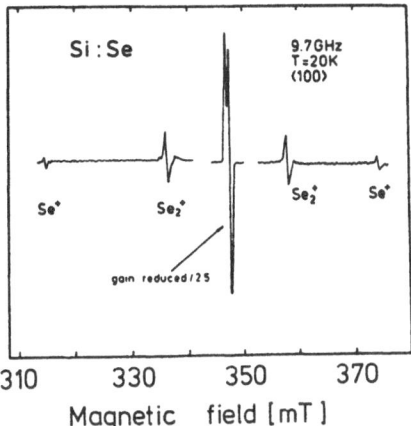

Fig. 3.18. EPR spectrum of selenium pairs (Se$_2^+$) and isolated Se$^+$ impurities in silicon. $\mathbf{B}_0 \parallel [100]$. The hyperfine splitting is due to the magnetic isotope ^{77}Se with 7.6% abundance. The strong central lines are due to the nonmagnetic isotopes (X–band). (After [3.20])

intensity ratio of the central pair line due to the nonmagnetic Se isotopes to the hf split pair lines is accordingly ($\approx$ 7:1). Thus, the hf interaction proves the presence of two equivalent impurities interacting with the unpaired electron. The hf splitting of the pair center is approximately half that of the isolated defect. This is roughly understood, since the unpaired electron is distributed approximately equally between the two equivalent Se nuclei [3.20]. Whether or not the Se pair is composed of two substitutional or interstitial Se atoms cannot be decided from the EPR spectrum. In a recent ENDOR investigation it could be shown that indeed both Se$_2^+$ and S$_2^+$ defects are substitutional aggregate centers [3.21].

In general, the hf interaction is anisotropic and the spectrum depends on the angle. For simplicity we consider an isotropic g factor and axial symmetry for the hf interaction in the approximation $b \ll a$. Then the spin *Hamiltonian* becomes:

$$\mathcal{H} = \mathcal{H}_0 + \mathcal{H}' + \mathcal{H}'',$$

$$\mathcal{H}_0 = g_e \mu_B B_0 S_z - g_n \mu_n B_0 I_z + a I_z S_z, \tag{3.5.19}$$

$$\mathcal{H}' = a(I_x S_x + I_y S_y),$$

$$\mathcal{H}'' = 2b I_{z'} S_{z'} - b(I_{x'} S_{x'} + I_{y'} S_{y'}).$$

x', y', z' is the principal axis system of the hf tensor, and x, y, z is the laboratory system with $\mathbf{B}_0 \parallel z$. The eigenfunctions of $\mathcal{H}_0$ are $|m_S\rangle |m_I\rangle$. $\mathcal{H}''$ does not cause energy shifts in first order. $\mathcal{H}'$ must be transformed into the laboratory frame x, y, z in order to calculate the diagonal matrix elements for the first order solution. With

$$\left. \begin{array}{l} I_{x'} = I_x \cos(\theta) - I_z \sin(\theta), \\ I_{y'} = I_y, \\ I_{z'} = I_z \cos(\theta) + I_x \sin(\theta), \end{array} \right\} \tag{3.5.20}$$

and similar expressions for $S_{x'}$, $S_{y'}$, $S_{z'}$ with θ the angle between (z', z), one obtains

$$E^{(1)} = g_e \mu_B B_0 m_S - g_n \mu_n B_0 m_I + \underbrace{\{a + b[3\cos^2(\theta) - 1]\}}_{A(\theta)} m_S m_I. \tag{3.5.21}$$

This is the simplest solution for the anisotropic hf interaction. Very often, however, b is not small compared to a. A more precise solution, which is obtained similarly to that for the g factor (Sect. 3.3), replaces $A(\theta)$ in (3.5.21) by ([3.2])

$$A(\theta) = \sqrt{A_\parallel^2 \cos^2(\theta) + A_\perp^2 \sin^2(\theta)}. \tag{3.5.22}$$

In general, there is a g tensor and not just an isotropic g factor, and a hf interaction which may be of the same order as the *Zeeman* interaction. The best way to solve the spin *Hamiltonian* is a numerical diagonalization. However, if the hf interaction is still small compared to the *Zeeman* interaction, a simple expression can be derived for the angular dependence, if the g tensor and the hf tensor have the same principal axis system x, y, z:

$$E = g_e \mu_B B_0 m_S + K m_S m_I, \tag{3.5.23}$$

with

$$g = \sqrt{g_{xx}^2 l^2 + g_{yy}^2 m^2 + g_{zz}^2 n^2}, \tag{3.5.24}$$

$$K = \frac{1}{g}\sqrt{A_{xx}^2 g_{xx}^2 l^2 + A_{yy}^2 g_{yy}^2 m^2 + A_{zz}^2 g_{zz}^2 n^2}. \tag{3.5.25}$$

l, m, n are the direction cosines between the magnetic field orientation and the principal axis system. Equation (3.5.22) follows from (3.5.25) for axial symmetry and an isotropic g factor. As an example of an angular dependent hf structure, Fig. 3.19 shows the EPR spectra of $Tl^0(1)$ and $Tl^0(2)$ centers in KCl. They are both axial centers, with the z–axis of both the g and hf tensor along a <100> orientation. Figure 3.19a shows the EPR spectrum for $B_0 \parallel [100]$; Fig. 3.19b the angular dependence for rotation of the magnetic field in a {100} plane. In order to facilitate the discussion, Fig. 3.20 shows the models for the two centers as derived from the EPR spectra, and a theoretical analysis of the hf data [3.22]. $Tl^0(1)$ centers are Tl^0 atoms next to one chloride vacancy (Fig. 3.20a), while $Tl^0(2)$ centers have two adjacent vacancies

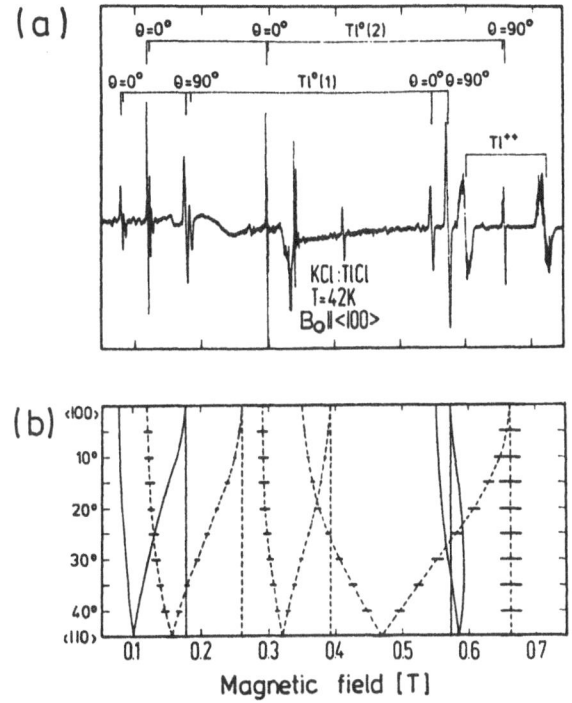

Fig. 3.19. a EPR spectra of $Tl^0(1)$ and $Tl^0(2)$ defects in KCl for $B_0 \parallel [100]$. The angles indicate the center orientation with respect to the orientation of B_0. **b** Calculated angular dependence in a {100} plane of the EPR spectra of $Tl^0(1)$ and $Tl^0(2)$ centers. *Solid lines:* $Tl^0(1)$; *dashed lines:* $Tl^0(2)$. The *horizontal bars* indicate the transition probabilities for $Tl^0(2)$. The intensity variation of $Tl^0(1)$ centers is only small. (After [3.22])

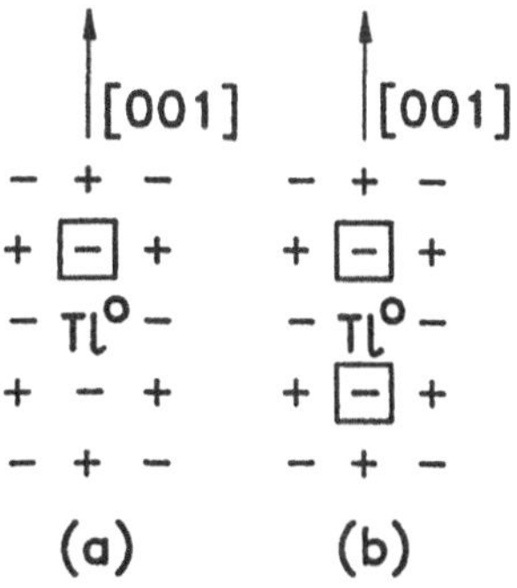

$$
\begin{array}{cc}
\uparrow [001] & \uparrow [001] \\
-\ +\ - & -\ +\ - \\
+\ \boxed{-}\ + & +\ \boxed{-}\ + \\
-\ Tl^0\ - & -\ Tl^0\ - \\
+\ -\ + & +\ \boxed{-}\ + \\
-\ +\ - & -\ +\ - \\
\text{(a)} & \text{(b)}
\end{array}
$$

Fig. 3.20. Center models for $Tl^0(1)$ centers **a** and $Tl^0(2)$ centers **b**. The *squares* symbolize a halide vacancy. (After [3.22])

(Fig. 3.20b). Both centers are produced by room temperature X–irradiation of KCl doped with Tl^+. In addition, Tl^{++} defects are created. Because of the axial center symmetry there are three center orientations present simultaneously in the crystal. For $B_0 \parallel [100]$ one center orientation is parallel and two are perpendicular to B_0. This is indicated by 0° and 90° in Fig. 3.19a above the lines. Upon rotating the magnetic field in a {100} plane, the line of those centers with orientation perpendicular to the rotation plane remain in the same field position (vertical lines in Fig. 3.19b). For $B_0 \parallel [110]$, the angle between B_0 and the z–axis of two centers is 45°, therefore, the lines of the two center orientations coincide. The angular dependences in Fig. 3.19b are typical "(100)–patterns", which both centers have. Tl has the two isotopes ^{205}Tl and ^{203}Tl with $I = 1/2$ and slightly different g_n values (^{205}Tl: $g_n = 3.2754$, 70.5% abundant and ^{203}Tl: $g_n = 3.2445$, 29.5% abundant). The typical hf doublet splitting for $I = 1/2$ is seen in the spectrum. The slight difference in g_n also shows up (see, for example, the lowest field lines in Fig. 3.19a). Since both $g_\parallel$ and $g_\perp$ differ greatly from each other and from $g = 2$, and there is also a large anisotropic hf interaction, the spin *Hamiltonian* has to be numerically diagonolized in order to describe the measured angular dependences [3.22].

From EPR it is clear that all centers have (100) axial symmetry, and from the production mechanism it seemed most likely that a halide vacancy is nearby to cause the axial symmetry. However, it cannot be said from the EPR spectra alone whether the vacancies are empty or, for example, filled by a nonmagnetic impurity like oxygen. It cannot be decided which of the models of Fig. 3.20 belongs to either one or the other hf split EPR line pattern with (100) symmetry. This assignment was made on the basis of the theoretical interpretation of the measured g values and hf interaction constants [3.22]. It is also not possible to decide whether the Tl^0 is relaxed towards the vacancy and, if so, how much. This can only be done by ENDOR [3.23].

Figure 3.19a shows another feature of conventional EPR spectroscopy. Since the ionizing radiation producing these centers creates many defects simultaneously, it is not easy to assign all the EPR lines to particular centers unambiguously and to follow their angular dependence. Similar experiments with Pb doping in alkali halides yielded spectra which contain even more lines. It will be shown in Chap. 4 that with optical detection of EPR each defect can be measured selectively, which greatly facilitates the analysis.

3.6 Superhyperfine Splitting of EPR Spectra

The magnetic interaction between the unpaired electron and the magnetic nuclei of the surrounding lattice is called superhyperfine (shf) or ligand hyperfine (lhf) interaction. Its physical origin is, of course, the same as discussed in the previous section. For the discussion of defects in solids it is, however, convenient to deal with the shf interaction in a separate section. Because of the large number of nuclei and their special symmetry conditions, the shf splittings of the spectra are often the most powerful source of information on the defect structure from an EPR spectrum. If the interaction occurs with too many nuclei with comparable shf interactions, then no shf structure is resolved, and one is left with the broad EPR lines typical of many defects in solids. The line width problem will be discussed in the following section. Assuming, for simplicity, an isotropic defect and $S = 1/2$, then the spin *Hamiltonian* (3.5.1) must be extended to be

$$\mathcal{H} = g_e \mu_B \mathbf{B}_0 \cdot \mathbf{S} + \sum_i (\mathbf{I}_i \tilde{A}_i \mathbf{S} - g_{n,i} \mu_n \mathbf{B}_0 \cdot \mathbf{I}_i) \qquad (3.6.1)$$

The sum runs over all nuclei with which a shf interaction is measured. The sum can, of course, also include a central nucleus, which usually has a larger interaction than the neighbor nuclei. The spectrum would then show hf split lines due to the central nucleus, and each hf line would exhibit a further splitting due to the shf interactions (an example will be given below).

Each neighbor nucleus has its own shf tensor, the orientation of which depends on the symmetry of the defect with respect to its center. The size of the shf interactions for a given orientation of $\mathbf{B}_0$, with respect to the lattice, depends on the relative orientations of $\mathbf{B}_0$ and the shf tensors, as seen from (3.5.21) for the case of axially symmetric shf tensors. If the angles θ_i between the magnetic field orientation and the shf tensor axes z_i of several nuclei having the same shf constants are the same, then their shf splitting is the same (it is supposed that the z–axis of the shf principal axis system always has the largest interaction). These neighbor nuclei are said to be "equivalent". The occurrence of equivalent nuclei in the shf structure of solid state defects is a characteristic feature, and the consequence of the high point symmetry often encountered. The presence of equivalent nuclei shows up in first order as a splitting pattern with equally spaced lines, and with a characteristic intensity ratio between the shf split lines. Figure 3.21 shows the energy levels for the cases of one neighbor nucleus and those of two to four equivalent nuclei with $I = 1/2$, and the resulting intensity ratio of the shf lines. For simplicity it is assumed that the nuclear *Zeeman* energy can be neglected compared to the shf interaction energy. It is easily read from the figure that, for example, for four equivalent neighbors with $I = 1/2$, there are five shf lines with the relative intensity ratio 1:4:6:4:1.

As an example, Fig. 3.22 shows the EPR spectrum of Ni^+ centers in CaF_2 for $\mathbf{B}_0 \parallel [100]$. Ni^+ has $S = 1/2$ ($3d^9$ configuration). There are two spectra with five lines each, suggesting a shf interaction with four equivalent nuclei of $I = 1/2$. In CaF_2 this can only be with ^{19}F ($I(^{19}F)=1/2$)). Tetragonal symmetry of the g tensor follows from the angular dependence. The low field group of lines corresponds to $g_{\parallel}$ and the high field group of lines to $g_{\perp}$ ($g_{\parallel} = 2.569$, $g_{\perp} = 2.089$ [3.24]). For $\mathbf{B}_0 \parallel g_{zz}$ and $\mathbf{B}_0 \parallel g_{xx}$ or g_{yy} four F^- must be equivalent to give the observed shf structures of five equidistant lines of

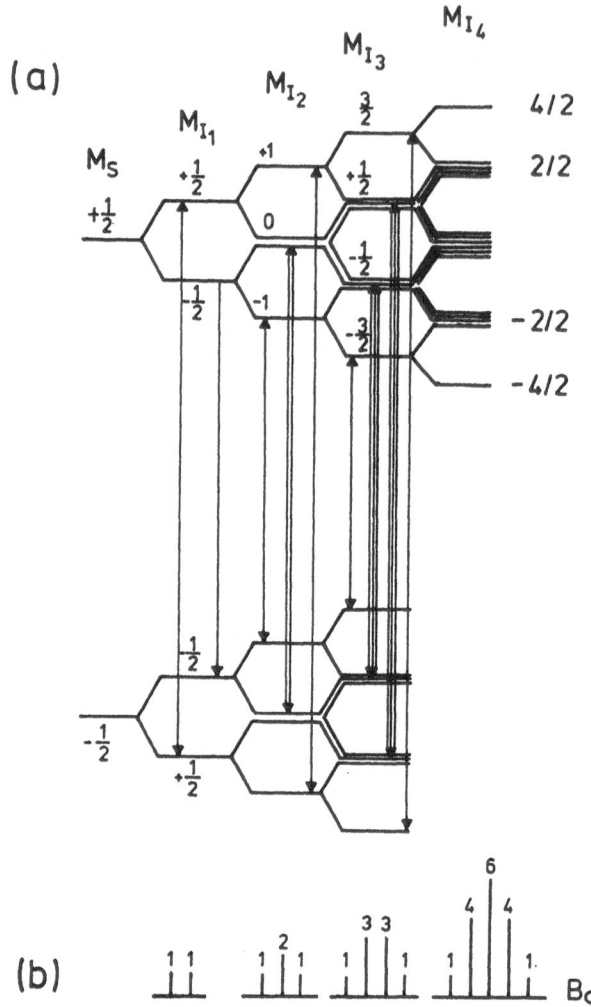

Fig. 3.21. a Electron *Zeeman* level scheme for an electron with superhyperfine coupling to one, two, three and four equivalent nuclei with $I = 1/2$ and the resulting EPR transitions. **b** Stick spectrum of the EPR superhyperfine lines due to a superhyperfine coupling to one, two, three and four equivalent nuclei with $I = 1/2$

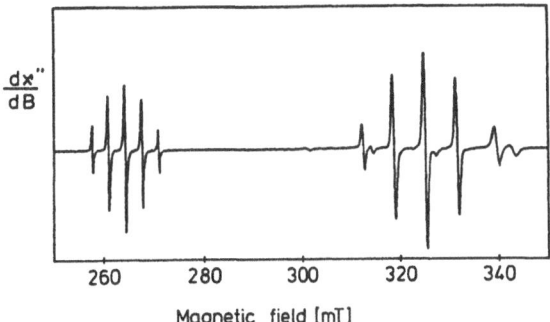

Fig. 3.22. EPR spectrum of tetragonal and Ni^+ centers ($3d^9$ configuration) in CaF_2 for B_0 parallel to a $<100>$ direction. $T = 20$ K. $\nu_{EPR} = 9490$ MHz. The small signals between the lines are due to another type of Ni^+ center. (After [3.24])

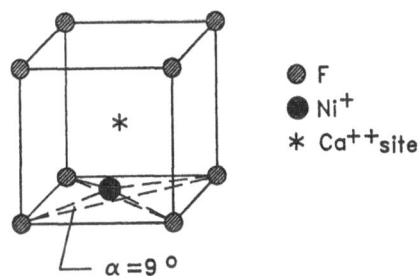

Fig. 3.23. Model of Ni^+ centers in CaF_2. (After [3.24])

the intensity ratio 1:4:6:4:1. At first sight this is surprising. The centers are produced by X–irradiation of CaF_2 doped with Ni^{2+} at room temperature. Ni^+ is formed by capture of an electron. Since Ni^{2+} substitutes for Ca^{2+}, one expects that Ni^+ is located at a body centered Ca^{2+} site, as indicated in Fig. 3.23. For this position one expects eight equivalent F neighbors with a shf splitting into nine equidistant lines. Apparently, such an octahedral site is not stable for a $3d^9$ electronic configuration (or a 3d hole configuration which would be equivalent). The Ni^+ is drawn towards a plane spanned by four F^- (Fig. 3.23). This result is later confirmed by ENDOR. From an interpretation of the measured shf interaction constants of the more distant lattice nuclei it is concluded that the angle α in Fig. 3.23, which describes the position of Ni^+ above the plane, is about $9°$ [3.25].

Figures 3.24a,b show the shf structure for two other Ni^+ centers in K_2MgF_4 doped with Ni^{2+}, which are also formed by X–irradiation. The magnetic field is parallel to the c–axis of the K_2MgF_4 crystal (Fig. 3.25), which has tetragonal symmetry. The center in Fig. 3.24a is produced by X–irradiation at 77 K (center I); the one in Fig. 3.24b is formed by X–irradiation

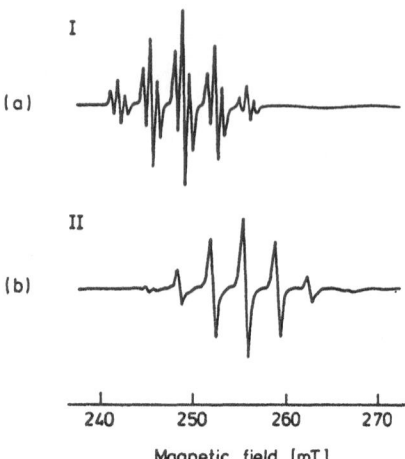

Fig. 3.24a,b. EPR spectrum of Ni^+ centers in K_2MgF_4 measured at 77 K for B_0 parallel to the **c**–axis of the crystal (X–band). **a** Center I, produced by X–irradiation of $K_2MgF_4:Ni^{2+}$ at 77 K **b** Center II, produced by X–irradiation of $K_2MgF_4:Ni^{2+}$ at 300 K and subsequent bleaching with visible light. (After [3.26])

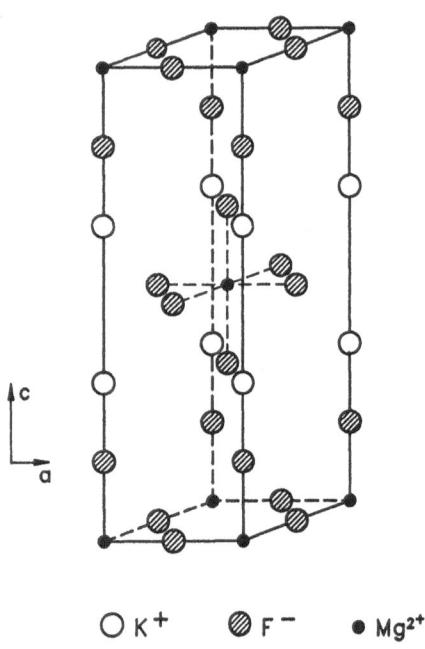

Fig. 3.25. Crystal structure of K_2MgF_4

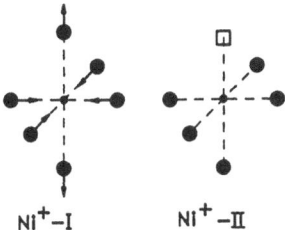

Fig. 3.26. Structure model of Ni^+ centers I and II in K_2MgF_4. (After [3.26])

at room temperature and subsequent bleaching with visible light (center II). Clearly, there is a marked difference in the shf structure. In center I, the unpaired electron shows a larger shf coupling with four equivalent ^{19}F nuclei similar to that of Ni^+ in CaF_2, and in addition, a smaller one with two further equivalent ^{19}F nuclei, causing a splitting of each of the five lines into a triplet with a relative intensity ratio of 1:2:1. The center II shows only the shf splitting due to four equivalent ^{19}F nuclei of about the same size as that for center I for this field orientation.

Figure 3.26 shows the structure models derived for the centers I and II from the shf structure. Only the nearest F^- neighbors are shown. In center I there is a larger interaction with four equivalent equatorial F^- and a smaller one with two axial F^-. This structure is due to a static *Jahn–Teller* distortion of the $3d^9$ configuration in which the orbital degeneracy of the d hole is lifted by a lattice distortion (for example [3.1]). In center II, it is also clear from the shf pattern for $B_0 \parallel c$ that there are four equivalent equatorial F^- neighbors, but apparently, the axial F^- neighbors are missing. One would conclude that there must be two F^- vacancies. However, for B_0 perpendicular to the c-axis and parallel to a $\langle 100 \rangle$ direction, all shf lines show an additional doublet structure with a 1:1 intensity ratio. This shows that one of the axial F^- must still be there, as in Fig. 3.26 [3.26]. Analogous Ni^+ centers were also found in $KMgF_3$[3.27].

The angular dependence of the EPR shf structure is usually rather complicated, especially if there are several center orientations present simultaneously. In Fig. 3.27, the angular dependence is shown for the Ni^+ center of type II in K_2MgF_4 discussed above. All centers are oriented along the c-axis, which makes the angular dependence simple. For $B_0 \parallel c$ there are five shf lines, for B_0 perpendicular to c there are seven lines; and for the intermediate orientations there is a bad resolution. The reason for this is, of course, that for other orientations of B_0, the four equatorial F^- neighbors which are equivalent for $B_0 \parallel c$ do not remain equivalent. For B_0 perpendicular to the c-axis and parallel to a <100> axis (Fig. 3.25) there are two sets of two equivalent F nuclei, which could give up to nine shf lines. How many lines are resolved depends on the ratio of their interaction constants. If the ratio is a small

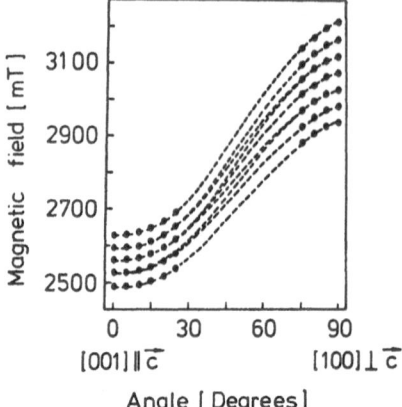

Fig. 3.27. Angular dependence of the EPR spectrum of Ni^+ centers of type II in K_2MgF_4 for rotation of the magnetic field in a $\{100\}$ plane from B_0 parallel to the c–axis to B_0 perpendicular to the c–axis. The *dashed lines* are calculated with the spin *Hamiltonian* parameters (X-band, $T = 77$ K). (After [3.26])

rational number, then one usually observes a resolution as in Fig. 3.27 for $B_0 \parallel [100]$. In fact, for this orientation there is this small additional doublet splitting as mentioned above, which is not depicted in Fig. 3.27. The center of gravity of the shf structure varies as a function of orientation according to the axial g tensor $(g_{[001]} = g_{\parallel} = 2.538, g_{[010]} = g_{\perp} = 2.116$ [3.26]).

For N nuclei with the same shf interaction, which may occur for a special field orientation, one can define a total nuclear spin NI, which leads to a $(2NI + 1)$-fold splitting into equally spaced lines of the EPR spectrum. The relative EPR signal heights are given by the statistical weights or degeneracies of the levels, see also Fig. 3.21. Table 3.1 gives the statistical weights of the states M_I for N equivalent nuclei with $I = 1/2$ and $3/2$. In general, the intensities of the $(2NI + 1)$ equally spaced lines are given by

$$I(M_I) = \frac{1}{(2I+1)^N} \sum_{j=0}^{k} (-1)^j \binom{N}{j} \binom{NI - M_I - j(2I+1) + N - 1}{N - 1},$$

(3.6.2)

where $M_I = \sum_{i=1}^{N} m_i$. k is the largest integer less than or equal to $(NI - M_I)/(2I+1)$. The last two terms in (3.6.2) are the binomial coefficients [3.2].

As a last example in this section, Fig. 3.28 shows the EPR spectrum of atomic hydrogen on interstitial sites (H_i^0 centers), on cation vacancy sites ($H_{s,c}^0$ centers), and on anion vacancy sites ($H_{s,a}^0$ centers) in KCl for B_0 parallel to a [100] crystal axis. The central hf splitting with the proton ($I = 1/2$) in the

Table 3.1. Statistical weights of the states M_I of N equivalent nuclei with spin I

M_I:	0	$\frac{1}{2}$	1	$\frac{3}{2}$	2	$\frac{5}{2}$	3	
$N=1$		1						
2	2		1					$I=\frac{1}{2}$
3		3		1				
4	6		4		1			
5		10		5		1		
6	20		15		6		1	

M_I:	0	$\frac{1}{2}$	1	$\frac{3}{2}$	2	$\frac{5}{2}$	3	$\frac{7}{2}$	4	$\frac{9}{2}$	5	$\frac{11}{2}$	6	$\frac{13}{2}$	7	$\frac{15}{2}$	8	$\frac{17}{2}$	9	
$N=1$		1		1																
2	4		3		2		1													$I=\frac{3}{2}$
3		12		10		6		3		1										
4	44		40		31		20		10		4		1							
5		155		135		101		65		35		15		5		1				
6	580		546		456		336		216		120		56		21		6		1	

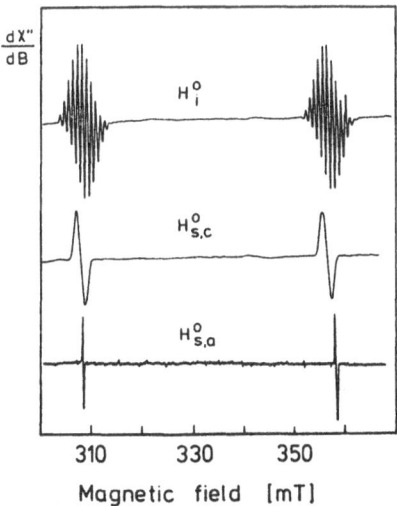

Fig. 3.28. EPR spectrum of three atomic hydrogen centers in KCl. ($\mathbf{B}_0 \parallel [100]$, X–band, $T = 77$ K). The hydrogen atoms occupy *interstitial sites* (H_i^0 centers), *cation vacancy sites* ($H_{s,c}^0$ centers) and *anion vacancy sites* ($H_{s,a}^0$ centers). The splitting caused by the hyperfine interaction with the proton is almost that of the free hydrogen atom. (After [3.30])

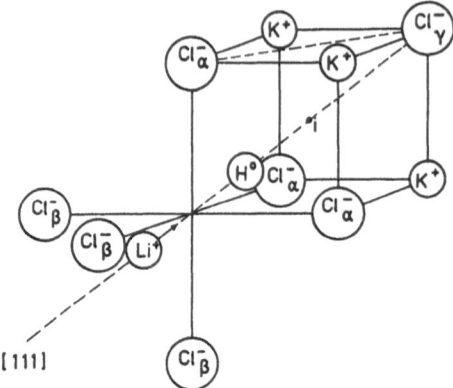

Fig. 3.29. Model of the $H_i^0(Li)$ center in KCl. i denotes the regular interstitial site. The position of H^0 and Li^+ are not drawn to scale. (After [3.31])

three cases is practically the same, and nearly that of the free hydrogen atom. Replacement of H^0 by D^0 results in three hf lines with the same shf structure and line widths, respectively ($I(D)=1$). This proves that one is dealing with hydrogen. Only for the interstitial site is a shf interaction with nearest neighbors resolved. Each proton hf line is split into 13 equally spaced lines with an intensity ratio of approximately 1:4:10:20:31:40:44:40:31:20:10:4:1. Inspection of Table 3.1 confirms that this is due to four equivalent nuclei of $I = 3/2$. The shf structure is due to a prominent interaction with the four tetrahedrally coordinated ^{35}Cl nuclei [3.28]. The substitutional sites cannot be inferred from the EPR spectrum; they can only be established by ENDOR experiments [3.29]. The reason for the lack of the resolved shf structure for these centers is not that there are no shf interactions large enough to be resolved. The reason is that there are too many shf lines superimposed, with rather similar shf splittings, with the result that their resolution is lost. This will be discussed systematically in the next section.

From the foregoing it is clear that a resolved shf structure in the EPR spectra gives one of the most important clues for a defect structure. In particular, if there are equivalent nuclei in highly symmetric defects, one can often deduce a rather precise structure model from the EPR spectrum.

The shf structure can be temperature dependent. For example, because of changes of the lattice parameters as a function of temperature, in EPR small changes of the interaction constants are usually not resolved. However, very clear changes of the shf structure can be observed because of dynamical effects. If, for example, a paramagnetic defect is on an off–center position at low temperature and experiences rapid thermally activated motion between the possible off–center positions at higher temperature, one will observe changes in the shf structure and g tensor. Figure 3.29 shows the structure of an

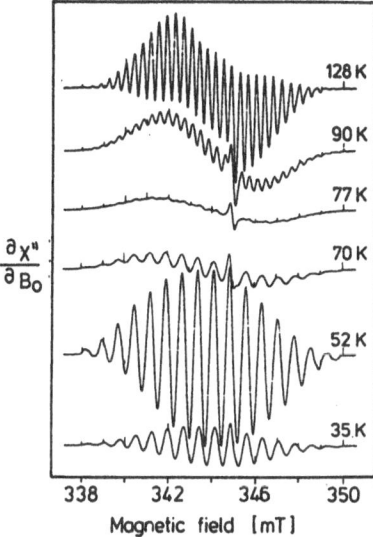

Fig. 3.30. High field hyperfine EPR line groups of $H_i^0(Li)$ centers in KCl for $B_0 \parallel [100]$ at various temperatures. (After [3.32])

interstitial atomic hydrogen center in KCl doped with Li^+ at low temperature. The hydrogen atom is displaced from the interstitial position towards the Li^+ impurity, which substitutes for K^+. Li^+ is also somewhat displaced along a [111] direction. One could say that both H^0 and Li^+ share a cation site. This low temperature model was originally derived from the EPR spectra (see below), and later established by ENDOR experiments [3.31]. Figure 3.30 shows the high field proton hf line for various temperatures for $B_0 \parallel [100]$. At low temperature (52 K), there is a shf structure of 16 equally spaced lines resolved, while at 128 K, 30 equally spaced lines can be distinguished with a splitting, which is about half of that observed at low temperature. For intermediate temperatures, a shf structure is not resolved any more. At low temperature, the g tensor is axially symmetric about [111]; there are eight center orientations in the crystal. At high temperature, the spectra become isotropic [3.32].

The shf structure at low temperature is due to the three equivalent Cl neighbors (Fig. 3.29), $(I(^{35}Cl)=I(^{37}Cl) = 3/2$, abundance of ^{35}Cl is 75.5%, that of ^{37}Cl is 24.5%), and the Li^+ cation $(I(^6Li)=1$, 7.4% abundance, $I(^7Li) = 3/2$, 92.6% abundance). Three equivalent Cl nuclei would lead to ten shf lines. The additional shf interaction with Li^+ of comparable size shows up in the observed number of 16 lines. At 120 K, a fast reorientational motion takes place, in which both Li^+ and H^0 move together like a molecule about the K^+ vacancy site. The shf structure is now due to six equivalent Cl^- neighbors; Cl_α and Cl_β are not distinguishable any more (Fig. 3.29). The interaction with

Li^+ is now isotropic. The Cl shf interaction parameters a and b at 128 K are almost half of the values measured at low temperature. At high temperature, the unpaired electron spends only half of the time at Cl_α, compared to the low temperature configuration. At intermediate temperatures, the frequency of reorientational motion becomes comparable to the shf interaction frequencies and, therefore, shf structure is not resolved any more. At 128 K, the motion is fast compared to the interaction frequencies, and one sees a motional average (for details of motional effects such as motional narrowing see [3.18]).

The interpretation of the shf structure, as discussed, was verified by a computer simulation of the EPR shf structure [3.32]. The interpretation that H^0–Li^+ behave like a molecule and have a separation of 1.44 Å could not be confirmed by ENDOR. The Li^+–H^0 separation came out to be 2.17 Å. The hydrogen atom is displaced from the interstitial site by only about 25% of the regular distance of $\sqrt{3}\, d_0$, d_0 being the lattice spacing [3.31]. However, at 128 K no ENDOR measurements were possible. Thus, the information on the dynamical behavior relies on the EPR spectra.

3.7 Inhomogeneous Line Widths of EPR lines

It is characteristic of EPR spectra of defects in solids that they exhibit a rather large line width which is typically of the order of 1–20 mT, and is often as large as 50 mT. The reason for the line widths is the large number of shf lines, which are all superimposed. The line width of each individual shf line is rather small, and mostly of the order of only 3×10^{-3} mT to 3×10^{-2} mT, or even less, depending on the transverse spin–lattice relaxation time T_2. This line width is called the homogeneous line width and it is seen as the ENDOR line width (Chap. 5). The shf broadened line width is called the inhomogeneous line width.

In order to illustrate the problem of the resolution of shf lines in Fig. 3.31a–c the integrated spectra of 3 different shf structures are calculated, whereby in each spectrum the single ("homogeneous") line width is assumed to be 0.3 mT. A *Gaussian* shape was assumed, the reason being that the single lines are also assumed to be inhomogeneously broadened by still smaller interactions. In Fig. 3.31a the interaction with four equivalent nuclei with $I = 3/2$ and a shf interaction of 3.6 mT is assumed; in Fig. 3.31b an addition to the shf interaction with another four equivalent nuclei with $I = 3/2$ and 0.72 mT interaction is taken into account, while in Fig. 3.31c a third set of four nuclei with $I = 3/2$ and an interaction of 0.15 mT is added. It is clear that in Fig. 3.31c all shf resolution is lost. Only three sets or "shells" of nuclei with decreasing interactions suffice to eleminate the resolution. The interaction decreases by a factor of five from shell to shell. In all three cases, it was assumed that the nuclei have a 100% abundance. If the interaction were to decrease less rapidly, the resolution would be lost for even fewer shells of in-

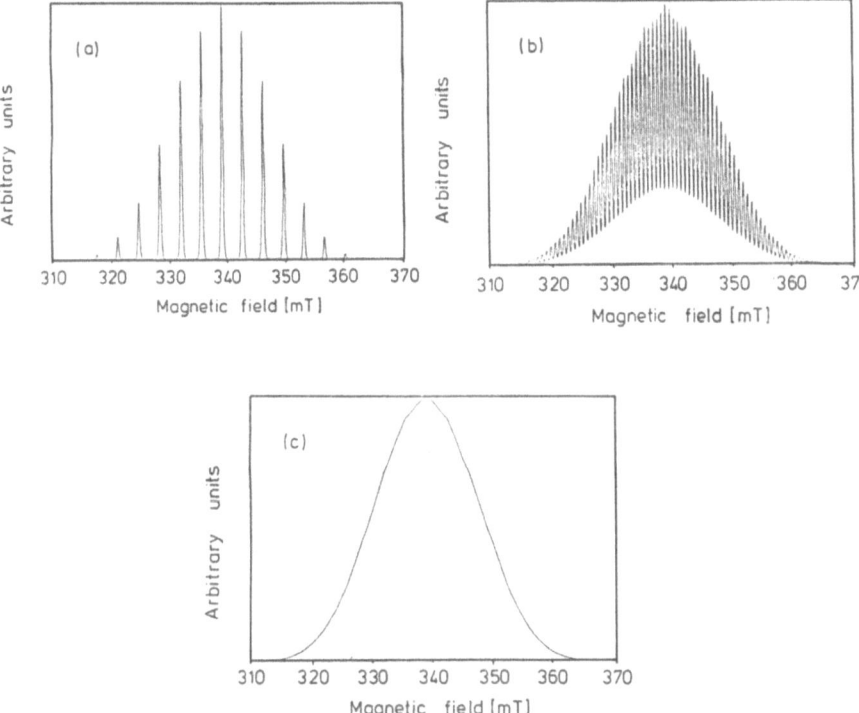

Fig. 3.31a–c. Simulation of EPR spectra with a superhyperfine structure for three different shells of interacting nuclei. In all spectra the line width of a single superhyperfine EPR line was assumed to be 0.3 mT and of *Gaussian* shape. **a** Simulated EPR spectrum for a superhyperfine structure due to four equivalent nuclei of $I = 3/2$ with a 100% abundance and an interaction of 3.6 mT. **b** Simulated EPR spectrum for a superhyperfine structure due to four equivalent nuclei of $I = 3/2$ (100% abundance) with an interaction of 3.6 mT, and four equivalent nuclei with $I = 3/2$ (100% abundance) with an interaction of 0.72 mT. **c** Simulated EPR spectrum of a superhyperfine structure due to the interaction with three shells of four equivalent nuclei with $I = 3/2$ (100% abundance) with the interactions 3.6 mT, 0.72 mT and 0.15 mT, respectively

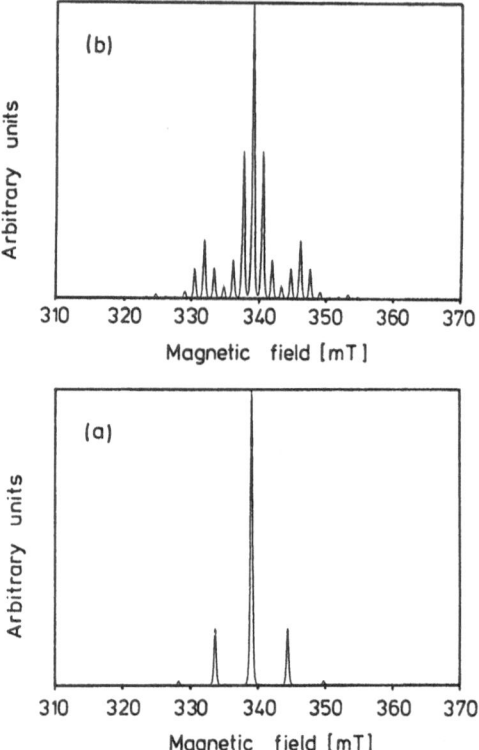

Fig. 3.32a,b. Simulation of the EPR spectrum with superhyperfine structure for two different cases of interacting neighbor nuclei with a small abundance of magnetic nuclei. The magnetic nuclei are assumed to have $I = 1/2$ and 4.7% abundance (as for example ^{29}Si). **a** four equivalent nuclei with an interaction of 10.7 mT. **b** four equivalent nuclei with 10.7 mT and 12 equivalent nuclei with 2.14 mT

teracting nuclei. The superposition of the homogeneous lines with *Lorentzian* shape (2.8.16), according to (3.6.2) results in a *Gaussian* line shape for the inhomgeneous EPR lines.

The line shape of a shf split EPR spectrum looks quite different if the neighbor atoms have nuclei of which only a small fraction are magnetic. An important lattice where this is the case is the silicon lattice, where only 4.7% ^{29}Si are magnetic nuclei with $I = 1/2$. Figure 3.32a,b show two simulations of EPR spectra for typical cases. In Fig. 3.32a, a substitutional paramagnetic defect is assumed which has four equivalent Si neighbors for $\mathbf{B}_0$ parallel to a cubic axis. The outermost shf lines (at 328 and 350 mT, respectively) are very weak compared to the central line (intensity ratio $\approx$ 1:100). A shf coupling to a second shell of 12 neighbors with a weaker interaction (by a factor of 5) splits each of the lines of Fig. 3.32a, where again, the outer lines are hardly

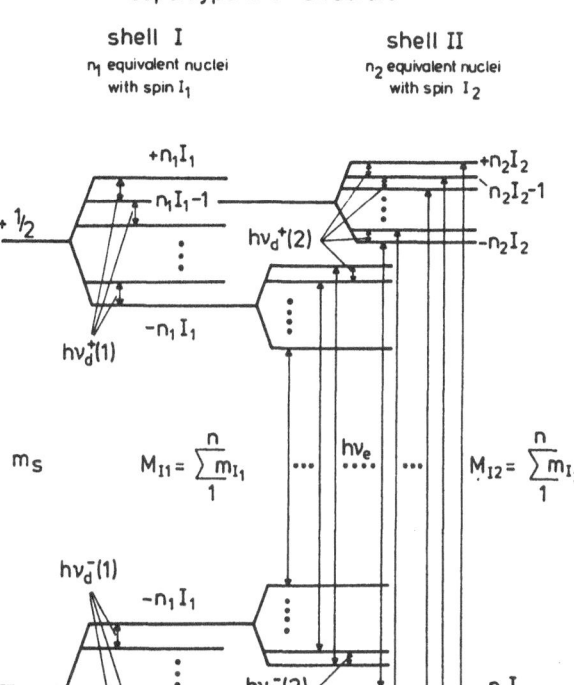

Fig. 3.33. Energy level scheme for an electron with superhyperfine coupling to several shells of equivalent nuclei. The EPR line is an *envelope* over all superhyperfine lines, which is usually not resolved because of the high number of superimposed lines

measurable. This makes a conclusive analysis of the shf structure particularly difficult.

The energy levels associated with the shf coupling to many shells of "equivalent" nuclei are shown schematically in Fig. 3.33, for $S = 1/2$ and 2 shells of equivalent nuclei, each of which contains n_1 and n_2 nuclei with the spins I_1 and I_2, respectively. Those nuclei are called to be members of a shell of nuclei, if their shf parameters are all the same, see Chap. 6 for a more rigorous definition. For a particular orientation of $\mathbf{B_0}$, not all members of a shell need to be equivalent in the sense discussed in the previous section. They are not equivalent if their shf tensor orientations relative to $\mathbf{B_0}$ are not the same.

For both electron *Zeeman* levels there are $N = (2n_1 I_1 + 1)(2n_2 I_2 + 1)$ sublevels. Therefore, there are up to N EPR transitions. For n equivalent nuclei with $I = 1/2$ there are 2^n EPR transitions. The number of EPR transitions rapidly increases with an increasing number of interacting nuclei, resulting in a loss of shf resolution. This is typical for solid state defects. Therefore, EPR is only sufficient in cases favorable for a structure determination.

As will be shown in Chap. 5, one of the great advantages of ENDOR is that the number of lines is dramatically reduced. In the case of Fig. 3.33, there are only four ENDOR transitions, provided that there are no quadrupole interactions present. If there is only one shell of n neighbors and $S = 1/2$, there are only two ENDOR transitions as compared to 2^n EPR transitions (for $I = 1/2$). Therefore, in ENDOR there is an excellent resolution of shf interactions, as will be discussed in Chap. 5.

4. Optical Detection of Electron Paramagnetic Resonance

Optical detection of EPR (ODEPR) differs from conventional detection basically in that a microwave–induced repopulation of paramagnetic *Zeeman* levels is indirectly detected by a change in some property of light, which is either absorbed or emitted by the defect under study. The light properties are polarizations or intensities, which are measured to detect EPR. These experiments are all double resonance experiments, one resonance being an optical resonance, the other one an EPR resonance. The optical detection of EPR has a number of interesting new features. One such feature is that, by virtue of the quantum transformation for detecting the signals from 10^{10} Hz to 10^{15} Hz, there is an enormous gain in sensitivity by several orders of magnitude. Thus, it becomes possible to study a very small number of defects. Originally, this sensitivity enhancement was used to study sparsely populated excited states of defects, and was, in fact, the major application of the optical detection of EPR. This aspect will not be discussed here in great detail. For an excellent review on the optical techniques of EPR, in particular the ODEPR of excited defect states, the reader is referred to the article by *Geschwind* [4.1]. It became evident only recently that the application of optical techniques to the detection of EPR is also very useful for the study of ground states of defects, especially in connection with materials science problems such as the structure determination of defects and their influence on bulk properties of solids. In this way, optical properties can be directly associated with particular defects and their structures. Properties connected with the energy levels of defects in the gap, such as electrical properties, which are of specific interest in semiconductor physics, can also be correlated with their EPR spectra.

Another interesting feature connected with ODEPR is the possibility of studying defects with a high selectivity. Often, in materials science, one of the problems for EPR is the simultaneous presence of many paramagnetic defects, which renders the spectra very complicated. An example of this was already discussed in Chap. 3 (Fig. 3.19). With optical detection, every defect can be investigated separately, except for the rare situation in which different defects (or slightly different defects) have identical optical properties (for an example of this see Sect. 4.4).

In ODEPR, the fact that magnetic sublevels can be selectively populated either by choosing appropriate experimental conditions, or by physical mechanisms, such as spin selection rules for radiative or nonradiative transitions, plays an important role. This is the basis of the application of different optical techniques to detect EPR. Basically, one can use the optical absorption, or the fluorescence or phosphorescence emission of a defect. Which technique is applied depends on the system and the kind of problem one wants to study. The absorption technique has proven to be particularly useful for the applications discussed in this book, and will, therefore, be dealt with in more detail than the emission techniques.

Optical absorption and emission bands of defects in solids are usually rather broad with typical half widths of 0.1–0.3 eV. This width is due to electron–phonon interactions. The *Zeeman* splittings of both ground and excited states are only of the order of 10^{-4} eV. The reason for the fact that one can observe ODEPR in the polarization of optical bands, is that the polarization properties of the absorption or emission bands are changed very little by the electron–phonon interaction. One can, therefore, use optical spectrometers with rather low optical resolution unless special features of very sharp optical transitions, such as zero–phonon lines, are of interest. It is due to this property of the electron–phonon interaction, that spin selection rules are not much affected by it, which makes the observation of ODEPR in the broad optical bands of defects in solids possible.

In the first section of this chapter, some basic features of optical transitions of defects in solids are briefly reviewed, which are necessary to understand the methods and examples given. Then the absorption detected ODEPR method of studying defect ground states and related techniques, such as the EPR excitation spectroscopy (so-called "tagging") used to correlate the optical transitions belonging to one defect with its EPR spectrum, as well as the measurements of the spin state and the longitudinal spin–lattice relaxation time are discussed. The several ways to detect EPR by optical emission will be discussed, again with the emphasis on studying defect structures. Specific aspects, such as spatial resolution of EPR, which is very important in materials science, and experiments with time resolution to study dynamical effects as well as to enhance the resolution of hyperfine structures, will briefly be described. A short discussion of the sensitivity of ODEPR is also given.

4.1 Optical Transitions of Defects in Solids

Many defects have localized energy states within the band–gap and possess optical absorption transitions (electrical dipole transitions) with photon energies below the band–gap energy E_g. These absorption transitions can be observed within the *optical window* of the crystal, where no band–gap transitions ($h\nu < E_g$) and no optical transitions caused by optical lattice phonons oc-

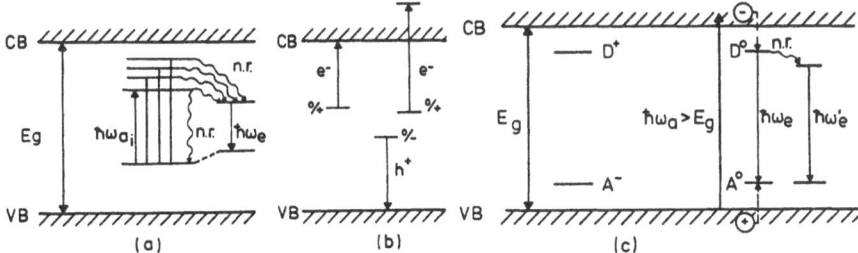

Fig. 4.1a–c. Schematic representation of optical transitions of defects in solids. **a** intracenter absorption and emission with *Stokes* shift. **b** Ionizing transitions to the conduction and valence bands and an intracenter transition into resonance states in the conduction band (semiconductors). **c** Donor–acceptor pair recombination luminescence $D^0 + A^0 \Rightarrow D^+ + A^- + \hbar\omega_e$ after creating an electron–hole pair by band–gap illumination and capturing of electrons and holes by the ionized donor D^+ and acceptors A^- to form the D^0–A^0 pair

cur. Wide–band–gap materials are insulators; narrow–band–gap materials are usually semiconductors. One can often observe optical emission after photo-excitation of the defect into one of the excited states, or after creating an electron–hole pair with light of energy above the band–gap.

The various possible transitions are shown schematically in Fig. 4.1. A defect in an insulator often has several absorption transitions from the ground state to unrelaxed excited states which show up as absorption bands (Fig. 4.1a). The excited states are also usually within the energy gap. Often, one can observe one emission band from the lowest relaxed excited state into the unrelaxed ground state. The relaxation is discussed below. Usually, the transitions from higher unrelaxed excited states into the lowest relaxed excited state from which the emission occurs, are nonradiative (indicated by n.r. in Fig. 4.1a), unless all excited states decay nonradiatively to the ground state. In many–electron systems, for example two–electron systems such as a ns^2 atomic configuration, the absorption transitions with high probability (allowed transitions) are singlet transitions, and usually one also observes a singlet emission. However, in addition, there can be a delayed emission, often at a smaller photon energy, from an excited triplet state, which is populated by intersystem crossing via the spin–orbit interaction. This triplet emission is spin–forbidden, and only partly allowed by slight admixtures of excited singlet states. The radiative lifetimes of these triplet emissions are much longer than those of the allowed singlet emissions. The transitions are often called phosphorescence emissions. They are observed in many molecular crystals [4.2,3], and also in ionic crystals [4.4,5] and semiconductors [4.6].

Defects can not only have the intracenter transitions just discussed, but also ionizing transitions into both the conduction and valence bands, respectively (Fig. 4.1b). These electron or hole emission processes are mostly observed in semiconductors. The photon energy at the onset of the process is determined by the energy levels of the defect in the gap. The ionization cross sections are usually smooth functions of the photon energy and look like broad absorption regions [4.7]. They can be measured until the band edge is reached.

Recently, a number of defects were studied, in which the excited states of the intracenter transitions were well within the conduction band and resonant with conduction band states (for example, Ref. [4.8]).

In semiconductors, an important emission is that due to donor–acceptor pair recombination. The mechanism is illustrated in Fig. 4.1c. On the left side, the energy levels of ionized acceptors A^- and ionized donors D^+ are indicated. Upon creation of an electron hole pair after illuminating the crystal with light of energy exceeding the band–gap energy, the acceptors capture a hole from the valence band, and the donors correspondly capture an electron from the conduction band. The recombination process $D^0 + A^0 \Rightarrow D^+ + A^- + \hbar\omega$ is often observed as recombination luminescence. Since the acceptors and donors often have their energy levels very close to the conduction and valence bands, respectively, (shallow donors and acceptors), this luminescence is observed near the band–gap energy. However, such a luminescence can also originate in deep levels, and is then observed at much lower photon energies. The luminescence of a deep donor D^+ can also be excited by light of sub–band–gap photon energy by raising an electron from the valence band directly into the ionized donor level. Sometimes there is an indirect nonradiative transfer of the electron from a shallow donor to a deeper donor (Fig. 4.1c), where the recombination luminescence then occurs. This nonradiative transfer can be spin–dependent, which has consequences for the ODEPR observation (Sect. 4.9).

Upon creation of electron–hole pairs, there is the possibility that electrons and holes form a bound excitonic state. Excitons can be trapped at impurities or defects, and often give rise to a luminescence with a photon energy near to the band edge, or there can be an excitonic absorption [4.9]. If the exciton is trapped at a paramagnetic defect, exciton emission or exciton absorption can also be used to optically detect the EPR spectrum of the paramagnetic defect.

4.2 Spectral Form of Optical Transitions of Defects in Solids

The intracenter absorption and emission transitions are mostly observed as rather broad bands. The reason is that in ionic and semiconductor crystals, most defect states are sensitive to the position of nearby atoms or ions, so that the form of the absorption and the energy position of the emission depend on the vibrations of the surrounding ions. This is conveniently discussed in

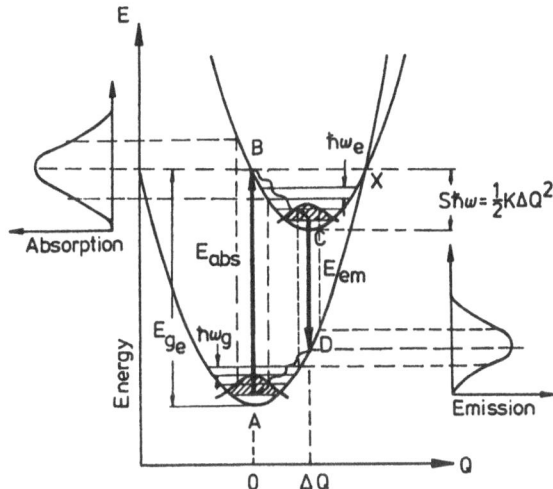

Fig. 4.2. Configuration coordinate (CC)–diagram of optical absorption and emission for strong electron–phonon coupling

the approximation of the configuration coordinate diagram (CC–diagram), where the lattice vibrations coupled to the defect are represented by a single localized mode with a *configuration coordinate* Q [4.4,5]. This is, of course, an approximation which will not be fully adequate. However, it allows one to describe the major features of the observed band shapes, and the energy position and occurrence of radiative emission processes.

The electronic states are coupled to the vibrations of the surrounding lattice neighbors. With the assumption that the electronic energy of the system depends only on one mode of displacement Q of the surroundings (the configuration coordinate), one can draw a curve which shows how the electronic energy depends on Q. Usually, a linear coupling is assumed for the vibrational energy, which results in a parabolic energy curve (harmonic approximation). The ground state becomes

$$E_g(Q) = \frac{1}{2}KQ^2. \tag{4.2.1}$$

K is a force constant and the energy is measured relative to the electronic energy of the ground state. The configuration coordinate for the equilibrium position of the ground state is set to $Q^g = 0$ (Fig. 4.2). In the excited state, the equilibrium configuration coordinate Q^e must be different, because the charge distribution is different after an electrical dipole transition has occurred, according to the selection rule that the orbital angular momentum must change ($\Delta l = \pm 1\hbar$). The lattice will relax and adjust to a new equilibrium position according to the new charge distribution, which is represented by

ΔQ in Fig. 4.2. The energy of the excited state is

$$E_e(Q) = E_{ge} + \frac{1}{2}KQ^2 - AQ,$$

$$E_e(Q) = E_{ge} - \frac{A^2}{2K} + \frac{1}{2}K(Q - \frac{A}{K})^2. \qquad (4.2.2)$$

It is assumed that the curvature of the parabolas in the ground and excited states are the same, that is, that the vibrational frequencies are the same. This is, of course, generally not the case. Here $A/K = \Delta Q$, the change in the configuration coordinate in the excited state (Fig. 4.2). Thus, A measures the difference in electron–lattice coupling between the ground and excited states. It is common to describe the difference in electron–lattice coupling by another dimensionless constant, the *Huang–Rhys* factor S, which is defined as:

$$S = \frac{A^2}{2K}\frac{1}{\hbar\omega} = \frac{1}{2}K\Delta Q^2 \frac{1}{\hbar\omega}. \qquad (4.2.3)$$

The quantity $S\hbar\omega$ is indicated in Fig. 4.2. S is a measure of the electron–phonon coupling, and of the lattice relaxation in the excited state and the displacement of the excited state parabola.

The vibrational states are quantized. The vibrational energy in each electronic state is given by $(n+\frac{1}{2}\hbar\omega)$, where n is the vibrational quantum number. The vibrational frequencies ω_e and ω_g are assumed to be the same in the simple model discussed here. In general, however, $\omega_g \neq \omega_e$.

The wave function for the vibrational state will depend on the electronic coordinate r and the nuclear coordinate Q. It can be written as the product $\phi_i(r_i, Q) = \psi_i(r_i, Q)\chi_{in}(Q)$, in which the electronic wave function ψ_i depends on both r_i and the configuration coordinate Q, where χ_{in} are the vibrational wave functions of the oscillator state n in the electronic state i.

Transitions can occur between the ground state and the excited electronic states according to

$$\hbar\omega_{nm} = E_{ge} - \frac{A^2}{2K} + m\hbar\omega - n\hbar\omega, \qquad (4.2.4)$$

where m depends on the vibrational quantum number in the excited state, and n is that of the ground state. In Fig. 4.2 only one transition for $n = 0 \Rightarrow m \neq 0$ is shown for absorption, and for $m = 0 \Rightarrow n \neq 0$ for emission. The transition probability is then proportional to the square of the electric dipole matrix element

$$P_{ge} \propto |\langle\psi_e(r_i, Q)\chi_{em}(Q)|e\mathbf{r}|\psi_g(r_i, Q)\chi_{gn}(Q)\rangle|^2. \qquad (4.2.5)$$

At very low temperature $\chi_{gn}(Q) = \chi_{g0}(Q)$ and the harmonic oscillator function has the maximum for $Q = Q_0^g$ (which is set to $Q_0^g = 0$ in Fig. 4.2).

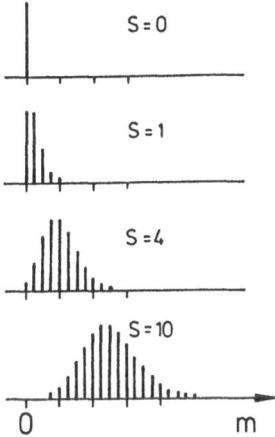

Fig. 4.3. Relative intensities of different electronic vibrational lines (stick diagram) for various values of the *Huang–Rhys* factor S

In the *Franck–Condon* approximation, it is assumed that a variation of the coordinate Q is small compared to that of electronic coordinates during an optical transition, and, therefore, Q in ψ_g is replaced by an average value; for small T this is approximately $Q_0^g = 0$. In the diagram of Fig. 4.2, this implies that the optical transition occurs as a "vertical" transition, as indicated. The transition probability then becomes

$$P_{ge} \propto |\langle \psi_e(r_i, Q_0^g)|er|\psi_g(r_i, Q_0^g)\rangle \langle \chi_{em}(Q)|\chi_{gn}(Q)\rangle|^2, \qquad (4.2.6)$$

$$P_{ge} \propto P_{eg}|\langle \chi_{em}|\chi_{gn}\rangle|^2. \qquad (4.2.7)$$

P_{eg} is the electronic transition probability, and is the same for all vibrational states. It determines the overall intensity (oscillator strength) of the transition, whereas the second vibrational matrix element contains the information on the shape of the bands, since it determines the intensity distribution over the various possible photon energies $h\nu_{nm}$ (*Condon* approximation, e.g., [4.7,9]. The vibrational matrix elements can be calculated analytically [4.10]. To illustrate the essential features, the result for $T = 0$ is quoted:

$$P_{m0} \propto |\langle \chi_{em}|\chi_{g0}\rangle|^2 = \frac{\exp(-S)\, S^m}{m!}, \qquad (4.2.8)$$

where S is the *Huang–Rhys* factor.

As an example for the resulting *Poisson* distribution, Fig. 4.3 shows the band shapes as "stick diagrams" for $S = 0$, 1, 4, and 10, where the single P_{m0} transitions are represented as "sticks". The total transition probability is independent of S, since $\Sigma|\langle \chi_{em}|\chi_{gn}\rangle|^2 = 1$. Therefore, the total intensity of the absorption band is independent of temperature. In Fig. 4.3, all shapes should have the same integrated intensity. They are, however, drawn to have the same maximal intensity [4.11] for a clearer comparison of the band shapes.

A special case is that for $S = 0$. The excited state parabola is not displaced with respect to the ground state parabola. There are identical harmonic oscillator functions in both states within that model. The transition is, thus, purely electronic and appears as a sharp line, the so–called *zero phonon line* (ZPL). For $S > 0$ the sideband transitions $0 \Rightarrow m$ involve the creation of phonons in the excited state. In practice, these sideband transitions appear as broad bands due to the wide spectrum of lattice vibration frequencies which can be generated. For smaller values of S the band shape is asymmetric; for larger ones it becomes symmetrical and of *Gaussian* shape (Fig. 4.3). For large values of S, the vertical lines intersect the excited state parabola in a region of almost linear character; the *Gaussian* shape for $T = 0$ is "reflected" at the excited state parabola (Fig. 4.2). After an absorption excitation into a higher vibrational state $\chi_{em}(Q)$, the system relaxes very fast by emission of phonons into the relaxed excited state for $m = 0$, the ground state level of $\chi_{em}(Q)$. The radiative decay into the ground state parabola can end at $\chi_{gn}(Q)$, with a high value of n leading to a broad *Gaussian* emission band (Fig. 4.2) because of the lateral displacement of the ground and excited state parabolas, $E_{em} < E_{abs}$ for $S > 0$. The energy shift between the peaks of the absorption and emission bands is called *Stokes* shift. If the excited state and ground state parabolas are identical, then absorption and emission bands are mirror images of each other. However, this is normally not the case.

Some further important features of the band shapes follow from (4.2.7). At low temperature their centroids occur at $m \approx S$. Thus, the centroid energy is E_{ge} and occurs at $Q^g = 0$. Its energy separation from the ZPL is $S\hbar\omega$. The second moment of the band shape is $S\hbar^2\omega^2$. The band width is, therefore, approximately $\sqrt{S}\hbar\omega$. The intensity of the ZPL relative to the whole band is $\exp(-S)$. It cannot be observed for $S \geq 6$.

At finite temperatures, higher vibration quantum states are occupied in the ground state parabola according to *Boltzmann* statistics. A line shape calculation must average over the thermal occupation of the ground state vibrational levels. It turns out that the line shape changes, but the total intensity of the absorption band is independent of temperature [4.10]. The ZPL intensity decreases like $\exp[-S\coth(\hbar\omega/2kT)]$, and the line width W increases according to

$$W(T)^2 = W(0)^2 \coth(\hbar\omega/2kT). \tag{4.2.9}$$

The centroid of the band shape remains at E_{ge}. S and $\hbar\omega$ can be determined from the temperature dependence of the line shape.

A further important result is that if there is coupling to several modes Q and a spectrum of vibrational frquencies ω_i, the ZPL remains sharp, while all the phonon assisted transitions are broadened. For small S, however, several

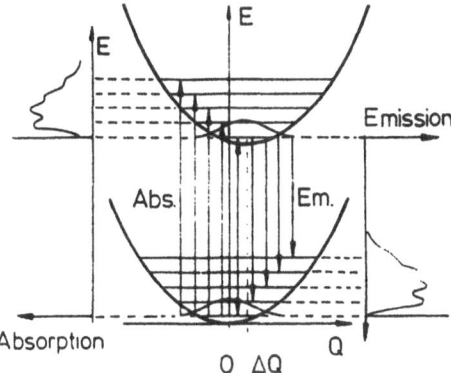

Fig. 4.4. Configuration coordinate (CC)–diagram of optical absorption and emission for intermediate electron–phonon coupling

phonon transitions may be seen as sharp lines, the so–called phonon–replica lines.

If the coupling is weak ($S < 1$) the ground and excited state parabolas are hardly displaced relative to each other, and the dominant spectral feature is the ZPL. Typical examples are transitions within the $4f^n$ configuration in rare earth ions in solids, and for $3d^n$ configurations of transition metals in semiconductors.

In intermediate coupling ($1 < S < 6$) the ZPL is resolved, but the multi-phonon structure is the dominant feature of the spectrum. The CC–diagram for such a case is shown in Fig. 4.4. An example of an absorption band is shown in Fig. 4.5 for a transition (the so–called R_2 band) of F_3 centers in various alkali halides; see Sect. 3.4 for F_3 centers. It is seen that S increases from LiF to NaCl, where the ZPL is hardly recognizable [4.12].

For strong coupling ($S > 10$) only the broad envelope of the band is seen. Typical band widths are 0.1–0.3 eV. For further details on the theory, the reader is referred to [4.4,5] and [4.13,14].

When a ZPL can be observed, the application of uniaxial stress and magnetic and electric fields can cause a splitting of the ZPL, from which structural information, such as the defect symmetry, can be achieved, see for example [4.4,5], and further references therein; for semiconductors also [4.15–17].

Defects with a strong electron–phonon coupling may not show a luminescence, or may show only a very weak one. In the simple theory of the CC–diagram, a nonradiative de–excitation occurs when the excited state energy reached in a *Frank–Condon* absorption transition (point B in Fig. 4.2) lies above the intersection of ground and excited state energy curves (point C in Fig. 4.2). Here, the excited defect can be de–excited through the intersection, directly to the ground state under phonon emission. In the approximation of

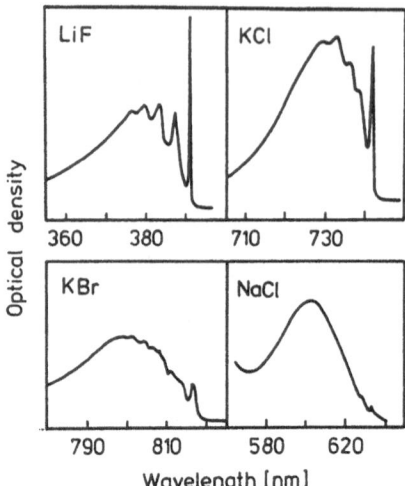

Fig. 4.5. The absorption spectrum of the R_2 band of F_3 centers level in alkali halides (4.2 K). (After [4.12])

linear coupling and equal vibration frequencies for ground and excited states, this occurs when $E_{em} < \frac{E_{abs}}{2}$ (*Dexter–Klick–Russel*-rule) [4.18,19].

In the discussion above, it was assumed that the electronic states are non-degenerate. If, however, there is an orbital degeneracy as a result of some high symmetry (for example an impurity p–electron in an octahedral crystalline environment), then the system will be unstable with respect to a distortion which lowers the symmetry and removes the electronic degeneracy. This is called the (static) *Jahn–Teller*-effect [4.20,21]. The major effect on the optical spectra of defects is that optical transitions to *Jahn–Teller* states may give bands with a multi–peaked structure (for example see Sect. 4.4). As a consequence of the dynamic *Jahn–Teller*-effect, certain defect properties, like the spin–orbit splitting or the response to external perturbations, can also be modified by the so–called *Ham* reduction factors. For further details see [4.22,23], and also Chap. 7.

4.3 EPR Detected with
Magnetic Circular Dichroism of Absorption Method

The detection of ground state EPR by measuring the microwave–induced change of the magnetic circular dichroism of the absorption (MCDA) has long been known from color center physics [4.24,25]. The method was neglected, however, because of the possibility of detecting ground state EPR by conventional means. Instead, the aim was to study optically excited states of

F centers. It was only recently that its usefulness was realized for problems arising in materials science, such as the investigation and structure determination of defects in semiconductors and in inorganic crystals, with applications in laser physics. It is mainly this aspect which will be discussed in this and the following sections. The technique is known as the MCD–technique or, more precisely, the magnetic circular dichroism of absorption (MCDA) technique.

The MCD of the absorption is the differential absorption of right and left circularly polarized light, where the light is propagating along the direction of an externally applied static magnetic field $\mathbf{B}_0$. One is, therefore, concerned with the circularly polarized transitions of the *Zeeman* effect. As a measure of the MCDA the quantity

$$\epsilon = \frac{\omega d}{2c}(k_r - k_l) \qquad (4.3.1)$$

is defined, where k_r and k_l are the absorption coefficients for right and left circularly polarized light, respectively. d is the thickness of the crystal and ω is the light frequency. Only the energy dependent absorption constants $\alpha_{r,l}(E)$ are measurable. With the relation

$$k(E) = \frac{\hbar c}{2E}\alpha(E), \qquad (4.3.2)$$

it follows with $E = \hbar\omega$ that

$$\epsilon = \frac{d}{4}(\alpha_r - \alpha_l). \qquad (4.3.3)$$

In the experiment, one measures the intensities $I_{r,l}$ which pass the sample,

$$I_{r,l} = I_0 \exp(-\alpha_{r,l}d), \qquad (4.3.4)$$

where I_0 is the light intensity incident on the sample. It follows from (4.3.4) that

$$\alpha_{r,l} = \frac{1}{d}\ln(\frac{I_0}{I_{r,l}}). \qquad (4.3.5)$$

From (4.3.5) one obtains a relation which depends only on the measurable quantities ΔI and I_a, where I_a is the average intensity, $I_a = \frac{1}{2}(I_r + I_l)$, and ΔI is the difference in intensity, $\Delta I = I_r - I_l$. Since $\Delta I \ll I_a$, that is $d(\alpha_r - \alpha_l) \ll 1$, one obtains the simple relation

$$\epsilon = \frac{\Delta I}{4I_a}. \qquad (4.3.6)$$

The quantity ϵ can be measured very precisely with the help of a stress modulator and lock–in techniques. This will be discussed in Chap. 9.

According to the selection rules for circularly polarized electric dipole transitions $(x \pm iy)$ in a magnetic field, the MCDA signal ϵ contains two parts [4.25],

$$\epsilon = \epsilon_p(P) + \epsilon_d(B_0), \qquad (4.3.7)$$

where ϵ_p is the paramagnetic part, which depends on the spin polarization P of the ground state, while ϵ_d is the diamagnetic part, which is proportional to $\mathbf{B_0}$ and arises from the unresolved *Zeeman* splittings in the optically excited states. For the detection of EPR we are only concerned with the paramagnetic part. For $S = 1/2$ one obtains for the paramagnetic term ϵ_p, since $P = (\frac{n_- - n_+}{n_+ + n_-})$ where n_+ and n_- are the occupation numbers of the $m_S = \pm 1/2$ states.

$$\epsilon_p(P) \propto P = (\frac{n_- - n_+}{n_+ + n_-}) = \tanh(g_e \mu_B B_0 / 2kT). \qquad (4.3.8)$$

For $S > 1/2$, P is given by the *Brillouin*–function (2.7.16).

The paramagnetic part ϵ_p is temperature and field dependent according to (4.3.8). By measuring ϵ_p as a function of temperature, one can easily distinguish it from nonmagnetic circular dichroisms of the sample. ϵ_d is not temperature dependent. The principle of EPR detection with the MCDA method is easily seen from (4.3.8). The equilibrium ground state spin polarization P can be changed by microwave–induced magnetic dipole transitions, as discussed in Chap. 3, provided the transition rate exceeds that of the spin–lattice relaxation and the light intensity is weak enough, so that the populations are not appreciably influenced by the optical transitions (for optical pumping effects see Sect. 4.8). The microwave transitions reduce ϵ_p. This reduction or decrease of ϵ_p is monitored to detect the EPR transitions. In principle, one can obtain a 100% effect if the EPR transition is fully saturated, i.e., $n_+ = n_-$.

ϵ_p is often very small. The detection limit is approximately 10^{-5} for an optical density of the absorption band of ≈ 1 (Chap. 9). Accordingly, the EPR–induced changes in ϵ_p are often difficult to measure. Therefore, samples which have a large (diamagnetic) dichroism due to crystal structure (e.g., due to birefringence), or which posess large stress–induced dichroisms, are not suitable for this technique to measure EPR.

To illustrate both the method and the experiment in more detail, let us discuss the MCDA of an atomic s $\Rightarrow$ p transition, such as that which occurs in a free alkali atom. The level scheme is shown in Fig. 4.6a. In a magnetic field the ground state *Kramers* doublet is split into *Zeeman* levels $m_S = \pm 1/2$, the excited p states are split by the spin–orbit interaction into $j = 1/2$ and $j = 3/2$ states with the magnetic substates m_j. The absorption transitions for right and left circularly polarized light are indicated by arrows, according to the quantum mechanical selection rules with their relative transition probabilities [4.26]. The equilibrium occupation of the ground state levels n_+ and n_- for finite temperatures is indicated. For $T = 0$ only $m_S = -1/2$ is

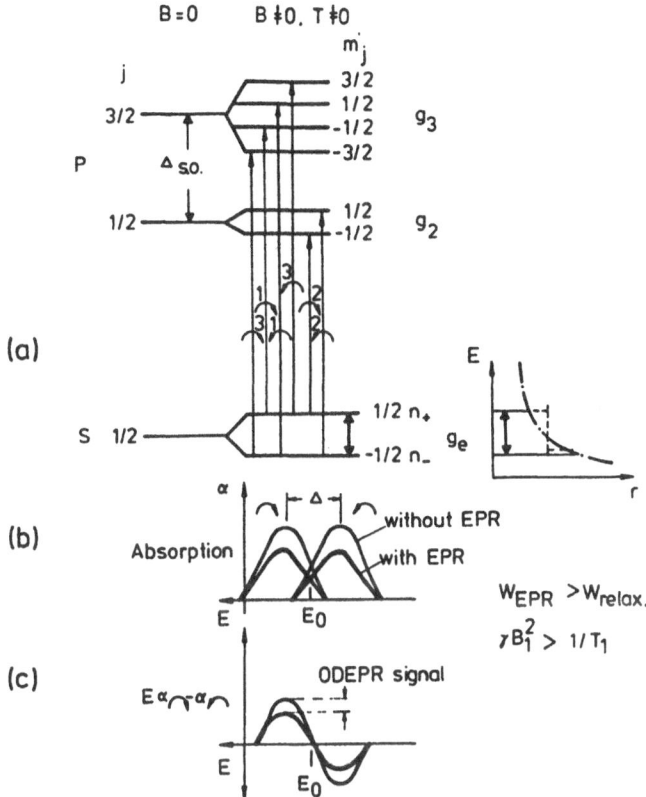

Fig. 4.6a–c. Simple atomic model to explain the magnetic circular dichroism method of the absorption (MCDA) to detect EPR. **a** Level scheme and circularly polarized optical absorption transitions. The *numbers* represent the relative transition matrix elements. **b** Absorption band for right and left circularly polarized light assuming strong electron–phonon coupling. **c** Magnetic circular dichroism and its change due to electron paramagnetic resonance transitions

occupied. If one neglects the small *Zeeman* splitting in the excited state, then one observes an absorption band with right polarization to $j = 1/2$, and one with left polarization to $j = 3/2$, both with equal intensities (the former with intensity 2, the latter with intensities $3 - 1 = 2$).

The spin–orbit splitting $\Delta_{s.o.}$ is often small compared with the absorption band width. In this case, it is convenient to measure the dichroism ϵ, which is the difference of the two absorptions α_r and α_l, and can be approximated by a derivative,

$$\epsilon(\omega) = \frac{\omega d}{2c}(k_r - k_l) \approx \frac{\omega d}{2c}\frac{dk}{d\omega}\Delta\omega$$

$$\approx \frac{Ed}{2\hbar c}\frac{dk(E)}{dE}\Delta E. \tag{4.3.9}$$

With (4.3.2) and the assumption, that the optical band shape is *Gaussian* according to

$$\left.\begin{array}{c} \alpha(E) = \alpha_0 e^{\frac{-(E-E_0)^2}{W^2}} \\[2mm] W_{1/2} = 2\sqrt{ln(2)}W \end{array}\right\}, \tag{4.3.10}$$

one obtains for the extrema of $\epsilon(E)$ (Fig. 4.6c)

$$\epsilon_{extr} = \mp\sqrt{\frac{ln(2)}{2e}}\frac{\alpha_0 d}{W_{1/2}}\Delta E, \tag{4.3.11}$$

with

$$\Delta E = 2\left[\frac{\mu_B B_0(g_2 + 5g_3)}{3} - \frac{\Delta_{s.o.}}{3}\tanh\left(\frac{g_e\mu_B B_0}{2kT}\right)\right]. \tag{4.3.12}$$

g_2 and g_3 are g factors of the excited state, α_0 is the absorption coefficient in the peak of the band, $W_{1/2}$ its half width, $\Delta_{s.o.}$ is the spin–orbit splitting of the excited $j = \frac{1}{2}$ and $j = \frac{3}{2}$ states [4.27]. The first term in (4.3.12) is the diamagnetic part, the second term the paramagnetic part of the MCDA. For the observation of EPR transitions, it is thus convenient to measure the change of the MCDA at the wavelength of the extrema of the MCDA. For $\Delta_{s.o.} \gg W_{1/2}$ there are two separate bands for ϵ, one with positive, the other with negative sign.

In order to measure EPR with the MCDA method it is not necessary to understand the optical transitions in detail, as long as a paramagnetic MCDA signal can be measured. Figure 4.6 is indeed a special case, which is approximately observed for F centers in alkali halides. In general, the MCDA spectra look quite different (see examples below).

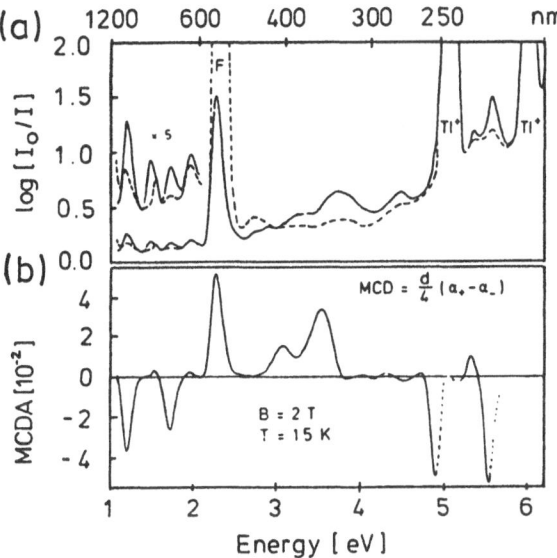

Fig. 4.7. a Optical absorption spectrum of KCl doped with Tl⁺ after X–irradiation at $-40°C$ (*dashed line*) and after F center bleaching at room temperature (*solid line*). **b** Magnetic circular dichroism of the absorption (MCDA) after F center bleaching. (After [4.28])

For a two–level system one can describe the paramagnetic MCDA in a more general way in the following expression:

$$\epsilon_p(E) \propto \alpha_0(E_0)d \left[\frac{\sigma_r(E) - \sigma_l(E)}{\sigma_r(E) + \sigma_l(E)} \right] \left(\frac{n_+ - n_-}{n_+ + n_-} \right). \qquad (4.3.13)$$

σ_r, σ_l are the cross sections or transition probabilities for right and left circularly polarized light, respectively.

The cross sections $\sigma_r(E)$ and $\sigma_l(E)$ are the sums over all transitions with right and left circular polarization, respectively, arising from the two magnetic sublevels, the small energy difference of which (10^{-4} eV) is neglected compared to the optical transition energies. $\alpha_0(E_0)$ is the absorption constant at the peak of the absorption band, d is the crystal thickness. From (4.3.13) it is seen that there is no ϵ_p if the ground state polarization is zero, or if there is no difference between $\sigma_r(E)$ and $\sigma_l(E)$. This difference depends on the properties of the excited states. As a general rule, the difference is large for a large spin–orbit interaction at the defect, such as that observed for heavy impurity atoms. The difference is also influenced by the crystal field splitting in the excited states. If both are large, one observes separated bands for right and left polarized light as a function of photon energy (for example, Fig. 4.7b). If α_0 is too small, then ϵ_p is also only a small effect, and hard to measure.

For a generalization to a multi–level ground state system for $S > 1/2$, the last term in (4.3.13) is to be replaced by the spin polarization P, which is then determined by the *Brillouin* function rather than the *Langevin* function, where the latter determines the last term in (4.3.13) for $S = 1/2$. Microwave transitions between fine–structure levels will also change the MCD of the absorption (see below for the example for Fe^{3+} centers in InP).

In (4.3.11) and (4.3.13), α_0 enters and, therefore, so does the oscillator strength of the optical transition. The oscillator strength is usually not known, and mostly very hard to calculate in a reliable fashion. Therefore, the measurement of the MCDA cannot be used to determine the concentration of defects quantitatively, unless the oscillator strength is known. The EPR spectra measured via MCDA cannot be calibrated with some other defect in a known concentration either, since the size of the spin–orbit interaction in the excited state enters in a crucial and usually unknown way. Although the MCDA method is very sensitive and selective (see below), it has the disadvantage that a quantitative determination of defect numbers is not directly possible, unless the oscillator strength and the spin–orbit splittings are known, or the MCDA can be changed in a controllable manner by some other processes. This is possible, for instance, if, by a charge transfer transition, the MCDA of the defect under study can be enhanced or decreased in a way in which the number of defects in the charge state from which the MCDA is built up or decreased, can be determined. If, on the other hand, the number of defects is known, the MCDA measurement can yield valuable information on the properties of the excited states.

In some cases both conventional EPR and ODEPR can be measured in the same sample. Then, by calibrating the conventional EPR signal (according to Sect. 2.9), one can calibrate the MCD of the absorption. This is valuable if the MCDA is used to investigate the spatial distribution of defects (Sect. 4.5).

The EPR spectrum of $Tl^0(1)$ centers detected with the MCDA method, which could also be measured with conventional EPR (Fig. 3.19 in Chap. 3), will be discussed as an example. Figure 4.7a shows the absorption spectrum of a Tl^+ doped KCl crystal after X–irradiation at $-40°C$ (dashed line), and subsequent F center bleaching at room temperature with light of about 550 nm wavelength (solid lines). There are numerous absorption bands due to several defects (in the UV, two Tl^+ bands from the Tl^+ doping are also seen). The MCDA in Fig. 4.7b shows fewer transitions than the absorption. Some of the defects created must be diamagnetic. Figure 4.8 shows the ODEPR spectrum measured in the absorption band (MCDA band) at 1040 nm. The absorption at 1040 nm has a negative paramagnetic MCDA. The MCDA, measured there for a fixed wavelength at 1.5 K as a function of the magnetic field under microwave irradiation (24 GHz), shows an almost linear field dependence according to (4.3.11,12) for this field range between 0.7–1.5 T, and the negative MCDA increases in magnitude with increasing field, except for several field values, where it changes abruptly. These field values are the resonance

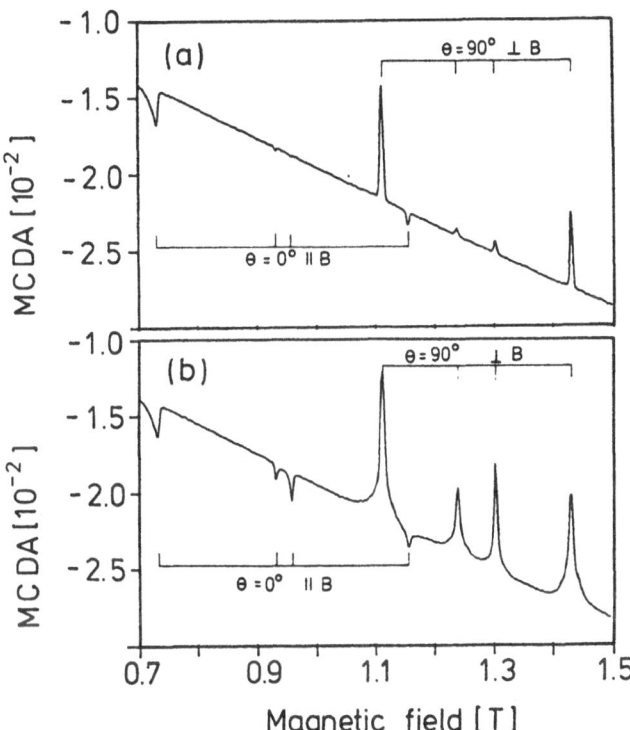

Fig. 4.8. a EPR spectrum of $Tl^0(1)$ centers in KCl detected as microwave–induced change of the magnetic circular dichroism of the absorption at 1040 nm, measured at low microwave power (24 GHz, 1.5 K). The linear change in the magnetic circular dichroism is due to the *Langevin* function (4.17). **b** Same as in **a** for high microwave power. The forbidden transitions $\Delta m_I = \pm 1$ are stronger compared to those in **a**. (After [4.28])

fields of EPR transitions. The g factors and hf splittings observed correspond exactly to those of $Tl^0(1)$ centers measured by conventional EPR (Fig. 3.19). The lines are due to centers with their axes parallel and perpendicular to the magnetic field.

In Fig. 4.8a it is seen that for the "perpendicular" centers the MCDA decreases in magnitude, as is expected from the foregoing discussion, while for the "parallel" centers the MCDA increases. The explanation is that for the parallel centers, there is a positive MCDA at 1040 nm superimposed on the stronger negative one due to the perpendicular centers, so that in total there is a negative MCDA. Thus, if this positive MCDA is decreased in a resonant microwave transition, the total negative MCDA increases in magnitude, as observed. Between the Tl hf doublet lines ($I = 1/2$) there are weak transitions seen in Fig. 4.8a, which are much stronger in Fig. 4.8b, where the same measurement is shown, except that the microwave power is increased by 30 db. They are due to so–called forbidden EPR transitions, in which the EPR selection rule $\Delta m_I = 0$ is broken. The transitions belong to $\Delta m_S = \pm 1$ and $\Delta m_I = \pm 1$ transitions, where the Tl nuclear spin is also changed. From the measurement, it is clear that the absorption band at 1040 nm belongs, indeed, to $Tl^0(1)$ centers [4.28]. This is of particular importance since the band at 1040 nm is used to pump the so–called "Thallium Laser" which emits at 1500 nm (in KCl) [4.29]. It is this measurement by ODEPR, which establishes that the laser active centers are indeed $Tl^0(1)$ centers, and not Tl perturbed $F_A(Tl)$ centers, as was speculated before these experiments [4.29]. The small difference in hf splittings due to the Tl isotopes ^{205}Tl and ^{203}Tl (both have $I = 1/2$) is not resolved. It could be resolved in conventional EPR (Fig. 3.19). It is a common observation that in ODEPR the individual line widths are broader than in conventional EPR.

The very high sensitivity of the method compared to conventional EPR and conventional optical absorption spectroscopy, can be demonstrated by experiments on the singly ionized mid–gap defect EL2 in semi–insulating as–grown GaAs single crystals, which were grown by the liquid encapsulated *Czochralski* (LEC) method. Figure 4.9a shows the optical absorption spectrum in the near infrared region for photon energies below the gap–energy of 1.52 eV. Only a weak band at 1.18 eV caused by an intracenter transition of the diamagnetic mid–gap defect EL2 is detectable. However, the MCDA reveals the existence of further absorption bands caused by paramagnetic defects, which turn out to be the singly ionized state of the mid–gap EL2 defect [4.30]. The ODEPR spectrum is measured at 1350 nm at 24 GHz (Fig. 4.10 shows a quartet structure, which is due to the hf splitting of ^{75}As which has $I = 3/2$ and is 100% abundant). In Fig. 4.10 the trivial MCDA effect due to the $\tanh(g_e \mu_B B_0 / 2kT)$ is already subtracted. Only the microwave–induced changes of the MCDA are shown as the ODEPR spectrum. In the conventional EPR spectrum the hf structure is hardly recognized. The signal–to–noise ratio observed for a defect concentration of $\approx 5 \times 10^{15}$, which is the one of

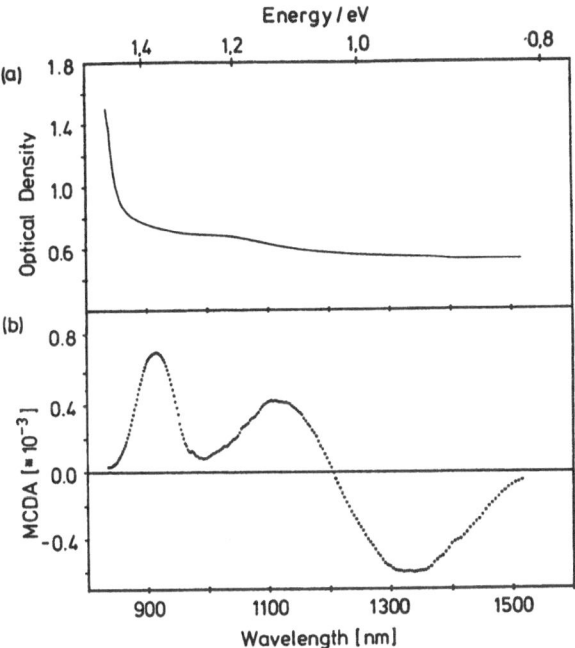

Fig. 4.9. a Optical absorption of as–grown semi–insulating GaAs at 1.4 K (crystal thickness 0.3 mm). **b** Magnetic circular dichroism of the absorption of **a**. $T = 4.2$ K, $B = 2$ T. (After [4.30])

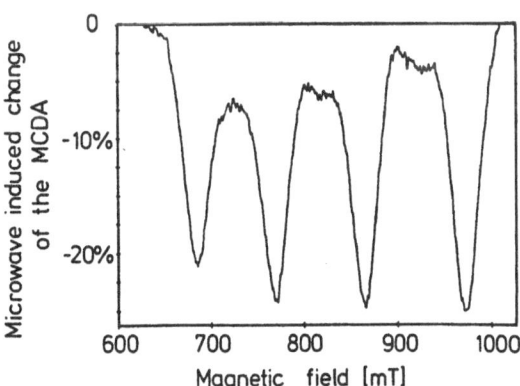

Fig. 4.10. Optically detected EPR spectrum of paramagnetic EL2 defects in semi–insulating as–grown GaAs measured as microwave–induced decrease of the MCDA (Fig. 4.9b) at 1350 nm. The spectrum shows a hf quartet structure due to ^{75}As ($I = 3/2$) (K–band, $T = 1.5$ K) (After [4.30])

Fig. 4.10, is not much greater than one. The weakness of the conventional EPR spectrum is caused by the quartet splitting and the large line width of each line (≈ 30 mT). The signal–to–noise ratio of the ODEPR spectrum is about two orders of magnitude higher. The defect structure is revealed by ODENDOR to be an As antisite–As interstitial pair defect [4.31]. As seen from Fig. 4.9a,b, a sensititivy enhancement is also found for the detection of the optical absorption bands of this defect, which cannot be measured in a conventional absorption experiment. Thus, the MCDA is also measurable for very weak absorptions as a consequence of the fact that one applies a sensitive modulation spectroscopy. Therefore, one also measures ODEPR spectra, even if the absorption spectrum does not show clear absorption bands.

Note that the change of the MCDA of each of the four hf lines in Fig. 4.10 is approximately 20–24%. Since each *Zeeman* level is split into four hf levels, each hf transition can theoretically cause a decrease of the MCDA by 25% for a saturating microwave transition. Apparently, this is reached here. However, this implies, that each hf line is a homogenous line and does not contain any shf interactions. As will be shown in Chap. 5, further shf interactions can be resolved by optically detected ENDOR using the same MCDA method. Thus, the ODEPR effect is larger than expected from a simple model of the m_I states and their statistical population, as discussed in Chap. 3. At present, there is no clear answer yet as to how to explain this observation. Also, the ODEPR effects of many other defects in both semiconductors and ionic crystals were found to be higher than expected from a statistical distribution of m_I states [4.31–33]. Further investigations are necessary to understand the reasons.

4.4 MCDA Excitation Spectra of ODEPR Lines (MCDA "Tagged" by EPR)

By conventional EPR and optical spectroscopy it is very difficult to unambiguously correlate the optical absorption spectra of a defect with its EPR spectrum, since they are both measured separately, and the presence of several defects can never be excluded. By ODEPR a direct correlation is possible, which is very important for all defect applications, where the optical properties are decisive, such as solid state laser material or photon detector material, for example X–ray detectors and X–ray storage phosphors.

One can measure a kind of excitation spectrum of the ODEPR lines. In Fig. 4.6a simple scheme was discussed to explain the MCDA method to measure the ground state EPR of a defect. Such a scheme applies to all optical transitions of a defect, i.e., the EPR spectra appear in the MCDA of all absorption transitions, provided, the condition (4.3.13) is fulfilled, in particular that $\sigma_r(E) - \sigma_l(E) \neq 0$ for the absorption transitions used for the experiments. The ground state polarization is, of course, the same for all transitions. Therefore, one can set the EPR resonance conditions to a particular EPR

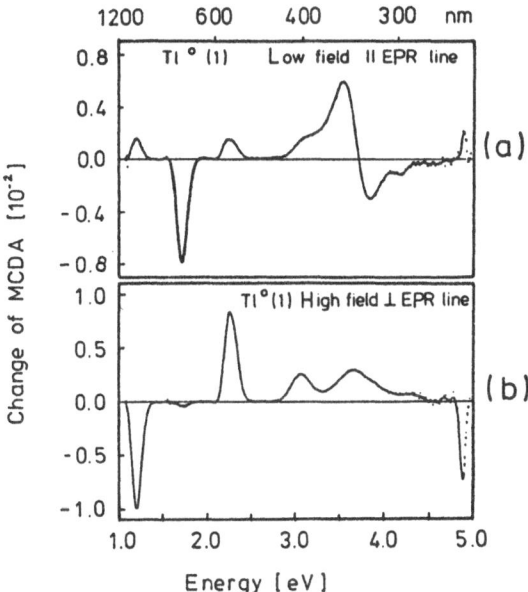

Fig. 4.11. a Magnetic circular dichroism at 1.4 K of the "parallel" $Tl^0(1)$ centers in KCl as tagged by the corresponding low field EPR transition. **b** Magnetic circular dichroism at 1.4 K of the "perpendicular" $Tl^0(1)$ centers as tagged by the corresponding high field EPR transition (K–band, $T = 1.5$ K). (After [4.28])

line, vary the optical wavelength and monitor the EPR signal intensity as the microwave–induced change of the MCDA. Thus, from the total MCDA of a sample, one can measure in this way that part which belongs only to the EPR line. Therefore, with this method one distinguishes the optical properties of different defects and also of different defect orientations, if the EPR spectra are anisotropic. One thus has the possibility of studying the polarization of optical transitions for different defect orientations. Such a measurement is particularly useful when a superposition of optical absorptions caused by several defects is present. That situation is typical for radiation damage or impurity problems. In Fig. 4.7b the total MCDA for $Tl^0(1)$ centers in KCl was shown. The measurement of the wavelength dependent change of the MCDA under resonance conditions for an EPR line of $Tl^0(1)$ centers with their axes parallel to $\mathbf{B_0}$, "parallel centers", (Fig. 4.11a), and of those with their axes perpendicular to $\mathbf{B_0}$, "perpendicular centers", (Fig. 4.11b), represent the MCDA excitation spectra of the two $Tl^0(1)$ center orientations. This kind of excitation spectrum was called "MCDA tagged by EPR", since only that MCDA appears which belongs to a particular EPR line [4.28]. In the case of $Tl^0(1)$ centers in KCl, an inspection of Fig. 4.11 shows that the optical absorption at 1040 nm for parallel centers has a positive MCDA, while that of

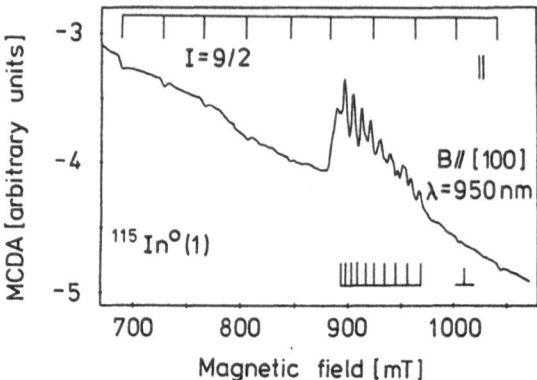

Fig. 4.12. EPR spectrum of axial In⁰(1) centers in KCl detected as microwave–induced change of the magnetic circular dichroism of the absorption at 950 nm for $B_0 \parallel [100]$. The ten In hf lines ($I = 9/2$) of parallel centers ($\theta = 0°$) and of perpendicular centers ($\theta = 90°$) are seen (K–band, $T = 1.5$ K). (After [4.32])

the perpendicular centers is negative. Furthermore, the transition at 725 nm is almost fully polarized perpendicular to the center axis. There is almost no transition probability for the electrical vector parallel to the center axis (Fig. 4.11b). In total eight absorption bands can be identified for the Tl⁰(1) centers [4.28]. Of course, the measurement yields only the MCDA spectra to be corrected by a factor representing the strength of the ODEPR effect. The photon energy range for the optical absorption transitions is taken from the measurement, however, not the details of the absorption band shapes. This would require an understanding of how the MCDA is derived from the excited state properties of the defect.

To illustrate the power of the method to detect hidden optical absorption bands, Figs. 4.12,13 show results for In⁰(1) centers in KCl. In⁰(1) centers are isoelectronic centers analogous to the Tl⁰(1) centers. The ODEPR spectrum resolves the hf interaction for parallel and perpendicular centers with ¹¹³In and ¹¹⁵In isotopes, which both have I = 9/2 and almost the same nuclear g factor (Fig. 4.12, the small difference in hf splitting is not resolved). From the tagging on the EPR lines for both orientations the photon energy of the peaks of the optical absorption bands can be determined. The positions are marked by arrows in the absorption spectrum, which one measures after production of the defects with X–irradiation of the In doped KCl crystals at room temperature (Fig. 4.13). Two transitions are hidden in the shoulder of optical absorptions due to other defects. Since there is no emission from the In⁰(1) centers either, their absorption could not have been determined otherwise [4.34].

Recent experiments also show that the zero–phonon lines and sharp spin–forbidden transitions can be measured as tagged MCDA spectra, and unam-

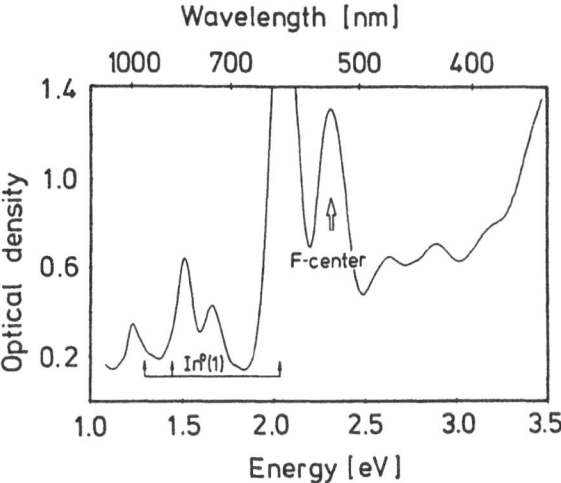

Fig. 4.13. Optical absorption spectrum of KCl doped with In$^+$ after X–irradiation at room temperature. The *arrows* mark the peak energy position of the F center absorption and of three absorption bands due to the In^0(1) centers. The energy position of the In^0(1) bands is determined by "tagging" the MCDA by In^0(1) ODEPR lines. (After [4.34])

biguously assigned to a particular defect. Figure 4.14 shows the absorption spectrum of high resistivity GaAs doped with V, measured with high resolution in a *Fourier* transform optical absorption spectrometer. The spectrum is attributed to V^{3+} defects, where V^{3+} substitutes for Ga. The broad feature is a $^3A_2 \Rightarrow {}^3T_1$ transition into excited *Jahn–Teller* states. The sharp features are the ZPLs and spin–forbidden transitions [4.35]. The optically detected EPR spectrum of the V^{3+} defects measured in the broad band (Fig. 4.15a) agrees with the one previously measured conventionally and has an unresolved hf splitting due to ^{51}V ($I = 7/2$) which is responsible for the unusual line shape. The MCDA tagged by this EPR spectrum (Fig. 4.15b, "ODMR–MCDA") shows the same ZPLs and spin–forbidden sharp transitions as the conventionally measured absorption spectrum, including some additional features. In this way it is verified that the structures in the optical absorption spectrum do indeed belong to this one defect V^{3+}, which is not possible on the basis of optical measurements alone. The crystal also contains V^{2+} defects and V–related complexes [4.36]. Figure 4.15c shows details of the MCDA spectrum around the ZPL at 1.008 eV, which cannot be detected conventionally.

In the tagged MCDA spectrum at (1.37 ± 0.02) eV, the onset of the ionization transition V$^{3+} \Rightarrow$ V^{2+} + hole (in the valence band) is observed in agreement with the known energy position of the V$^{3+/2+}$ level at $E_c - 0.15$ eV. This is a remarkable observation because it implies that the ODEPR spectrum can be observed in a ionizing transition, in this case to the valence band.

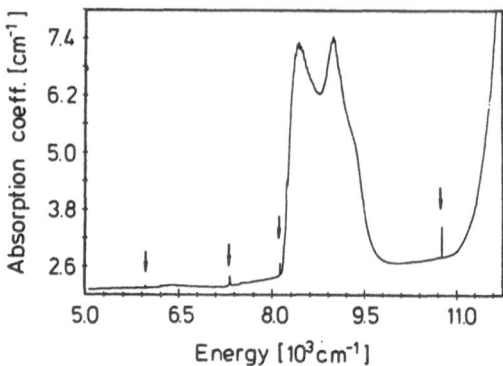

Fig. 4.14. Optical absorption spectrum of a semi–insulating GaAs:V sample measured with a *Fourier* transform spectrometer. Zero phonon lines are observed close to 5900 cm^{-1}, at 7333, 8131 and 10773 cm^{-1}. All transitions correspond to V^{3+} internal transitions. (After [4.35])

Since this first observation [4.36], this phenomenon has been observed several times for a number of transition metal ions (Ti, Mn, Fe) in GaAs and GaP [4.36,37]. From the foregoing discussion, such an observation is not expected, since localized states are always assumed in the mechanism to explain ODEPR. Possibly, during the time needed to induce an optical transition, which is of the order of 10^{-15} s, defect–induced states resonant with the bands exist, which allow the MCDA and ODEPR observation before lattice phonons destroy them, and before the electrons or holes become thermalized within the bands. However, a theory for explaining the recent observations is not yet available. The fact that the electrons and holes are thermalized after the transition is seen by the observation that no optical pumping was possible (Sect. 4.8).

From a practical point of view, the possibility of observing the EPR in ionizing transitions is very important, since such defect transitions are always present in semiconductors, while there may not be intracenter transitions. It seems, therefore, that the MCDA method of investigating defects in semiconductors can be applied in a very general way.

Finally, it should be mentioned that different EPR spectra can belong to indistinguishable MCDA spectra, and, thus, cannot be differentiated by the tagging experiment. Figure 4.16 shows the structure of the laser active Pb$^+$ center in KMgF$_3$, which can be excited at 640 nm, and exhibits tunable laser action between 850–900 nm [4.38]. Pb^{2+} is doped and upon X–irradiation at room temperature, Pb^{2+} captures an electron to form Pb$^+$ and attracts a F$^-$ vacancy along a <110> direction. It was found by ODEPR and ODENDOR experiments that the Pb^{2+} replaces K$^+$, and not Mg^{2+} as first expected. Therefore, there must be an additional K$^+$ vacancy for charge compensation.

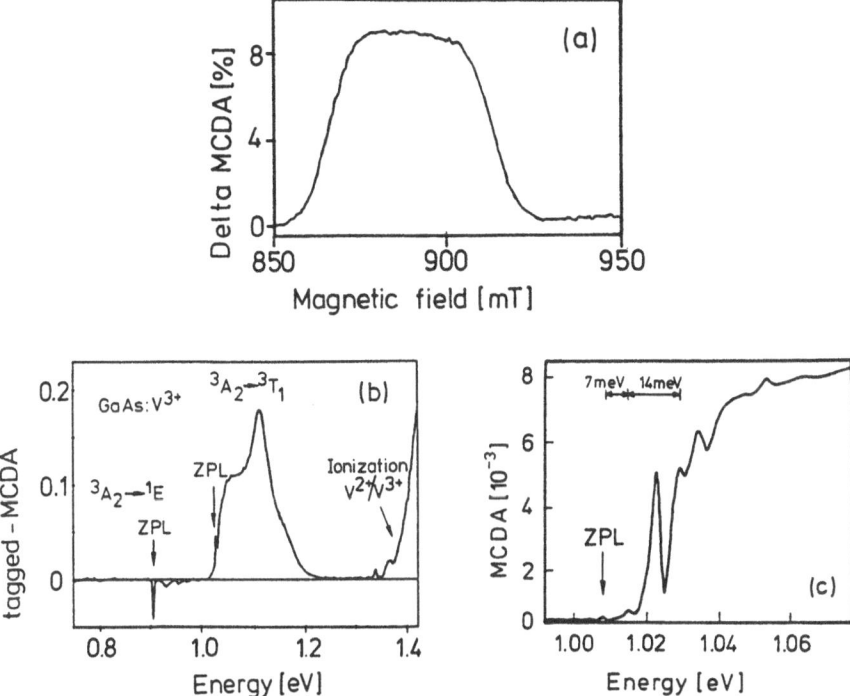

Fig. 4.15. a ODEPR spectrum of V_{Ga}^{3+} centers in high resisistivity (p–type) GaAs measured with the MCDA method at 1.1 eV ($g = 1.96 \pm 0.01$, $\nu_{EPR} = 24.41$ GHz, $T = 1.5$ K). **b** Magnetic circular dichroism of the absorption of V_{Ga}^{3+} centers in high resistivity GaAs (p–type) at $T = 1.6$ K tagged by the EPR spectrum of **a**. **c** Part of the MCDA spectrum of V_{Ga}^{3+} centers in GaAs measured with high spectral resolution. (After [4.36])

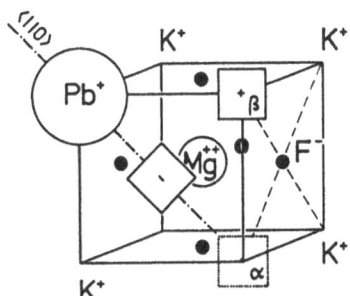

Fig. 4.16. Model of the laser active Pb^+ centers in $KMgF_3$. Vacancies are represented by *squares*. The two K^+ vacancy positions α and β are indicated. (After [4.33])

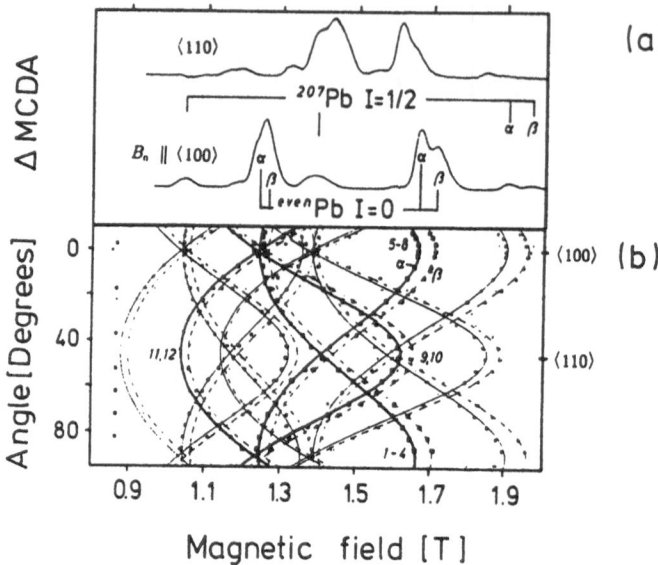

Fig. 4.17. a ODEPR spectrum of the laser centers Pb$^+$ in KMgF$_3$ measured with the MCDA method (K–band, $T = 1.5$ K). **B$_0$** ||<110> (upper trace) and **B$_0$** ||<100> (lower trace). **b** Angular dependence of the spectra of **a** for rotation of the magnetic field in a {100} plane. (After [4.33])

From the spectrum it follows that this vacancy is situated at the two possible positions indicated in Fig. 4.16. The ODEPR spectrum measurement and its angular dependence clearly show two defects with <110> symmetry, which are only slightly different (α and β lines in Fig. 4.17). The hf structure with the isotope ^{207}Pb ($I = 1/2$, abundance 22.6%) is resolved. However, the MCDA bands cannot be distinguished for the two EPR spectra. The optical properties are determined only by the Pb$^+$–F$^-$ vacancy part of the defect, and not influenced in a measurable way by the different positions of the two K$^+$ vacancies [4.33].

4.5 Spatially Resolved MCDA and ODEPR Spectra

In materials science it is often necessary to have a nondestructive method to measure the spatial distribution of defects. In semiconductor physics, for example, one wants to know the spatial distribution of different charge states of defects across a wafer on which microelectronic devices are to be fabricated. Such measurements are not possible with conventional EPR. However, with optical detection, such experiments are possible within the limitations of the

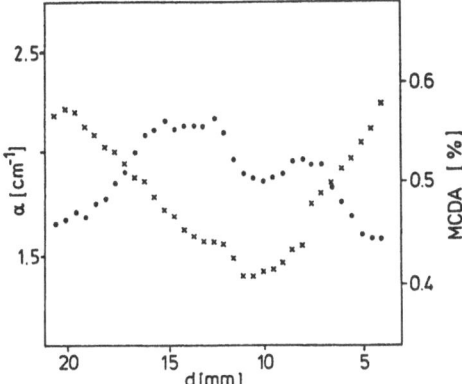

Fig. 4.18. Spatial distribution of mid–gap EL2 defects (*crosses*) and paramagnetic EL2 defects (*dots*) measured across half a diameter of a semi–insulating GaAs wafer. The center of the wafer is at $d = 4$ mm. The spatial resolution was about 300 μm. The mid–gap EL2 was measured as the 1.18 eV intracenter absorption band, the paramagnetic EL2 as the MCDA of the absorption at 1350 nm ($B = 1$ T, $T = 1.5$ K). (After [4.39])

cross section of the light beam. One integrates, of course, over the thickness of the sample.

The first example of such a measurement is shown in Fig. 4.18, where the spatial distribution of the two charge states of the EL2 defect in semi–insulating GaAs, is investigated across half the diameter of a GaAs wafer of 0.3 mm thickness. The light beam of a halogen lamp was focused and limited in diameter by a pin–hole, such that the spatial resolution of about 300 μm was achieved. The charge state of the mid–gap EL2 (dots in Fig. 4.18) is monitored by its intracenter absorption band at 1.18 eV (Fig. 4.9a), the charge state of the paramagnetic EL2 by the MCDA at 1350 nm (1 T, 1.5 K), see Fig. 4.9b for the MCDA, (crosses in Fig. 4.18). From Fig. 4.18 it is seen that the two charge states of the same defect are not equally distributed. In places where there is more mid–gap EL2, there is less paramagnetic EL2, and vice versa. This points to the role of acceptors which ionize the mid–gap EL2, and to their distribution in the wafer, which must be similar to that of the paramagnetic EL2 [4.39,40]. Figure 4.19 shows the distribution upon converting the optical absorption and the MCDA into defect concentrations, after performing a calibration both of the MCDA and the optical absorption [4.40]. It is seen that there is, indeed, a significant anti–correlation between the two charge states of the EL2 defect, and that the distribution of the defects in both charge states (triangles in Fig. 4.19) is not homogeneous.

This kind of mapping experiment can be improved in several ways. One is to use a focused laser beam to improve the resolution. The higher the spatial resolution, the less light passes through the sample, and, therefore, the signal–

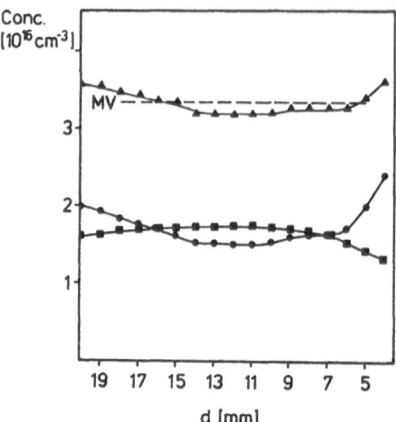

Fig. 4.19. Spatial distribution of mid–gap EL2 defects and paramagnetic EL2 defects of the wafer of Fig. 4.18 after calibrating both the absorption band of mid–gap EL2 and the MCDA of paramagnetic EL2. (After [4.40])

to–noise ratio drops. With higher light intensity levels one can improve the resolution, but this is feasible only if the defects are not light sensitive, which was found to be the case for the EL2 defects. Also, normal lasers have a light output with a rather high noise of up to 1%. This is disadvantageous for the MCDA method (Chap. 9).

One of the interesting possibilities of this mapping technique is that after identification of the MCDAs of several defects, one does not necessarily need to spatially resolve the ODEPR spectra, but can simply measure their MC-DAs at characteristic wavelengths, where there is no MCDA superposition of different defects. This mapping technique allows the simultaneous nondestructive investigation of the distribution of several defects. One can thus study their interactions upon external influences on the crystal, such as heat treatments.

This mapping can be done in two dimensions, and one can establish topographic pictures of the distribution of defects. Such investigations using the MCDA and absorption technique have just started to be realized [4.41]. One must take care, however, that when a large sample is cooled to very low temperatures for the MCDA experiments, it does not exhibit stress dichroisms, because it is not easy to reliably correct for this effect, especially if it has the same order of magnitude as the magnetic circular dichroism. And low temperatures are necessary for sensitivity reasons (Sect. 4.12). For the influence of nonmagnetic dichroisms on the results, see Chap. 9.

4.6 Measurement of Spin–Lattice Relaxation Time T_1 with MCDA Method

The measurement of T_1 for a solid state defect using conventional EPR is problematic. Any cw method relying on the saturation of the EPR signal has the difficulty that the saturation factor contains the product $T_1 T_2$ (Sect. 2.8) and T_2 is not known. In an inhomogeneous line one can saturate a single spin packet and observe its recovery, for example, with a microwave pulse (hole burning). However, spin diffusion may influence the result [4.1].

With the MCDA method, T_1 can be measured in a rather simple way. One drives the ground state polarization P from thermal equilibrium by either rapidly changing the magnetic field or the temperature, or by applying a (saturating) microwave pulse and observing the return of the MCDA to the equilibrium value. The advantage is that the MCDA only depends on the longitudinal magnetization, and, therefore, only on T_1. The most convenient disturbance of the equilibrium population is achieved by a microwave pulse. The shortest T_1 measurable depends on the speed with which the disturbance can be applied. Neither the magnetic field, nor the temperature can be easily varied fast enough, although T_1 times as short as 10 μsec have been measured using pulsed heaters [4.42]. The spin system may also be driven from equilibrium by optical pumping (Sect. 4.8). This was applied, for example, to F centers in alkali halides and rare earths in alkali fluorides [4.42,43].

Microwave switches are very fast (in the ten nanosecond range). The limitation for a measurement of a fast T_1 is the MCDA detection, which uses a stress modulator, the frequency of which is of the order of 10^4 Hz. Therefore, T_1 as short as 10–100 μsec may be measured with this method. In order to assess the relaxation process one needs to measure T_1 as a function of T and B (Chap. 2). An effective field variation is achieved by using several microwave bands like X, K, and Q band, which can be used in the same MCDA apparatus. For very high fields and microwave frequencies, respectively, quasi-optic arrangements can be used rather than cavities (Chap. 9).

For inhomogeneous lines the same problems exist with spin diffusion, as mentioned for the hole burning technique with conventional EPR, if the disturbance is achieved by the microwave pulse. This is not so, if one uses optical pumping, or a temperature or magnetic field change instead, since one then looks at the entire ground state population and not only on spin packets.

At 1.5 K, the usual temperature for measuring ODEPR and MCDA, T_1 is rather long for those defects for which one can measure an EPR signal. If T_1 becomes too short, no sufficient MCDA change can be achieved with the microwave power available in a cavity (Chap. 9). A measurement of the time evolution of the MCDA after driving it from thermal equilibrium has still an application other then the measurement of T_1. Often, it is not clear whether the total MCDA measured is due to only one defect, or to a superposition of

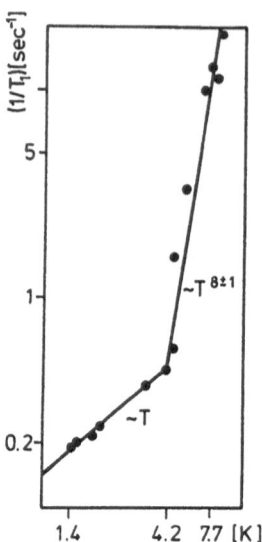

Fig. 4.20. Spin–lattice relaxation rate T_1^{-1} as a function of temperature of paramagnetic EL2 defects in semi–insulating GaAs measured at 960 mT by observing the recovery of the MCDA signal after a saturating microwave pulse. The transition from the direct process to a *Raman* process between 4 and 5 K is clearly seen. (After [4.44])

several MCDAs, of which only some EPR spectra are seen, so that the tagging method does not tell the whole story. By such a time evolution experiment as a function of the wavelength, it can be investigated as to how many time constants are involved in the MCDAs. One single defect should normally show a simple exponential return to equilibrium, unless there are cross relaxation effects or complicated relaxation schemes.

A recent example for a measurement with the MCDA method of T_1 as a function of temperature is shown in Fig. 4.20 for the paramagnetic EL2 defects in semi–insulating GaAs. At low temperature, T_1^{-1} is proportional to T, while at higher temperature, T_1^{-1} is proportional to $T^{(8\pm1)}$. The relaxation mechanism changes from a dominant direct process at low temperature (between 4–5 K) to a *Raman* process. [4.44,45].

4.7 Determination of Spin State with MCDA Method

The knowledge of the charge of a defect is quite important. Connected with the charge state is the spin state. For many defects it is clear that $S = 1/2$, but for many–electron systems such as for transition metal ions ($3d^n$), the spin state can be derived from the EPR spectra only if a fine–structure or crystal field splitting is observed. This splitting can be too small to be resolved. Then the measurement of an EPR spectrum does not allow one to conclude on the spin and charge state of the defect. It may be determined from ENDOR, if ENDOR measurements are possible (Chap. 5).

However, there is an alternative way: by using the MCDA method. The following is exact for orbital singlet states. According to (4.3.7), the MCDA is the sum of a diamagnetic part $\epsilon_d(B)$ and a paramagnetic part $\epsilon_p(P)$. The diamagnetic part is proportional to the magnetic field B, and usually small compared to the paramagnetic part ϵ_p, which, in turn, is proportional to the spin polarization P. For a general value of the spin, the polarization is given by the *Brillouin* function $B_s(x)$

$$P = g_e \mu_B S\, B_s(x) \tag{4.7.1}$$

$$x = \frac{g_e \mu_B B_0}{kT} \tag{4.7.2}$$

$$B_s(x) = \frac{1}{S}\left[\left(S + \frac{1}{2}\right) \coth\left(S + \frac{x}{2}\right) - \frac{1}{2}\coth\frac{x}{2}\right] \tag{4.7.3}$$

where g_e is the electronic g factor of the defect.

According to (4.3.7), P, which contains the information on the spin state S, cannot be directly measured by a MCDA measurement because of the diamagnetic term ϵ_d. However, since ϵ_d is linear in B, one can measure the total MCDA ϵ for several magnetic fields and temperatures, and form the following ratio, for example, for B_1 and T_1:

$$R_{exp} = \frac{\epsilon(B_1, T_1) - \epsilon(B_1, T_2)}{\epsilon(B_2, T_1) - \epsilon(B_2, T_2)} \tag{4.7.4}$$

$$= \frac{\epsilon_p(B_1, T_1) - \epsilon_p(B_1, T_2)}{\epsilon_p(B_2, T_1) - \epsilon_p(B_2, T_2)}. \tag{4.7.5}$$

Equation (4.7.5) follows from (4.7.4) since the diamagnetic term ϵ_d cancels out. Since, according to (4.3.13)

$$\epsilon_p = CS\, B_s(x), \tag{4.7.6}$$

where C is a proportionality constant which contains $g_e \mu_B$ and the right and left circular polarized absorption cross sections, and, thus, information on the nature of the excited states. As will be seen below, C does not enter the spin determination. From (4.7.4) and (4.7.5) follows

$$R(S) = \frac{B_s(S, B_1/T_1) - B_s(S, B_1/T_2)}{B_s(S, B_2/T_1) - B_s(S, B_2/T_2)}. \tag{4.7.7}$$

Therefore, a measurement according to (4.7.4) can be compared to the calculated $R(S)$ varying S. For the correct value of S both must be equal: $R_{exp} = R(S)$ [4.36].

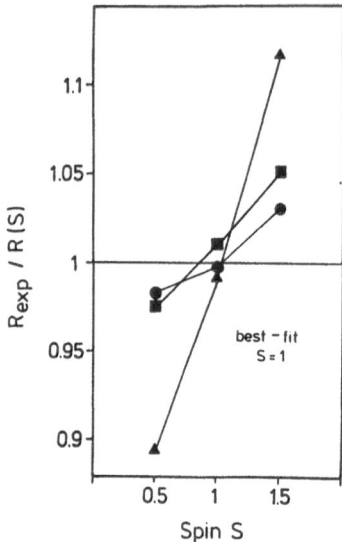

Fig. 4.21. Ratio of R_{exp} and $R(S)$ as a function of S for V_{Ga}^{3+} centers measured at 1.1 eV in high resistivity (p–type) GaAs for the following pairs of magnetic field values B_1 and B_2 (in T): 1.3 and 0.5 (*triangles*), 1 and 0.8 (*circles*), 0.8 and 0.5 (*squares*); at temperatures at $T_1 = 1.65$ K and $T_2 = 4.2$ K (see text). (After [4.36])

The application of the method is illustrated for V^{2+} and V^{3+} defects in GaAs. From the ODEPR spectrum (Fig. 4.15a) the g factor of V^{3+} is known to be $g = 1.96 \pm 001$. A determination of (4.7.4) at 1.1 eV (Fig. 4.15b) and comparison to (4.7.7) for varying S, gives Fig. 4.21 for three pairs of B_1 and B_2 at temperatures T_1 and T_2. It is seen that $S = 1$. This is in agreement with an earlier spin determination by ENDOR [4.46,47]. Figure 4.22a shows the ODEPR spectrum of V^{2+} defects and Fig. 4.22b the MCDA spectrum in a GaAs sample, which contains both V^{2+} and V^{3+} defects $(g(V^{2+}) = 2.07 \pm 0.02)$. Comparison with Fig. 4.15b shows that at 1.25 eV there is no MCDA due to V^{3+} defects, only that of V^{2+} defects. The MCDA of V^{2+} is rather broad and featureless throughout the spectral range of Fig. 4.22b. Applying the same method, one finds $S = 1/2$ (Fig. 4.23) [4.36]. The $3d^3$ configuration of V^{2+} has a low spin configuration in agreement with theoretical predictions [4.48].

The method outlined above is strictly valid only for an orbital singlet state. V^{2+} has, however, an ^2E ground state, which has a twofold orbital degeneracy. A splitting of the twofold degeneracy is not observed (the EPR line is isotropic). If a splitting was present, however, one would have to take into account that, besides the ground state, a nearby excited state could contribute to the ground state spin polarization, its amount depending on the magnetic field. Figure 4.24 shows the level scheme for the assumption of

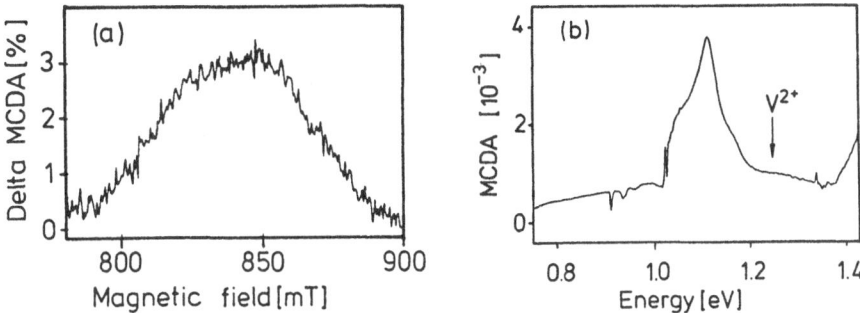

Fig. 4.22. a ODEPR spectrum of V_{Ga}^{2+} centers in high resistivity (n–type) GaAs measured at 1.25 eV with the MCDA method ($g = 2.07\pm0.02$) ($\nu_{EPR} = 24.42$ GHz, $T = 1.5$ K). **b** Magnetic circular dichroism of high resistivity (n–type) GaAs:V ($T = 1.6$ K, $B = 1$ T). The *Fermi* level is pinned to the $V^{2+/3+}$ level. (After [4.36])

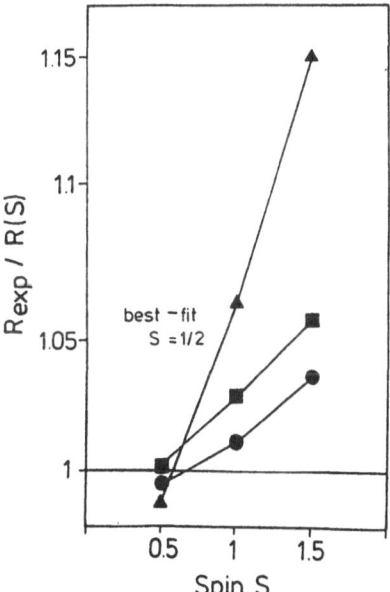

Fig. 4.23. Ratio of R_{exp} and $R(S)$ as a function of S for V_{Ga}^{2+} centers measured at 1.25 eV in high resistivity (n–type) GaAs for the following pairs of magnetic field values B_1 and B_2(T): 1.3 and 0.5 (*triangles*), 1 and 0.8 (*circles*), 0.8 and 0.5 (*squares*), and temperatures $T_1 = 1.65$ K and $T_2 = 4.2$ K (see text). (After [4.36])

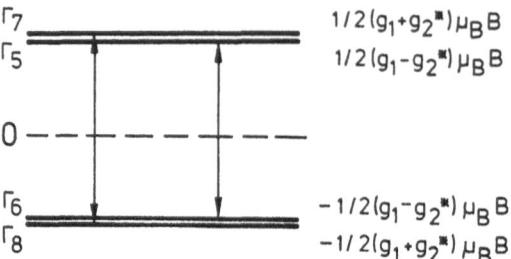

Fig. 4.24. Energy levels of a ^2E state with spin–orbit splitting of the doubly degenerated states. For details see text. (After [4.51])

a small splitting, which is described by a small g factor $g_2^* \ll g_1$. In tetrahedral symmetry, which is realized here, g_2^* is determined by the spin–orbit splitting and crystal field splitting ($g_2^* \approx \lambda/\Delta$). It is shown in Ref. [4.49] that for tetrahedral symmetry the magnetization is given by:

$$M = \frac{\mu_B}{2} \left[g_1 \tanh \left(\frac{g_1 \mu_B B}{2kT} \right) + g_2^* \tanh \left(\frac{g_2^* \mu_B B}{2kT} \right) \right], \qquad (4.7.8)$$

where g_2^* is small compared to g_1 and angular dependent. $g_2^* = 0$ for $\mathbf{B}_0 \parallel <111>$ and $g_2^* = -4\lambda/\Delta$ for $\mathbf{B}_0 \parallel <001>$ (λ is the spin–orbit interaction constant, Δ the crystal field splitting). The spin polarization was measured for several crystal orientations and no deviation from the spin value $S = 1/2$ was found. If there is a splitting in the case of V^{2+} in GaAs, it must be very small. In general, for an orbitally generated state one cannot simply use (4.7.7) to determine the spin state. One must calculate the spin polarization according to the details of the system, and then perform analogue measurements as outlined above.

The determination of the spin state with the MCDA method can also be used in a different kind of application. One is to apply the method for all optical wavelengths, i.e., to perform a sort of "tagging" experiment for the spin determination, and the other is to distinguish between diamagnetic and paramagnetic MCDAs with the aim of discriminating between superimposed MCDAs for different defects.

Recent measurements of Fe^{3+} defects substituting for In in InP, will be discussed to illustrate the ideas. Fe^{3+} is in a $3d^5$ state, and its ODEPR spectrum measured in the MCDA has a corresponding fine–structure splitting. The spin determination according to (4.7.7) yields $S = 5/2$. Figure 4.25a shows the measurement of the total MCDA (curve a). Between 0.76 and about 0.86 eV, a rich fine–structure is resolved due to excitonic transitions at the Fe^{3+} ($Fe^{3+} \Rightarrow Fe^{2+}$ + bound hole) with phonon replicas (Fig. 4.25b). Several of these lines were also observed by calorimetric absorption spectroscopy [4.50]. The strong MCDA setting in at about 1.15 eV is due to ionizing transitions. The onset is marked by an arrow in Fig. 4.25a. Figure 4.25c shows

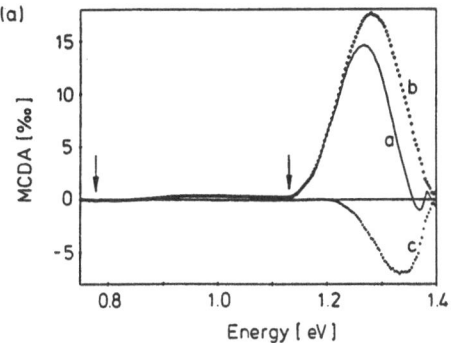

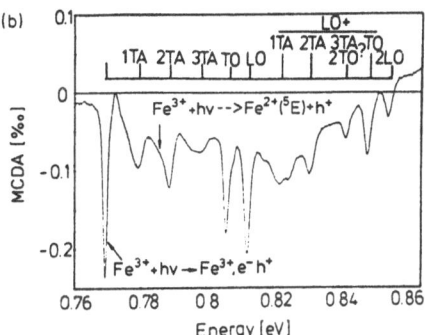

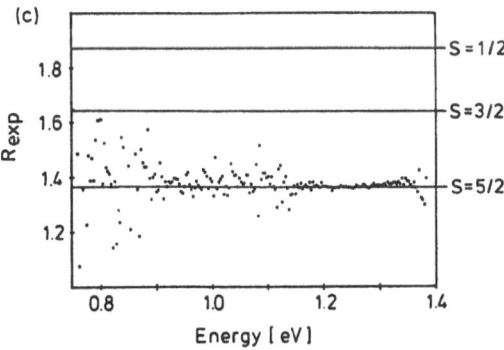

Fig. 4.25. a Magnetic circular dichroism of the absorption of InP:Fe ($B = 1$ T, 1.5 K). *Curve a*: total MCDA; *Curve b*: paramagnetic part of the MCDA due to Fe^{3+} centers; *Curve c*: diamagnetic part due to an unknown defect. **b** measured with higher resolution around 0.8 eV. **c** Determination of R_{exp} as a function of photon energy and comparison with various values of S. Clearly, Fe^{3+} has $S = 5/2$. (After [4.51])

R_{exp} according to (4.7.4), measured as a function of photon energy, and $R(S)$ for $S = 5/2$ and $3/2$ as horizontal lines.

It is seen that S equals $5/2$, as was expected throughout the total spectrum. The noise at low photon energy is, of course, due to the weak MCDA effect there. Such a spin–tagging experiment can be useful if there are several MCDAs superimposed due to different spin states of defects which can be discovered in this way.

Figure 4.25a (curve b) shows the paramagnetic part of the MCDA which can be determined by measuring $\epsilon(B_1, T_1) - \epsilon(B_1, T_2)$, i.e., the MCDA at different temperatures and the same field. It does not agree with curve a between 1.2 eV and 1.4 eV, since there a diamagnetic MCDA (curve c in Fig. 4.25a) is superimposed. Curve b was scaled such that the paramagnetic part and the total MCDA agree at 1 eV, where only a paramagnetic MCDA exists which is known from EPR tagging. The diamagnetic MCDA is due to yet another unknown defect (note that the diamagnetic MCDA from the *Zeeman*–effect in the excited state of Fe^{3+} would be smaller by orders of magnitude and buried in the noise of the experiment) [4.36,51]. Therefore, when applying the MCDA method, it is useful to check the field and temperature dependence of the MCDA throughout the photon energy range.

4.8 EPR of Ground and Excited States Detected with Optical Pumping

Optically detected paramagnetic resonance can also be measured by detecting microwave–induced changes of the optical emission. There are several ways to achieve this, which will be discussed in the following sections. In principle, one can take the level scheme of Fig. 4.6 and reverse the sense of the optical transitions from absorption to emission. The initial states would then be the relaxed excited states, and the final states the unrelaxed ground states of the defect system. Instead of measuring the magnetic circular dichroism of the absorption, one measures the magnetic circular polarization of the emission (MCPE). The condition for seeing a resonance in the excited state would be that the magnetic sublevels are occupied differently (e.g., by nonradiative transitions from unrelaxed excited states). The microwaves would then change the distribution of these occupations, and would have to do this within the radiative lifetime of the excited states. A different occupation of the magnetic sublevels can, of course, also be due to spin–lattice relaxation, as in the ground state. However, this requires that the spin system is able to thermalize in the excited state within a time which is short compared to the radiative lifetime. Especially at low temperature, where the radiative lifetime is long enough, so that the microwave transition rate is sufficiently fast to cause occupation changes within that radiative lifetime, T_1 is often very long and thermalization is not possible within the radiative lifetime.

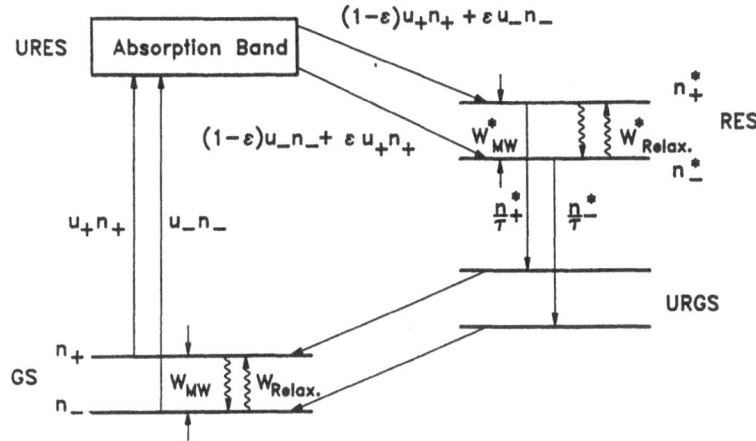

Fig. 4.26. Schematic representation of the optical pumping cycle of a *Kramers* doublet

A selective feeding of the magnetic sublevels can also result from diffe-rent radiative probabilities for the different magnetic sublevels. One takes advantage of this effect mainly when investigating the excited triplet states and donor–acceptor pair recombination luminescence, to be discussed in Sects. 4.9,10.

A different population of magnetic sublevels of ground and excited states can also be achieved by optical pumping and spin memory effects. This will be discussed in this section.

The optical pumping cycle essentially consists of the following four pro-cesses: absorption, that is the transition into the unrelaxed excited states (URES); nonradiative transitions from the unrelaxed excited states into the relaxed excited states (RES); radiative transitions (emission) from the relaxed excited states into the unrelaxed ground state (URGS); and finally, the sy-stem returns in nonradiative transitions into the relaxed ground state (GS), see Fig. 4.26 for a *Kramers* doublet. The spin polarization of the ground state can be changed by going through the optical pumping cycle, if there is no perfect *spin memory*. The term *spin memory* refers to a connection between the initial ground state magnetization M_0 in thermal equilibrium, and to the magnetization M in a later state, to which the system is carried by optical pumping. Spin memory is said to be positive, if the ground state magne-tization maintains the same sign, and negative, if it is reversed by optical pumping.

If, for example, by means of the selection rules for the optical absorption, only spins of the $m_S = -1/2$ state (a *Kramers* doublet) are pumped, they can return without spin reversal into their original state, in which case they would have perfect spin memory. If the spins returned into the $m_S = +1/2$

state, they would have perfect negative spin memory. Usually, there is only a certain probability that they return without a spin reversal. If this probability is only 50%, then there is a complete loss of spin memory [4.25]. It is often found that the spin memory loss is rather small, in spite of the many electron–phonon interactions involved in the relaxation process in the excited states. The orbital–lattice interaction does not flip the spins. However, by spin–orbit interaction, spin flips can occur in radiative and nonradiative transitions. A detailed picture of what happens depends very much on the details of the system under study, and is thoroughly investigated for F centers in alkali halides [4.1,25] (see also [4.52] for a recent review).

An application of spin memory effects in optical pumping is the observation of the ODEPR of both the ground and relaxed excited states of a defect. There are possibilities to detect the EPR in both absorption and emission by either monitoring the MCDA or MCPE. This is discussed for a *Kramers* doublet in a simplified scheme [4.25]. In Fig. 4.26 u^+ and u^- are the optical pumping rates from the states $|+\rangle$ and $|-\rangle$, respectively, for a given sense of the circularly polarized light. They are proportional to light intensity and matrix elements. n_+ and n_- are the populations in the ground state, n_+^* and n_-^* are the populations and τ is the radiative lifetime of the relaxed excited states.

The parameter ε is the spin–mixing parameter, which describes the loss of spin memory upon going from the URES ("absorption band") with non-radiative transitions to the RES. $\varepsilon = 0$ means perfect spin memory ($\varepsilon = 1$ perfect negative spin memory), $\varepsilon = 0.5$ describes total loss of spin memory.

For the populations n_+, n_-, n_+^* and n_-^* of Fig. 4.26, it is a straightforward matter to write down the rate equations in an analogous way as was done for the *Zeeman* doublet in Chap. 2. For W_{Relax}, see (2.7.8) and (2.7.20). The solution for the stationary state is easily obtained, see also [4.25] and [4.1]. T_1 is the spin–lattice relaxation time of the GS, T_1^* that of the RES. The ground state polarization P_g is obtained to be

$$P_g = \left(\frac{n_+ - n_-}{n_+ + n_-}\right) = \frac{P_{es} - (T_p/T_1) \tanh{(g_e\mu_B B_0/2kT)}}{[1 + (T_p/T_1)]}, \qquad (4.8.1)$$

where

$$P_{es} = \left(\frac{u^- - u^+}{u^- + u^+}\right) \qquad (4.8.2)$$

is the so–called "saturation polarization", and

$$T_p^{-1} = \varepsilon(u^+ + u^-). \qquad (4.8.3)$$

Thus, the ground state polarization can be influenced by the optical pumping. If $\varepsilon = 0$, that is complete spin polarization, T_p goes to infinity and P_g reaches $-\tanh(g\mu_B B_0/2kT)$ which is the equilibrium value. For $\varepsilon = 0$ there is no optical pumping effect expected. However, for $\varepsilon \approx 0$, T_p is finite. If the pumping light is sufficiently intense, then $T_p \ll T_1$, and $P_g \Rightarrow P_{es}$. The saturation polarization may be positive or negative, depending on whether u^+ or u^- prevails. Under saturating pumping conditions the ground state polarization is given by the *spin memory* and the optical transition rates, rather than T_1. Thus, optical pumping may be used to drive P_g from thermal equilibrium, and to measure T_1, after switching off the pumping light, by observing the recovery of the polarization to the thermal equilibrium value (Sect. 4.6).

Inversion of the ground state polarization by optical pumping has been observed, for example, for F centers in KCl [4.25] and for Cr^{3+} in Al_2O_3 [4.53].

The polarization of the excited state for a fixed polarization of the pumping light (circular or linear polarized light) depends on the assumption that $\tau/T_1^* = 0$ (i.e., T_1^* is long compared to the radiative lifetime, which is usually the case).

$$P_e = \left(\frac{n_+^* - n_-^*}{n_+^* + n_-^*} \right), \tag{4.8.4}$$

$$= \frac{(1 - 2\varepsilon)[u^- - u^+ + (u^+ + u^-)\tanh(g_e\mu_B B_0/2kT)]}{\{[(u^+ + u^-) + (u^+ - u^-)\tanh(g_e\mu_B B_0/2kT)] + 4\varepsilon u^+ u^- T_1\}}. \tag{4.8.5}$$

From (4.8.5) a very interesting behavior is seen for two extreme cases. One is that there is only weak pumping with unpolarized light, for which $u^+ = u^- = \frac{1}{2}u$. One obtains

$$P_e = \left(\frac{1 - 2\varepsilon}{1 + \varepsilon u T_1} \right) \tanh \left(\frac{g_e\mu_B B_0}{2kT} \right). \tag{4.8.6}$$

For weak pumping $\varepsilon u T_1 \ll 1$. Then one obtains a polarization in the excited state

$$P_e = (1 - 2\varepsilon)\tanh \left(\frac{g_e\mu_B B_0}{2kT} \right). \tag{4.8.7}$$

The ground state polarization is pumped into the relaxed excited state diminished by the factor of $(1 - 2\varepsilon)$. If there is a fast relaxation in the excited state, then (4.8.7) is to be modified by considering the effect of T_1^* (for details see [4.25]). For strong pumping ($\varepsilon u T \gg 1$), $P_e \Rightarrow 0$ if one assumes one single "overall" spin–mixing parameter ε as done so far. Things change by assuming more realistically different spin–mixing parameters for the transitions during the pumping cycle.

There are two ways to use the effect of optical pumping to measure the EPR of the relaxed excited state. One is to use the MCDA method in connection with (4.8.1) and (4.8.5). A resonant microwave transition between the

Zeeman levels of the RES will change their populations if the transition rate is large enough compared to the inverse of the radiative lifetime. The radiative lifetime must be of the order of at least 100 ns to $1\,\mu s$ for the available microwave powers at low temperature. This population change is equivalent to an increase in ε in the pumping cycle, which, in turn, changes P_g in (4.8.1) by changing the pumping rate via decreasing T_p in (4.8.3). Thus, if one monitors P_g by a continuous MCDA measurement (as discussed in Sect. 4.4) while pumping with u^+ or u^- light, one can detect the EPR in the RES by measuring the changes in the ground state polarization induced by a change of the optical pumping and effective spin–mixing parameter ε. In this way, one can measure both the ground state EPR and the excited state EPR of a defect. Thus, the excited state EPR is detected by its influence on the ground state polarization P_g, which, in turn, is monitored by MCDA. The variation ΔP_g of P_g can be calculated from (4.8.1) by differentiating [4.1]:

$$\Delta P_g = \frac{\partial P_g}{\partial T_p} \Delta T_p = \frac{(-1/T_1)}{[1 + (T_p/T_1)]^2} \left[P_{es} + \tanh \left(\frac{g_e \mu_B B_0}{2kT} \right) \right] \Delta T_p. \quad (4.8.8)$$

ΔP_g has a maximum value for $T_p = T_1$. Experimentally, the pumping light intensity is chosen such as to fulfill this condition. It is shown for F centers in alkali halides, that it is advantageous experimentally to pump with modulated circularly polarized light, quickly alternating between u^+ and u^- with the frequency ω_m [4.54]. For the ground state polarization one then obtains

$$P_g = \frac{1}{(1 + \varepsilon u T_1)} \tanh \left(\frac{g_e \mu_B B_0}{2kT} \right). \quad (4.8.9)$$

For strong pumping, P_g can be diminished, as well as changing the effective ε, which in turn can be monitored as a change in the MCDA.

The excited state polarization is also dependent on the modulation frequency ω_m [4.55,56]. The other way to observe both ground and relaxed excited state EPR is by monitoring the paramagnetic part of the magnetic circular polarization of the emission MCPE. Since

$$\text{MCPE}_p = \left(\frac{I_l^e - I_r^e}{I_l^e + I_r^e} \right) = \beta_e P_e, \quad (4.8.10)$$

where β_e describes the emission probability (selection rules, matrix elements for the excited states). $I_{r,l}^e$ is the emitted intensity of right and left polarized light. For pumping with unpolarized light, for example, P_e is given by (4.8.5). By a resonance transition in the RES, ε is effectively increased, therefore, P_e is decreased and so is the MCPE in (4.8.10), which is monitored to detect the EPR of the RES. On the other hand, a change in P_g due to a resonance in the ground state also decreases P_e, since less ground state polarization is pumped to the relaxed excited state. Therefore, the resonance of the ground state can also be monitored in the MCPE.

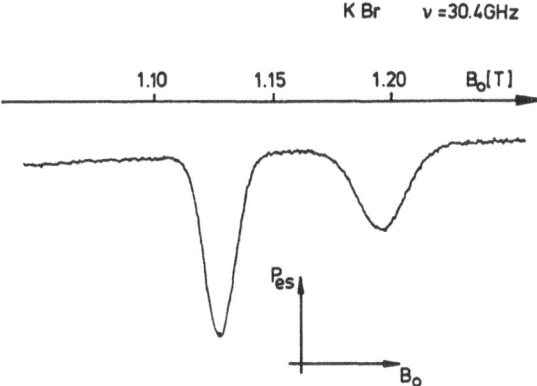

Fig. 4.27. EPR spectrum of the ground state and relaxed excited state of F centers in KBr with the MCDA method and saturated optical pumping ($\nu_{EPR} = 30.4$ GHz). (After [4.58])

Because of the pumping cycle and spin memory effects, in principle, both GS and RES EPR can be obtained in either the MCDA or the MCPE. When it comes to a specific system, the details of the experiments to be performed depend on the specific properties of the defect system.

For example, there may be two different spin–mixing parameters ε^+ and ε^- for the two magnetic sublevels of the GS [4.56,57] and another $\varepsilon_{r,l}$ for the relaxation process to the RES, as was found for F centers in alkali halides when studying the optical pumping cycle in detail. These details determine which kind of optical pumping procedures (polarization, intensity, modulation frequency, etc.) must be chosen. Particularly, P_e is not zero for saturating pumping of the RES assuming different ε^+ and ε^-. Therefore, direct ODEPR measurements of the RES are usually possible via monitoring the MCPE.

The methods described above are not suitable for semiconductors unless the excited states of the defects are within the band gap and there is a closed pumping cycle. If they are resonant in the bands, or if there are only ionizing transitions, then there is a complete loss of *spin memory* when the electrons or holes thermalize in the bands. For recent reviews and further references on ODEPR for this optical pumping technique in ionic crystals see [4.1,52].

To illustrate the obtainable results, Fig. 4.27 shows the EPR spectrum of F centers in KBr in the GS (low field line, $g = 1.984$) and RES (high field line, $g = 1.873$) [4.58] measured with the MCDA technique under the saturated optical pumping conditions. Both spectra are well separated due to their different g factors. They both have an unresolved superhyperfine structure.

In the field of laser physics one is interested in being able to unambiguously correlate both the laser pump band (absorption band) and the laser emission band with a particular defect. The absorption is normally correla-

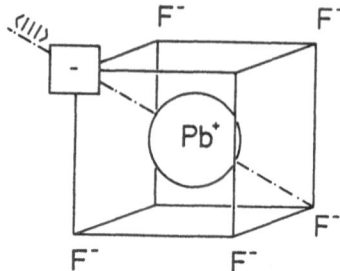

Fig. 4.28. Model of Pb$^+$(1) centers in alkali earth fluorides. (After [4.32])

ted to the emission with a measurement of the wavelength dependence of the excitation of the emission band. However, if there are overlapping absorption bands of several defects, this need not be unambiguous. Therefore, one would like to measure the EPR of the responsible defect both in the laser pump absorption band and the laser emission band. An example for this is shown in Figs. 4.28–31 for the Pb$^+$ centers in SrF$_2$, which are potential candidates for tunable laser centers [4.32]. To create these centers SrF$_2$ doped with Pb^{++} is X–irradiated at room temperature. Pb^{++}, which substitutes for a Sr^{++}, captures an electron and attracts a F$^-$ vacancy. The structure model (Fig. 4.28) was derived by ODENDOR experiments [4.33]. The optical absorption spectrum (Fig. 4.29a) is due to a superposition of many defects, the MCDA (Fig. 4.29b) is due to a superposition of two Pb$^+$ centers, one of which has a Pb^{++} impurity in the near neighborhood [4.32].

The one center to be discussed here (Fig. 4.28) has the "tagged MCD" spectrum of Fig. 4.29b, curve 3, which was measured for the ODEPR spectrum in Fig. 4.30, curve 1, belonging to the ground state of the Pb$^+$(1) centers. The spectrum shows the doublet hf splitting due to the magnetic isotope ^{207}Pb ($I = 1/2$), and a central line due to the nonmagnetic isotopes. In a wavelength region around 660 nm the emission band peaking at 1040 nm is excited (Fig. 4.31). The ground state EPR spectrum could also be measured in this emission as a microwave–induced change of the MCPE, whereby the system was pumped with alternating right and left circularly polarized light. The EPR spectrum is the same as that of the ground state and belongs to the same center (curve 2). The orientation of the magnetic field is slightly different compared to the MCDA experiment so that the central line starts to split according to the angular dependence. In this case, no EPR of the RES can be detected [4.32].

In Fig. 4.32, a section of the EPR spectrum of Tl^0(1) centers in NaCl is shown measured as a microwave–induced change of the MCPE (pumped with alterating polarizations). There are lines due to the ground state [lines (1) in Fig. 4.32] and the relaxed excited state of the center [lines (2) in Fig. 4.32]. If the modulation frequency of the pumping light becomes too fast, then the

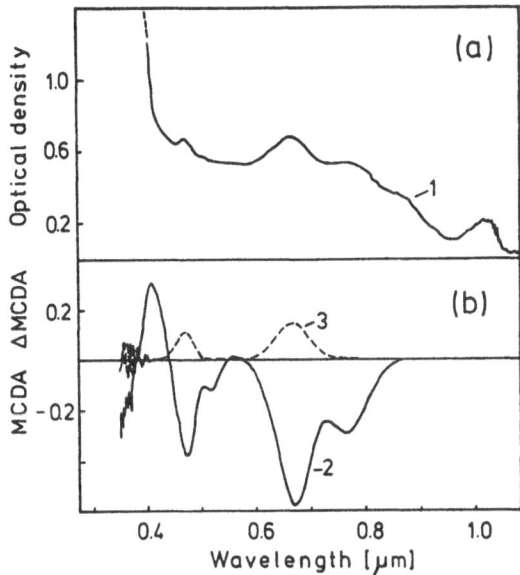

Fig. 4.29. a Optical absorption spectrum of SrF_2:Pb^{2+} after X–irradiation at 300 K measured at 10 K. **b** *Curve 2*: magnetic circular dichroism of the spectrum of **a**. *Curve 3*: magnetic circular dichroism tagged by the EPR spectrum of $Pb^+(1)$ centers ($B = 1.25$ T, $\nu = 24$ GHz, $\mathbf{B_0} \parallel [100]$). (After [4.32])

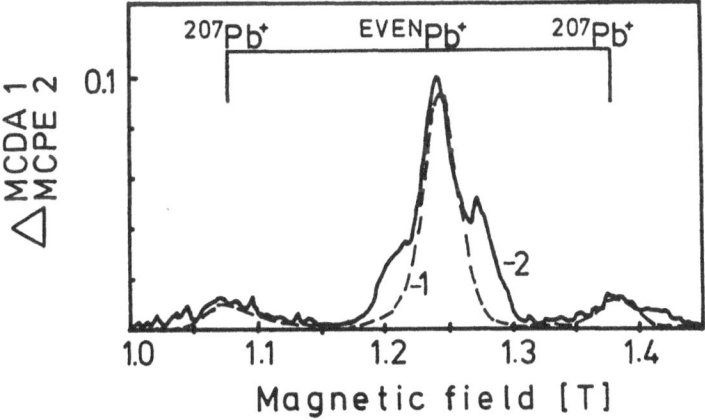

Fig. 4.30. EPR spectrum of the ground state of $Pb^+(1)$ centers in SrF_2 measured with the MCDA method (*curve 1*) and in the MCPE of the emission at 1.061 μm under optical pumping conditions (*curve 2*) ($T = 1.5$ K). Optical pumping was done with the unpolarized light of a 6 mW He–Ne laser. $\mathbf{B_0} \parallel [100]$ for *curve 1*; $\mathbf{B_0}$ was a few degrees off [100]. (After [4.32])

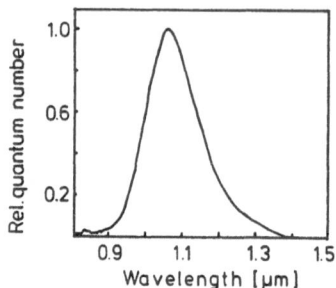

Fig. 4.31. Optical emission spectrum of Pb$^+$(1) centers in SrF$_2$ excited at 0.66 μm ($T = 10$ K). (After [4.32])

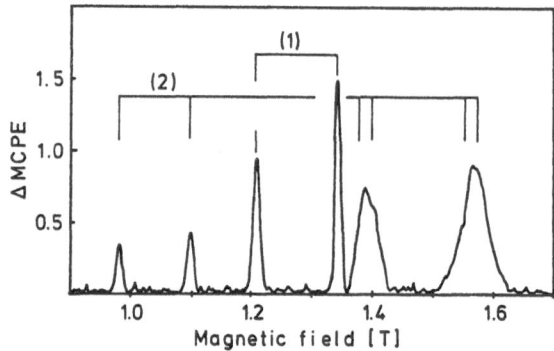

Fig. 4.32. EPR spectrum of the ground state and relaxed excited state of Tl^0(1) centers in NaCl measured under optical pumping conditions in the MCPE of the emission at 1.5 μm (**B**$_0$ ∥ [110], $\nu = 24$ GHz, $T = 1.5$ K). The optical pumping was done with modulated ($\nu_m = 1.2$ GHz) linearly polarized light of a ca. 500 mW Nd–YAG–laser at 1.065 μm. The lines (1) are from the ground state, the lines (2) are from the relaxed excited state (After [4.59])

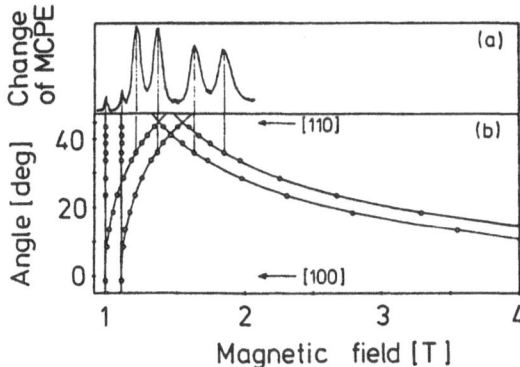

Fig. 4.33. a EPR spectrum of the relaxed excited state of $Tl^0(1)$ center in NaCl (1.6 K, $\nu = 24$ GHz) measured in the MCPE under similar conditions as in Fig. 4.32. The modulation frequency of the pumping light was higher so that the ground state EPR lines were not measured because of the longer spin–lattice relaxation time of the ground state. **b** Angular dependence of the EPR spectrum. The magnetic field was rotated in a {110} plane from [110] to {100}. (After [4.59])

GS resonance is not seen any more, only that of the RES, because T_1 is in the ground state too long. Figure 4.33 shows the EPR spectrum of the RES of $Tl^0(1)$ centers in NaCl, and the angular dependence of this axial center for rotation of the magnetic field in a {100} plane. The (100) symmetry of the defect is also present in the relaxed excited state [4.59].

Another example for the measurement of the EPR spectrum of the RES with the optical pumping technique of an impurity center is that of interstitial atomic hydrogen centers in alkali halides [4.60,61].

In semiconductors, optical pumping and *spin memory* effects were used to investigate the EPR of conduction electrons in the conduction–to–valence band emission. The selection rules for circularly polarized absorption transitions between the conduction band states and those of the valence band, which are characterized by a large spin–orbit splitting in analogy to the discussion of defects in Sect. 4.3, were used to produce a pumped spin polarization of the conduction electrons, and to monitor the microwave–induced change of the MCPE. Such experiments were successful, in particular, in the III–V semiconductors GaSb [4.62,63] and GaAs, where the valence band spin–orbit interaction energy is of the same order as that of the band gap energy. For details see [4.64] and further references therein.

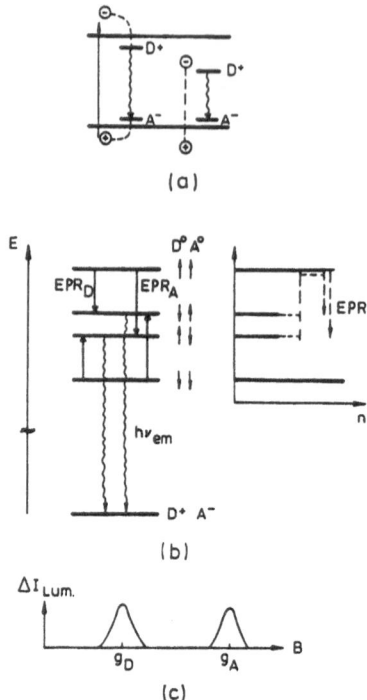

Fig. 4.34a–c. Schematic representation of the ODEPR measurement in the donor–acceptor pair recombination luminescence. **a** donor–acceptor pair recombination **b** level scheme for weakly coupled donor–acceptor pairs **c** EPR spectra from donors and acceptors

4.9 EPR Optically Detected in Donor–Acceptor Pair Recombination Luminescence

The donor–acceptor pair recombination luminescence was already discussed in Sect. 4.1; it is schematically shown once more in Fig. 4.34a. Donors D^0 can be described by $S = 1/2$, while for acceptors the angular momentum depends on the energy position relative to the valence band. If the acceptors are very shallow, that is, if their energy level is very close to the valence band, the wave function depends on the upper valence band states and $J = 3/2$. However, if they are deep enough, then $J = S = 1/2$. In order to have recombination luminescence, donors and acceptors must be at least slightly coupled with a slight overlap of their wave functions. In this case, the level scheme of Fig. 4.34b applies. For weak coupling, the energy levels of the donors D^0 and A^0 are determined only by their g factors and the magnetic field.

The necessary difference in population of the magnetic sublevels is usually not caused by spin–lattice relaxation processes but by the different probabi-

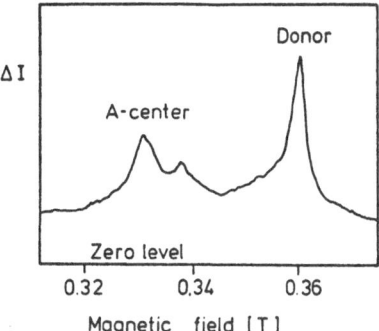

Fig. 4.35. ODEPR spectrum of donors and deep acceptors (A centers) in hexagonal InS for $B_0 \parallel$ c–axis measured in the donor–acceptor pair luminescence at 2.70 eV. (After [4.65])

lities for radiative transitions into the singlet ground state. The transition probability for the two triplet spin configurations into the singlet ground state is low compared to the two singlet–singlet transitions, where the donor and acceptor spins are antiparallel. Therefore, the spin system will have a stationary overpopulation in the two triplet states upon continuous excitation with band gap light. By inducing EPR transitions between the donor or acceptor magnetic sublevels, as illustrated in Fig. 4.34b, the overpopulation in the triplet states is reduced, and the recombination luminescence out of the singlet states is enhanced. Therefore, it is sufficient to monitor the change of the luminescence intensity to detect EPR (Fig. 4.34c).

One can thus see both the donor and acceptor EPR lines, provided the g factors are sufficiently different. Very often one observes, however, only the donor resonances since the p–type hole states of the acceptors experience a dynamical *Jahn–Teller* effect, which makes the resonance very difficult to observe. The method just described has the high sensitivity of the optical detection and requires only a rather simple spectrometer, which can be built by modifying a conventional EPR spectrometer (Chap. 9). Many experiments were performed with this technique, in particular in the II–VI semiconductors, which was reviewed by *Cavenett* [4.64].

As an example, Fig. 4.35 shows the ODEPR spectrum of donors and deep acceptors in hexagonal ZnS for **B** $\parallel$ c–axis. The donor is isotropic ($g = 1.886$). The acceptor is axially symmetric about one of the four <111> bond directions ($g_{\parallel} = 2.006, g_{\perp} = 2.052$). The spectrum is recorded as a microwave–induced change of the luminescence intensity at 2.70 eV. The acceptor is associated with the A center, known from conventional EPR to be a Zn–vacancy–Cl–complex (V_{Zn}–Cl), where Cl is next to the V_{Zn} along <111> [4.65]. If there are shallow acceptors characterized by $J = 3/2$, one may be

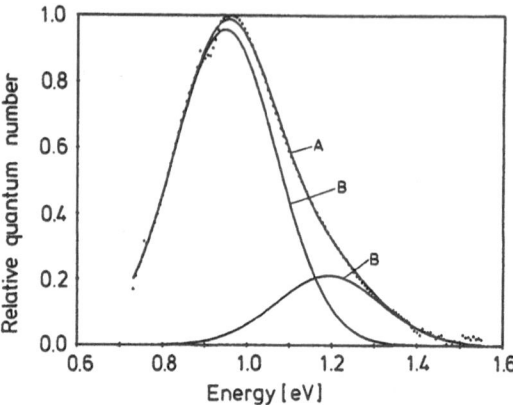

Fig. 4.36. Infrared luminescence spectrum of GaP:Zn after above–band–gap excitation ($T = 10$ K) (*curve A*). *Curves B* represent a decomposition into two symmetric *Gaussian* bands. (After [4.66])

able to observe an MCPE due to the selection rules for circularly polarized light. Such effects are seen in GaAs:Cr [4.64].

As in the case of the MCDA method, it is possible to directly correlate the emission band with a particular EPR line (i.e., a particular defect) by monitoring the EPR line while varying the wavelength of the emission. A recent example in III–V semiconductors is shown in Fig. 4.36, which shows two strongly overlapping luminescence bands in GaP:Zn (p–type) peaking at 0.95 and 1.20 eV, respectively, after excitation with light of energy exceeding the band–gap energy. In Fig. 4.37, the EPR spectrum measured in the 0.95 eV band is that of the PP_4 antisite defect, which shows the characteristic doublet hf splitting with the ^{31}P nucleus ($I = \frac{1}{2}$) of the P atom occupying a Ga site [4.66]. Figure 4.38 shows the ODEPR spectrum measured in the 1.20 eV luminescence, which is caused by a P antisite defect in a triplet state and the acceptor (middle line) that is participating in the luminescence process. The two low and high field lines are caused by the antisite defect. The splitting between the two lines of each doublet is caused by the ^{31}P hf interaction, while the field separation of the centers of the two doublets (between ca. 240–400 mT) corresponds to the separation caused by the fine–structure interaction. It was concluded that the structure model is a PP_3Y_p defect with an unknown atom on one of the four nearest neighbor P sites, and that the acceptor may be a nearby interstitial Fe^{3+} impurity [4.66].

When exciting the donor–acceptor pair recombination luminescence with light of energy above the band–gap, one excites only a rather small layer of the crystal because of the high optical absorption constant for the light of energy exceeding the band–gap. One usually excites the sample only in a layer of 1 μm, thus limiting the number of donor–acceptor pairs to be

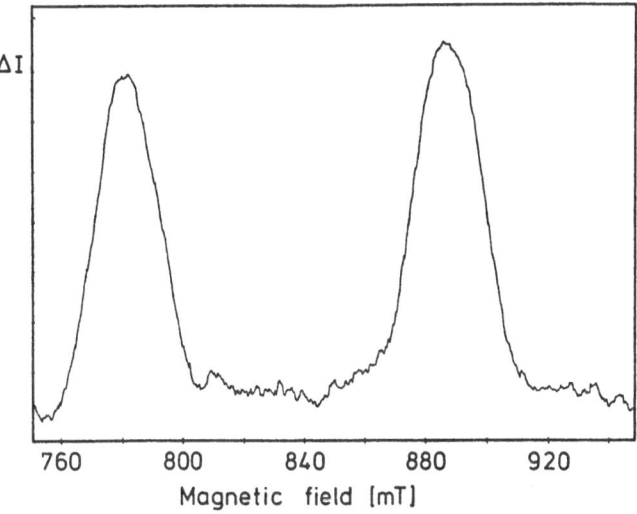

Fig. 4.37. EPR spectrum of the PP_4 antisite defect in GaP:Zn measured in the donor–acceptor pair recombination luminescence at 0.95 eV. $\mathbf{B_0} \parallel [111]$, $T = 1.5$ K, $\nu = 23.42$ GHz. The hyperfine splitting is due to $^{31}P(I=1/2)$. (After [4.66])

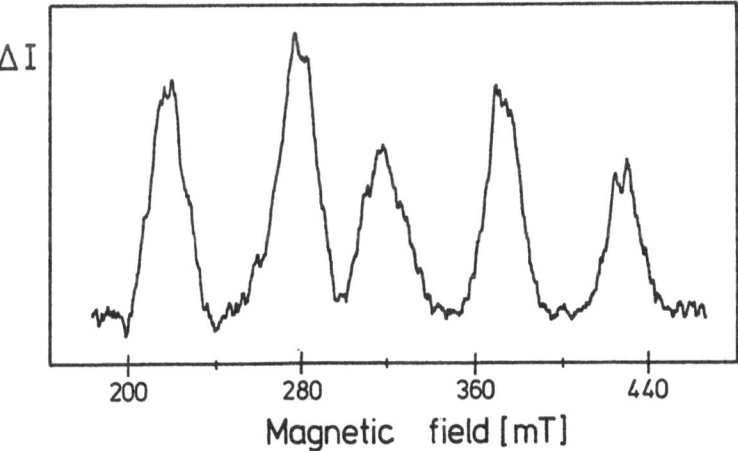

Fig. 4.38. EPR spectrum of the PP_3Y_p triplet antisite defect and acceptor in GaP:Zn measured in the donor–acceptor pair recombination luminescence at 1.2 eV. $\mathbf{B_0} \parallel <111>$, $T = 1.5$ K, $\nu = 9.15$ GHz. For the antisite defect, fine–structure and hyperfine structure is resolved. The line in the middle is due to the acceptor, probably interstitial Fe^{3+}. (After [4.66])

excited. The luminescence from deep donors can also be excited with sub-band–gap light by lifting electrons from the valence band into the D$^+$ states directly, and not via the conduction band. This has the advantage that one excites the whole sample. This can enhance the signal–to–noise ratio. On the other hand, by varying the photon energy of the exciting light, one can also determine the level position of the defect from the onset of the luminescence (or ODEPR) and correlate the EPR spectrum with the energy level position; see also Chap. 5 for further photo–excited experiments.

The luminescence method is the only successful method, so far, to study thin layer samples such as epitaxially grown III–V or II–VI semiconductor crystals.

It is a common observation that the ODEPR spectra observed with the donor–acceptor recombination technique are rather broad. This is usually also true for the emission bands. The emission energy is given by the difference of the energy levels of the donors and acceptors, and the *Coulomb* interaction between them:

$$E = E_g - E_D - E_A + \frac{e^2}{\varepsilon\, r}, \tag{4.9.1}$$

where r is the separation between acceptors and donors, ε is the dielectric constant. For different donor and acceptor pair distances one would expect a series of sharp lines, which was, indeed, observed in GaP [4.67]; also see the review by *Dean* [4.68]. However, the emissions are mostly broadened by exchange interactions and electron–phonon interactions. The ODEPR lines for different separations of donors and acceptors are also broadened by exchange interactions, as shown schematically in Fig. 4.39. This exchange interaction makes the spectra broad and featureless; no hf interactions are resolved, which would be the basis for a structure determination. Therefore, in many cases one can record donor ODEPR spectra, but is not able to say anything about the structure of the defects behind them. The larger the overlap between donor and acceptor wave functions, the larger the exchange interaction, and thus the shorter the radiative lifetime for their recombination luminescence. Close pairs of donors and acceptors have short τ_R values contrary to distant pairs. This points to a way to enhance the resolution of the spectra by performing time resolved experiments. One measures the microwave–induced change of ΔI_e within a certain time window of the emission to detect specifically separated pairs.

An easier way is to detect only distant pairs by performing the ODEPR experiment after all close pairs have already recombined, as was demonstrated by *Cox* et al. for ZnO:In [4.69]. The ODEPR spectrum, measured as microwave–induced change in luminescence intensity in a stationary experiment, showed only broad and featureless lines. Figure 4.40a shows the scheme used for the time resolved experiment. After the exciting light is shut off, there is a delay time before the microwave (hf in Fig. 4.40) and the

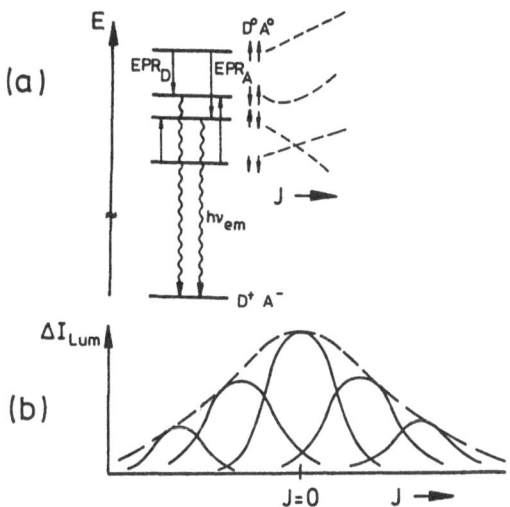

Fig. 4.39a,b. Schematic representation of the ODEPR measurement in the donor–acceptor pair recombination luminescence for exchange coupling between donor–acceptor pairs. **a** Level scheme with increasing exchange coupling J. **b** resulting broad EPR lines

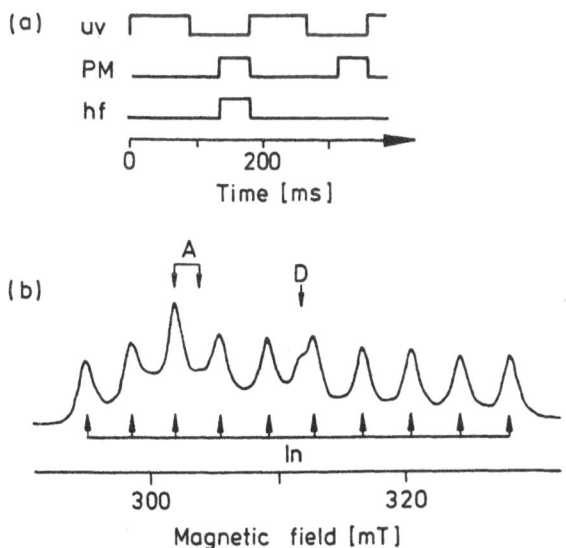

Fig. 4.40a,b. ODEPR measurement with time resolution in the donor–acceptor pair recombination luminescence of ZnO:In. **a** Pulse timing: laser beam (uv), microwave power (hf), photo–multiplier (PM) (high = on, low = off). **b** ODEPR spectrum for $\mathbf{B_0}$ perpendicular to the c–axis, $\nu = 8.53$ GHz. *Lines A, D,* and *In* correspond to Li acceptor, unidentified shallow donors and the In donors with resolved hf structure, respectively. (After [4.69])

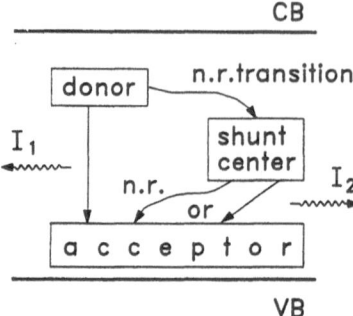

Fig. 4.41. Schematic representation of the influence of a shunt center in the recombination path of a donor–acceptor pair. I_1 is the donor–acceptor pair recombination luminescence, I_2 the recombination luminescence between the shunt center and the acceptor. n.r. means nonradiative transitions (see text). (After [4.70])

photo–multiplier are switched on. Thus, all fast recombination processes are not measured, only those with long radiative lifetimes. This time sequence is repeated and the signals are accumulated. Now the In donor structure is resolved (Fig. 4.40b). There are ten hf lines seen due to the In ($I = 9/2$, for both ^{114,125}In isotopes with nearly the same g_n factors). Sometimes, also with the appropriate choice of the microwave modulation frequency, a certain time window is realized in a crude way and a better resolution is achieved. So far not many experiments with time resolution have been performed. Because of sensitivity in thin expitaxial layers, the only way to observe ODEPR is via luminescence. Therefore, the development of time resolved techniques should be encouraged.

The correlation of an emission band with the EPR lines of a specific defect can be hampered by nonradiative processes, to other deep defects. This is schematically explained in Fig. 4.41, where, besides the donor and acceptor, a so–called "shunt" center is shown, to which the electron from the donor can be transferred in a nonradiative way. The shunt center acts in competition to the donor with respect to the luminescence. An increase in the competing recombination rate I_2 during resonance at the shunt center will cause a decrease in the emission I_1. Therefore, the resonance at the shunt center is seen as a decrease in intensity on the I_1 luminescence (so–called "quenching" signal), while it is seen as an increase in the I_2 emission ("enhancing" signal). If the shunt center is nonradiative, then the competing nonradiative path will result in only quenching signals in I_1. It is important to note that in either case, the quenching signals are not directly related to the emission band that is being monitored: they are due, instead, to the center that provides the competing decay path, Ref. [4.70] and further references therein for such observations.

Also, cross relaxations can be an efficient transfer mechanism of the EPR spectrum of one defect into other emission bands, as observed for the Mn^{2+}

ground state resonance in ZnSe [4.71]. It was pointed out by *Davies* that under certain conditions, energy transfer can also produce enhancement signals in the emission of a defect, which has nothing to do with the resonance seen in this emission [4.70]. Therefore, one must be cautious when correlating emission bands and EPR spectra. For a correlation, it may be useful to use sub–band–gap light for the excitation of an emission, and the ODEPR spectrum as discussed above.

In the case of inhomogenously broadened lines, an enhancement of the ODEPR signal can be achieved by applying an oscillating longitudinal magnetic field. This field sweeps the resonance through the spin packets of the inhomogeneously broadened line, and thus brings all spin packets to participate in the resonance transitions, shifting more populations into the radiative channel. An enhancement factor of ten and more is observed [4.72,73].

4.10 Optically Detected EPR of Triplet States

Triplet states are often found as excited states of two–electron systems such as ns^2 impurity configurations, or, as strongly coupled electrons and holes after a band–to–band excitation, if triplet excitons are formed, which may be free excitons or excitons bound to an impurity. In this latter case, they are interesting probes for the study of the defect structure of the impurity. Triplet states are very common in molecular crystals. Optically detected EPR of triplet states, therefore, has a long tradition. For a recent review see *Lynch* and *Pratt*, where specific reviews on the ODEPR of triplet states in many areas are also listed, including those on inorganic and semicondutor crystals [4.74]. The method is discussed here only very briefly.

Figure 4.42a shows the energy levels for a triplet system with fine–structure interaction for B_0 parallel to the fine–structure axes. In zero–field the $m_S = \pm 1$ and $m_S = 0$ levels are split by the fine–structure constant D (3.4.5). As discussed in Chap. 3, there are the two allowed fine–structure EPR transitions, $|0\rangle \Rightarrow |+1\rangle$ and $|-1\rangle \Rightarrow |0\rangle$. The triplet levels are usually occupied after an optical transition from a singlet ground state into a singlet excited state and subsequent intersystem crossing. A weak radiative transition to the singlet ground state is allowed for the $|+1\rangle$ and $|-1\rangle$ levels, but is not allowed for the $|0\rangle$ level. Therefore, the $|0\rangle$ level will have a higher population compared to the $|+1\rangle$ and $|-1\rangle$ states, as indicated in Fig. 4.42a, provided T_1 is larger than the radiative lifetime, in which case no thermalization of the electrons occurs. The levels $|+1\rangle$ and $|-1\rangle$ are weakly radiative due to admixtures of higher singlet states with the selection rule, that from the $|+1\rangle$ state σ_+ light is emitted, and from the $|-\rangle$ state σ_- light is emitted. EPR transitions will enhance the emitted light intensity by shifting populations from the $|0\rangle$ level into the $|\pm 1\rangle$ levels. If one observes only ΔI_e, then, for both fine–structure transitions, an increase in emitted light intensity is

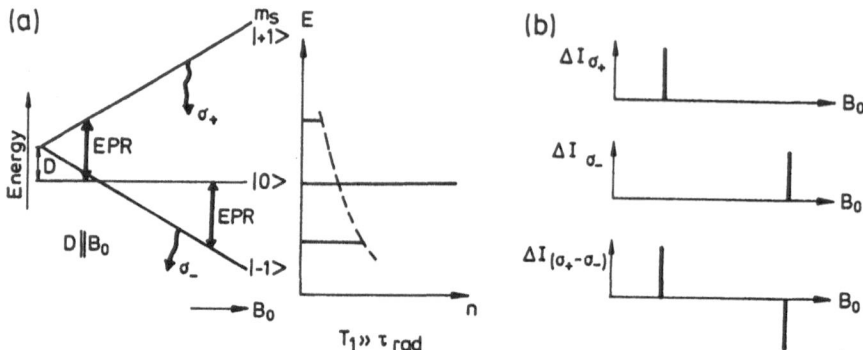

Fig. 4.42. a Level scheme of triplet states to illustrate the optical detection of EPR by microwave–induced changes of the population of radiative levels. B_0 is assumed to be parallel to fine–structure axis and T_1 is assumed to be large compared to the radiative lifetime τ_{rad}. **b** Microwave–induced intensity changes of σ_+ and σ_- light as well as of the MCPE as a function of the magnetic field

observed. Such a spectrum was already shown in Fig. 4.38 for the PP_3Y_p antisite defect in GaP, where, in addition, the ^{31}P hf interaction split the two fine–structure levels. Figure 4.42b shows what is observed by only measuring either σ_+ or σ_- light, or by measuring the MCPE. From the observation of the circularly polarized emission one can determine the sign of D. For a negative D, the low field line in the MCPE would be of σ_- character ($m_s = 0 \Rightarrow |-1\rangle$), the high field line would be of σ_+ character ($0 \Rightarrow |+1\rangle$). For $T_1 \ll \tau$ there is thermalization of the electron in the triplet state. Then, for the low field line one observes an increase in σ_+, and for the high field line a decrease in σ_-emission (for $D > 0$).

An axial triplet state defect investigated thoroughly with these techniques is the F center in CaO, a two–electron system in which an O^{2-} vacancy is occupied by two electrons. Upon optical excitation in the UV, the system goes into an excited triplet state and can be observed by ODEPR. The fine–structure axis is parallel to a [100] cubic axis. Figure 4.43 shows the ODEPR spectra for several modes of observation. It has to be considered that there are three center orientations present simultaneously due to the three cubic axes [4.72]. In a CaO crystal doped with 1% Mg, new F centers are formed, which differ from the F center, in that one or two of the Ca^{2+} ions are replaced by Mg^{2+} (F_A and F_{AA} centers). Figure 4.44 shows the ODEPR spectrum measured in the MCPE of the total emission for wavelengths longer than 665 nm ($B_0 \parallel [100]$), for which both F_A and F_{AA} centers emit. The structure of the two centers is established by resolving the ^{25}Mg ($I = 5/2$) hf structure, which is seen in Fig. 4.45 [4.73]. The line intensities of the six line hf structure relative to the lines due to the nonmagnetic isotopes, correspond to the abundance of ^{25}Mg of 10%. For F_{AA} centers, the intensity ratio is twice

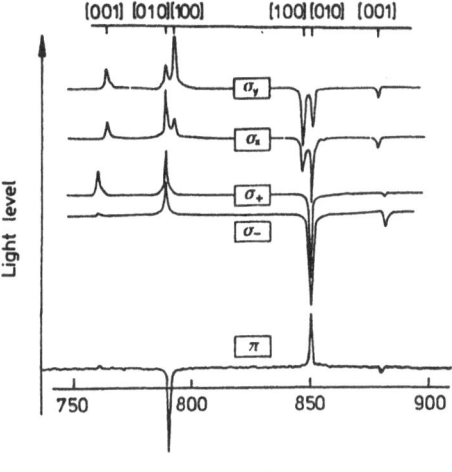

Fig. 4.43. EPR spectrum of F centers in CaO in the excited triplet state for $B_0 \parallel [100]$ measured as microwave–induced change of the triplet phosphorescence for several polarizations of the emitted light. The π–components were observed perpendicular to the field orientation. (After [4.72])

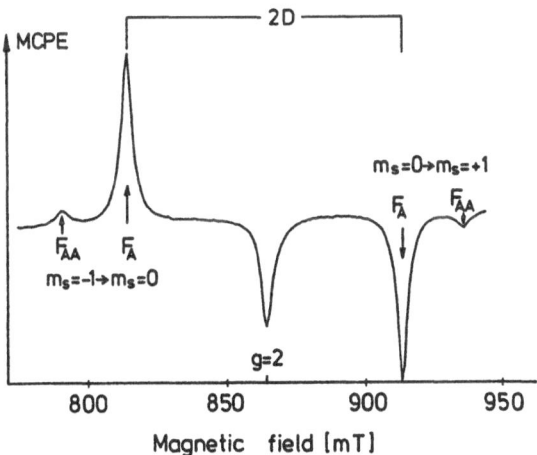

Fig. 4.44. Magnetic circular polarization of the emission of the excited triplet states at wavelengths longer than 665 nm under cw–microwave irradiation for $B_0 \parallel [100]$ in CaO:Mg^{2+} containing F, F_A and F_{AA} centers. $T = 1.6$ K, $\nu = 24$ GHz. (After [4.76])

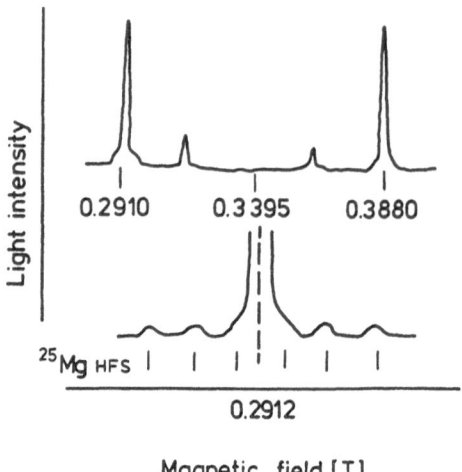

Fig. 4.45. EPR spectra of F_A centers in CaO:Mg^{2+} in the excited triplet state for $\mathbf{B}_0 \parallel$ [100] measured as triplet phosphorescence intensity change under cw–microwave irradiation (X–band). The lower trace shows the lower field fine–structure line with higher resolution, where the ^{25}Mg shf structure is resolved. (After [4.75])

that of the F_A centers, since there are two Mg^{2+} ions interacting. Because of the tetragonal symmetry of the center, the two Mg^{2+} must be opposite to each other along <100> [4.75].

The spectrum of Fig. 4.44 shows two unusual features. One is an off–set of the MCPE outside resonance, and the other is that the MCPE becomes zero for $g_e \approx 2$ (precisely at the g factor of the ground state of F_A^+ centers, which can be measured with convential EPR). These observations can be explained by the tunneling of electrons in the excited state (Fig. 4.46). From an excited F center an electron tunnels to an F_A^+ center, forming an F_A center in the excited triplet state. In the magnetic field the occupation of the two *Zeeman* levels of F_A^+ is different in thermodynamic equilibrium. The spin of the F center electron can either form an excited F_A triplet state, with spin up or down with spin occupations (which are determined by the spin polarization of the F_A^+ ground state), and an MCPE is observed accordingly (Fig. 4.44). When inducing a saturating EPR transition in the F_A^+ ground state, one equalizes the *Zeeman* spin populations and thus reduces the MCPE to zero [4.76].

With measurements of microwave–induced emission intensity changes many triplet excitons were also investigated. For a review in inorganic crystals see [4.77,52], for self trapped excitons in alkali halides and some oxides see [4.81].

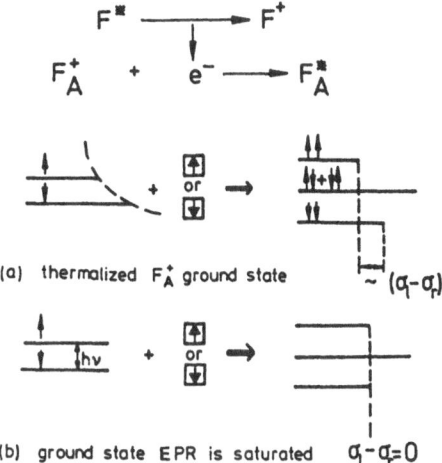

Fig. 4.46. Schematic explanation of the electron transfer from excited F centers to F_A^+ centers. (After [4.76])

4.11 ODEPR of Trapped Excitons with MCDA Method

The observation of excitons trapped at impurity centers is normally carried out by the investigation of their luminescence decay. The nature of the trapping impurity can be determined from ODEPR spectra only if the trapped exciton is a triplet exciton, and the mechanism discussed in the previous section applies for this investigation. It was recently observed, however, that excitons bound to shallow donors could also be investigated with the MCDA method. An example is that of GaP with the neutral donor Ge, where Ge substitutes for Ga. Figure 4.47 shows the MCDA spectrum. It is dominated by the ZPL and followed by a series of phonon replicas, which coincide with those investigated previously in photo–luminescence [4.77] from which the phonon replica assignments were taken. In the whole MCDA the ground state resonance of the neutral donor Ge could be measured ("tagging" experiment). This proves that the whole structured MCDA of Fig. 4.47 belongs to the Ge–bound exciton. It cannot be safely said from the previous emission experiment alone that all the emission lines belonged to the one donor bound exciton. Similar spectra were found for GaP:Si [4.78].

To explain the observation of an MCDA, Fig. 4.48 shows the level scheme for the exciton bound to the neutral donor, with the selection rules for circularly polarized transitions according to [4.79] for the emission. The two electron spins are paired in the excitonic ground state so that electron hole exchange and the electron coupling to the magnetic field are not important. The magnetic splittings are determined by the single hole. Similarly, a para-

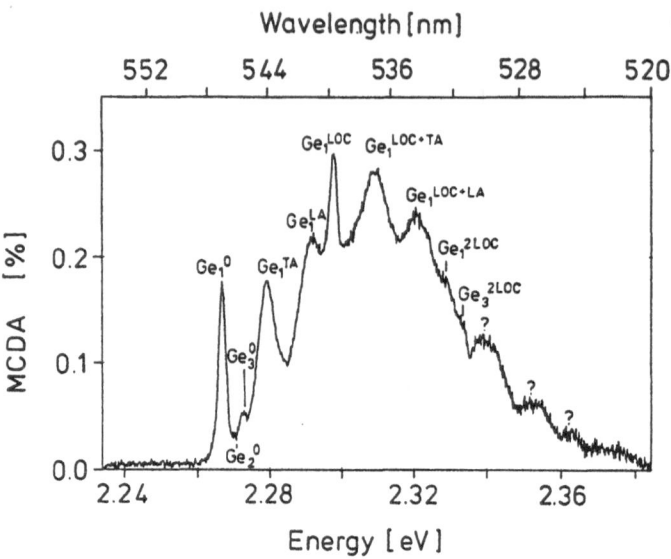

Fig. 4.47. Magnetic circular dichroism of the absorption of excitons bound to the neutral donor Ge in GaP. In the whole spectrum the ODEPR of the neutral donor could be measured. (After [4.78])

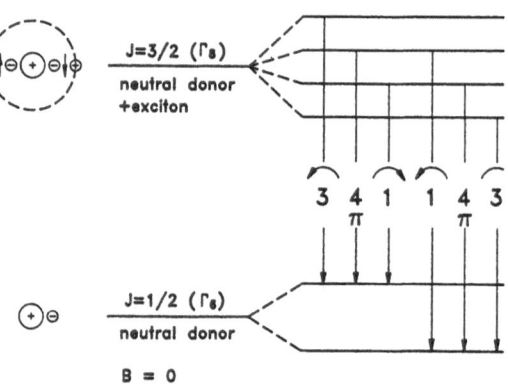

Fig. 4.48. Schematic representation of the no–phonon levels of an exciton bound to a neutral donor, the neutral donor itself, and the dipole allowed transitions in a magnetic field for different polarizations. The numbers give the relative transition probabilities. (After [4.79])

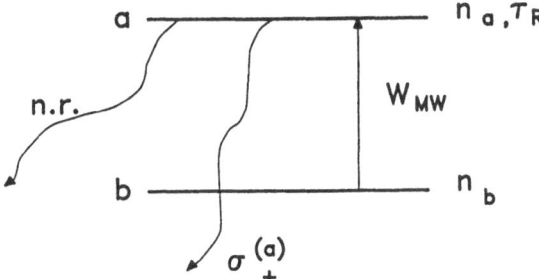

Fig. 4.49. Luminescent excited state levels and transitions for the discussion of ODEPR sensitivity (see text)

magnetic MCDA is expected, as discussed previously for the atomic model, because of the spin polarization of the ground state. The numbers in Fig. 4.48 again give the relative matrix elements for the differently polarized transitions, the senses of which are reversed in the MCDA experiment.

4.12 Sensitivity of ODEPR Measurements

In ODEPR, the sensitivity is basically limited by the shot noise rather than by kT, as in conventional EPR. Therefore, in principle, a very high sensitivity results, considering that intense light sources are available. In practice, however, the sensitivity depends very much on the system investigated and the technique used. There is, up to now, no systematic investigation of the ODEPR sensitivity which would require a calibration of the defect numbers measured, since the ODEPR techniques are not quantitative techniques, as discussed before.

For the detection of excited state EPR via luminescence, a rather straightforward estimate of the minimum number of spins detectable can be made following *Geschwind* [4.1]. Considering the simple scheme of Fig. 4.49, one obtains the luminescence signal S_e from the level occupied by n_a spins in the following way. The total number of photons emitted per unit time from level a is n_a/τ_R (τ_R is the radiative lifetime), of which a fraction η is collected at the detector. A time constant of one second is assumed for the experiment. η takes into account the solid angle seen by the detector, that is, the geometry of the optical system. Let α be the actual change of n_a upon resonance and E the efficiency of the photo–detector. One then has the signal S_e

$$S_e = \frac{\alpha n_a}{\tau_R}\,\eta E. \tag{4.12.1}$$

The signal is to be detected against the background of shot noise (all luminescence from level a) given by

$$N = \sqrt{\frac{n_a}{\tau_R}} \eta E. \tag{4.12.2}$$

The signal–to–noise–ratio then is

$$\frac{S}{N} = \alpha \sqrt{\frac{n_a}{\tau_R}} \eta E. \tag{4.12.3}$$

In the limiting case for S/N = 1, one needs the minimum number of spins in n_a

$$n_a = \frac{\tau_R}{\alpha^2 \eta E}. \tag{4.12.4}$$

For a realistic set of values of $\alpha = 0.1$, $\eta = 10^{-3}$, $E = 10^{-2}$ and $\tau_R = 3$ ms one obtains the low number $n_a \geq 3 \times 10^4$ spins, which must be compared with $n \approx 10^{11}$ spins/mT in conventional EPR. One is, therefore, able to study a small number of excited states, or a low concentration of defects in thin layers. The value of α depends, of course, on the nature of the EPR line. It is smaller for inhomogeneously broadened lines, since only a fraction of the spin packets is involved in the luminescence intensity change. According to (4.12.4) and the set of values assumed for α, η, E, and τ_R, a concentration as low as 10^{10} defects per cm^3 in an epitaxial layer of 1 μm thickness can be detected, whereby a light spot of 1×1 mm^2 is assumed, and excitation of the luminescence with light of energy near the band edge, so that the whole thickness of 1 μm is homogeneously excited. More realistically would be to assume a quantum efficiency (QE) for the emission of only 10% or less, which then yields 10^{11} centers per cm^3, with a homogeneous line to be detected ($\tau_R = \tau_L/QE$, τ_L being the luminescence decay time). α may be smaller, of course. A number of 10^{11}–10^{12} centers per cm^3 seems to be about the right order of magnitude, in view of recent observations, although, as mentioned above, there is no quantitative investigation yet.

It should also be possible to detect surface layer defects with the luminescence method, by illuminating a large enough area and working in ultra–high–vacuum conditions. No such experiments seem to have been done yet.

An estimate of the sensitivity for the MCDA method is less straight-forward. In principle, the sensitivity is also only limited by the shot noise. However, the quantities, like the oscillator strength of the absorption and the spin–orbit splitting in the excited states, enter in the decisive quantity $(\alpha_r - \alpha_l)$, which is detected as the MCDA signal.

In order to get an idea about the sensitivity, one can use the simple alkali atom model, as discussed in Sect. 4.3. In order to relate the absorption constant α_{max} at the peak of the absorption band to the concentration of defects, we assume the *Smakula* equation [4.80]

$$Nf = 1.3 \times 10^{17} cm^{-3} \frac{n}{(n^2 + 2)^2} \alpha_{max} W_{1/2}, \tag{4.12.5}$$

where N is the number of defects per cm^3, f the oscillator strength, n the refractive index and $W_{1/2}$ the halfwidth of the absorption band. For the MCDA we obtained (Sect. 4.3)

$$\epsilon_{extr} = \sqrt{\frac{\ln 2}{2e}} \left(\frac{\alpha_{max} d}{W_{1/2}} \right) \Delta E, \tag{4.12.6}$$

$$\Delta E \approx \frac{2}{3} \Delta_{s.o.} \tanh \frac{g_e \mu_B B_0}{2kT}, \tag{4.12.7}$$

where $\Delta_{s.o.}$ is the spin–orbit splitting of the excited states. The diamagnetic term of (4.3.12) is small and can be neglected for this estimate. Typical experimental conditions are $B_0 = 1$ T, $g \approx 2$, $T = 1.5$ K, for which $\tanh(g_e \mu_B B_0/2kT) = 0.445$ (this corresponds to an experiment in K–band). The value of ϵ_{extr} which can be achieved, depends on $\alpha_{max} d/W_{1/2}$ and the spin–orbit splitting $\Delta_{s.o.}$.

$$\varepsilon_{extr} = 0.10 \times \alpha_{max}[\text{cm}^{-1}] \Delta_{s.o.}[\text{eV}] d[\text{cm}] \times 1/W_{1/2}[\text{eV}] \tag{4.12.8}$$

The sample thickness is limited by the cavity size. For K–band it is ≤ 0.5cm. $W_{1/2}$ is usually of the order of 0.2 eV. For $d = 0.5$ cm, $W_{1/2} = 0.2$ eV we obtain for

$$\epsilon_{extr} = 0.23 \, \alpha_{max}[\text{cm}^{-1}] \, \Delta_{s.o.}[\text{eV}] \tag{4.12.9}$$

ϵ_{extr} can be determined experimentally to about $1 - 2 \times 10^{-5}$ (Chap. 9). This limits the lowest absorption constant detectable to a range of 2×10^{-5} cm^{-1}– 2×10^{-6} cm^{-1} for $\Delta_{s.o.}$ in a range of 0.1–1 eV.

The defect concentration which can be determined in this range of absorption constants depends on the oscillator strength. For $\alpha_{max} = 3 \times 10^{-5}$ cm^{-1} ($n = 3$, for example GaAs), and $W_{1/2} = 0.2$ eV according to (4.12.5) $Nf = 2 \times 10^{10}$ cm^{-3}. Assuming $f = 0.1$, one therefore needs of the order of 10^{11} defects/cm^3 to be able to detect the MCDA. To see an EPR signal, in practice, one will need about an order of magnitude more, 10^{12} cm^{-3}.

This is roughly of the same order of magnitude as needed for conventional EPR, depending on f and $\Delta_{s.o.}$, of course. The advantage is, however, that the sample size does not need to be as big as in conventional EPR, where the filling factor of the sample enters critically (Chap. 2). In the MCDA technique, a narrow laser beam would suffice, provided the defects under study can withstand a high light intensity without being destroyed. On the other hand, it was observed for GaAs, and also in other semiconductors and ionic crystals, that the microwave–induced decrease in MCD was as large as expected for homogeneous lines, although the lines were inhomogeneous (Sect. 4.3 and the discussion of the ODEPR of EL2 defects). If this happens, then there is a considerable sensitivity gain compared to conventional EPR, since in the latter it is the inhomogeneous line which enters critically in the sensitivity.

As already mentioned in Sect. 4.3, the reason for this is not understood at present. In the case of the EL2 defects, it seems that this is the major source of the observed sensitivity gain of about two orders of magnitude, compared to conventional EPR (the total line width of the paramagnetic EL2 EPR lines is $\Delta B \approx 4 \times 30$ mT= 120 mT, see Fig. 4.10).

5. Electron Nuclear Double Resonance

In the previous chapters the usefulness of EPR for the characterization of defect structures was demonstrated. EPR is usually a powerful tool, in particular, for preliminary first information about the main symmetry properties of the defect. If optical detection of magnetic resonance is applicable, additional knowledge about the nature of the defect arising from its optical properties can be linked to its spin resonance properties.

However, there are important questions about the details of the defect structure and the delocalization of its unpaired electrons or holes in the lattice, which can not be answered by EPR experiments. EPR mostly does not resolve these details. This chapter deals with ways to overcome this limitation of EPR.

5.1 The Resolution Problem, a Simple Model

The origin of the inhomogeneous line width of EPR lines was already discussed in Sect. 3.7. The discussion is resumed here because of its importance to explain the electron nuclear double resonance (ENDOR) method. First, a very simple situation is considered:

One electron with an isotropic g value g_e in an external magnetic field B_0 interacts with two different neighbor nuclei which both have nuclear spin $I_1 = I_2 = \frac{1}{2}$. There are two hyperfine interaction constants a_1 and a_2 which are both assumed to be isotropic. In a simple first order solution of the spin *Hamiltonian* (3.5.1) one obtains for the energy levels of this system

$$E = m_S g_e \mu_B B_0 + m_S m_{I1} a_1 + m_S m_{I2} a_2. \qquad (5.1.1)$$

For EPR transitions there is the selection rule $\Delta m_S = \pm 1$, $\Delta m_{I1} = 0$, and $\Delta m_{I2} = 0$. Obviously there are four possible EPR transitions because of $m_{I1} = \pm 1/2$, and $m_{I2} = \pm 1/2$. Each of these four EPR transitions may have a very small natural line width, about 0.004 mT. Adding just one further neighbor nucleus, again with nuclear spin $I = 1/2$, then one obtains $2^3 = 8$ possible EPR transitions. In the more general case, with N interacting neighbor nuclei with nuclear spin $I = 1/2$, the total number, N_{EPR}, of possible EPR transitions amounts to $N_{EPR} = 2^N$. This simple observation has a tremendous impact on the shape of the EPR spectrum. Figure 5.1 shows results

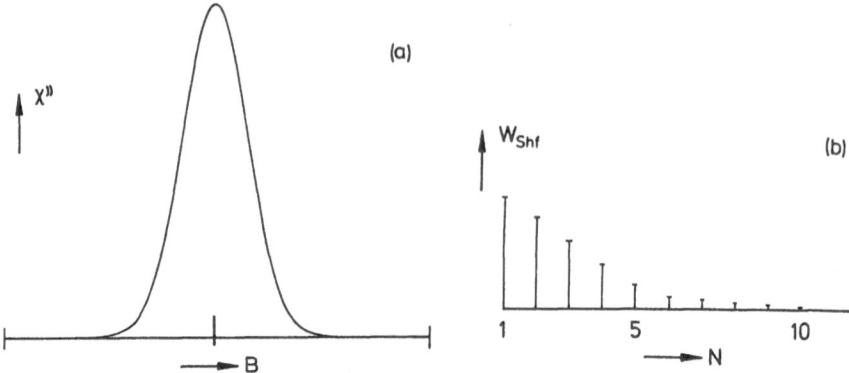

Fig. 5.1. a Model calculation of the EPR absorption spectrum for the interaction of a defect center with ten shells of neighbor nuclei. Each shell consists of four equivalent nuclei with $I = 1/2$, N is the shell number. **b** Relative shf interaction energies W_{shf} for the ten shells

of a model calculation. It is assumed that the electron of a defect center interacts with ten different shells of neighbor nuclei, where each shell consists of four nuclei with nuclear spin $I = 1/2$. The hyperfine interaction energies for the nuclei within a shell are assumed to be identical. See Sect. 6.1 for the definition of a neighbor shell. The hyperfine interaction energies W_{shf} vary among the neighbor shells as indicated in Fig. 5.1b by the length of the vertical lines (relative units). If one calculates the resulting EPR spectrum for this situation, one obtains the spectrum of Fig. 5.1a. (For the computer algorithm see appendix G). The resulting 2^{40} EPR transitions all overlap, giving a broad structure with nearly *Gaussian* shape, despite the fact that the natural line width of each single EPR transition was only 10^{-6} of the magnetic field scale shown in the figure. EPR spectra of this kind are typically found for delocalized unpaired electrons of color centers in ionic crystals, or for defects in many semiconductors.

A different situation is assumed for the calculation shown in Fig. 5.2. There, the hf interaction with the first neighbor shell is larger by a factor of ten, compared to the first shell in Fig. 5.1 (Fig. 5.2b). This prominent interaction now creates the nicely resolved EPR spectrum in Fig. 5.2a. One can immediately see from the shape of the spectrum, that the main interaction takes place with four equivalent neighbor nuclei with spin 1/2, one of the advantages of the EPR as discussed in Sect. 3.6. However, there is no information from the spectrum about the interaction with all the other neighbor shells, which is completely buried in the line width of each of the five resolved lines. Note that exactly the same information is buried in each of these lines. Whether or not an EPR spectrum exhibits resolution, and all further details of the shape of the EPR spectrum, are determined by the distribution of the

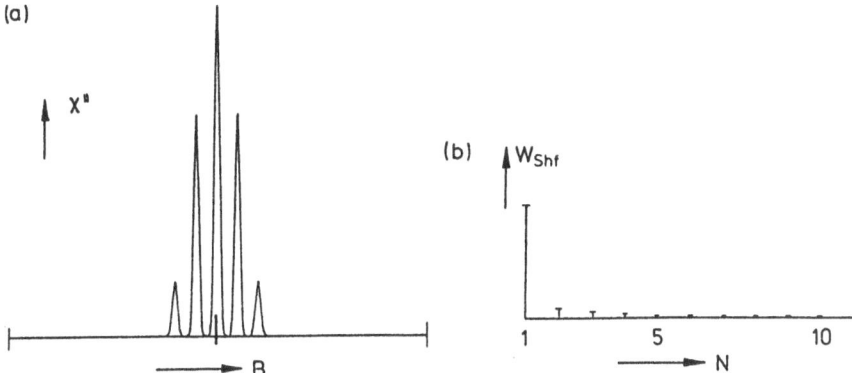

Fig. 5.2. a Model calculation of the EPR absorption spectrum as in Fig. 5.1, but the interaction with the first shell is increased by a factor of ten compared to the first shell in Fig. 5.1. b Relative shf interaction energies W_{shf} for the spectrum in a

hf interactions within the different neighbor shells. Little or no resolution is typical for EPR spectra of defects in solids.

A completely different situation exists if one solves (5.1.1) for nuclear magnetic resonance (NMR) transitions with the selection rules $\Delta m_S = 0$, $\Delta m_{I1} = \pm 1$ and $\Delta m_{I2} = 0$, or $\Delta m_S = 0$, $\Delta m_{I1} = 0$ and $\Delta m_{I2} = \pm 1$. It follows immediately from (5.1.1) that there are only two nuclear magnetic resonance (NMR) transitions for each neighbor nucleus, $m_S = 1/2$ and $m_S = -1/2$, respectively. With N neighbor nuclei with nuclear spin $I = 1/2$ and electron spin $S = 1/2$, the total number of lines in an NMR spectrum amounts to 2N.

The NMR spectra for the configurations discussed in Figs. 5.1,2 would contain only 20 different lines, with absolutely no problem of resolution. The NMR lines of the nuclei of each shell are the same in the assumed configuration. The interaction with each particular neighbor nucleus can now be studied individually and independently of the other neighbors.

5.2 Type of Information from EPR and NMR Spectra

So far, NMR was considered just as a possibility to overcome the resolution problem. But beside resolution enhancement, NMR provides some information which cannot be obtained by EPR experiments. On the other hand, some information provided by EPR cannot be obtained by the kind of NMR considered in the preceding section.

In EPR, the main interaction observed is usually the interaction of the electron spin with the external magnetic field. This provides information on the g matrix, and, consequently, on the symmetry of the defect. The fine-structure interaction, if present, again tells something about the symmetry features. The hyperfine interaction of the electron spin with the central nucleus or with near neighbor nuclei, if resolved, yields the symmetry of the shell containing the nucleus or the nuclei, and yields the nuclear spin and the number of interacting nuclei.

A different situation is met in NMR. The nuclear g values are smaller than the electron g value by about three orders of magnitude. Therefore, in most cases, the interaction of the nuclear spins with the external magnetic field is no longer the most prominent interaction. The nuclear g values are isotropic, and, therefore, there is no symmetry information from this interaction. However, very important new information arises from the possible measurement of the magnitude of the nuclear g value for each individual neighbor nucleus. This provides information about the chemical identity of the neighbor nuclei. Further information, usually not obtainable from EPR, is provided by the quadrupole interaction of neighbor nuclei with nuclear spin $I > 1/2$ with local electric field gradients. Knowledge about electric fields around the defect center may be helpful to clarify, for example, the charge state of the defect, or, in special cases, the lattice site of the defect. Additional information from a quadrupole interaction is the value of the nuclear spin responsible for this interaction. A direct spin–spin interaction between nuclear spins, analogous to the electronic fine–structure interaction, is an unmeasurably small effect in the cases considered here. However, indirect couplings between different nuclear spins, provided by the paramagnetic electron via large hyperfine interactions, are sometimes important. The resulting line splittings are usually not very valuable, although they may tell something about the number of nuclei taking part in a hyperfine interaction of interest.

The most important information obtained from NMR transitions is the hyperfine interaction of nuclear spins with the paramagnetic electron of the defect center. Exactly the same interaction is also observed in EPR, if resolved, but in a very different way. For example, the shape of the EPR spectrum is determined by the simultaneous interaction of all neighbor nuclei together, as can be seen from Fig. 5.2. In principle, an EPR spectrum provides the additional information about the spin of the interacting nuclei and the number of nuclei taking part in the different hyperfine interactions. This is usually too much simultaneous information with the well–known problems of resolution. The situation is completely different when measuring the same interactions by NMR transitions. In first order, one looks at each neighbor nucleus individually, as if all the other nuclei did not exist. The gain in resolution compared to EPR is tremendous. However, nearly all information about the number of interacting nuclei is lost. This is a severe deficiency, which sometimes makes it very difficult, if not impossible, to establish an unequivocal defect model. Some ways to overcome this problem are at least partially discussed below.

It is evident that the information principally contained in an EPR spectrum is very valuable if resolved. If not, the only way is to then attempt the tedious task of ENDOR spectroscopy described in the following two chapters.

5.3 Indirect Detection of NMR, Double Resonance

Unfortunately, the NMR transitions of defect neighbor nuclei can never be directly detected because of intensity problems. Due to the small NMR quanta involved, one would need center concentrations in excess of about 10^{19} cm^{-3}, using straight NMR techniques. The absolute majority of defects of interest have concentrations orders of magnitude below this value. It is, therefore, necessary to find ways to detect NMR transitions with much higher sensitivity. The most successful way, so far, is the application of quantum transformation processes in multiple resonance experiments. There is a basic principle in all of these techniques. By inducing EPR transitions at a sufficiently high transition rate, one induces a spin level occupation different from thermal equilibrium. This is the EPR saturation discussed in Sect. 2.7. Whether this is experimentally possible or not, depends on the spin–lattice relaxation time T_1 of the electron spin, which is usually strongly temperature dependent, and depends on the intensity of the microwave field. NMR transitions are then induced simultaneously by a sufficiently intense radio–frequency (rf) field. If one of the levels between which the NMR transition takes place is identical with one of the EPR levels, then the occupation of this EPR level is again modified, now towards thermal equilibrium. All one needs, in order to detect the NMR transition indirectly, is some method to detect the changes in occupation of the EPR levels and its changes. There are different ways to achieve this.

(i) The basis of an optically detected EPR experiment of the ground state is the measurement of its spin polarization by detecting the MCDA (e.g., (4.22) in Sect. 4.3). The spin polarization is diminished by a saturating, or, at least, partially saturating microwave EPR transition, which is measured as a decrease of the MCDA. In principle, the polarization can be driven to zero if the microwave transition rate is large enough compared with spin–lattice relaxations. This cannot be achieved if there is a hyperfine interaction, since the spin polarization can only be partly diminished due to the EPR selection rule. This is illustrated for the simple configuration of $S = 1/2$ and $I = 1/2$ in Fig. 5.3, where one of the two EPR transitions is indicated. In this case, for a saturating EPR transition, the polarization can only be diminished by roughly 50%, and then the MCDA can only be reduced to half its value in thermal equilibrium. The situation considered in Fig. 5.3 is once again sketched in Fig. 5.4a. The two possible EPR transitions are indicated by the arrows

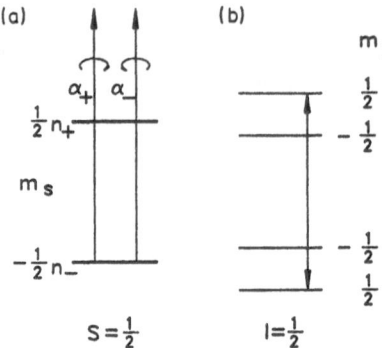

Fig. 5.3. a Schematic representation of the MCDA for $S = 1/2$ (see also Fig. 4.6). In principle the spin polarisation can be driven to zero with a saturating EPR transition between the two levels. **b** Energy level scheme for the interaction of the electron with one nucleus with $I = 1/2$. The MCDA can now only be reduced to half of its value in thermal equilibrium by a saturating EPR transition (see text)

between the levels A and D, and between B and C, respectively. The occupation of the levels A and B is nearly identical in the thermal equilibrium due to the small energy separation of these levels. The same is true for the levels C and D, respectively. In Fig. 5.4 the abscissa gives the relative occupation of the levels. The signal observed in the MCDA is proportional to $(N_B + N_A) - (N_C + N_D)$. Due to a saturating EPR transition between the levels A and D, the occupation of these levels is equalized, $N_D = N_A$ (Fig. 5.4b). In this case, the MCDA signal is only proportional to $(N_B - N_C)$, which amounts to roughly half the value considered in Fig. 5.4a. Now if an additional strong NMR transition is induced between the levels C and D, then the occupation of these levels is also equalized, see Fig. 5.4c. The saturating EPR transition is still present, with the result that now the occupation of the three levels A, D and C is identical, or at least nearly identical. However, the occupation of the level C, N_C, is now bigger than in Fig. 5.4b, whereas the occupation of the level B remains unchanged. As a result, the observed MCDA signal, still proportional to $(N_B - N_C)$, is additionally diminished, compared to Fig. 5.4b, by the NMR transition. Thus, the NMR transition is indirectly detected by the optical detection of the spin polarization as an additional decrease of the MCDA signal, compared to the saturating EPR transition.

For excited states, analogous arguments hold when monitoring the MCPE (Sects. 4.9,10). A special case is a closed optical pumping cycle (Sect. 4.8) with spin memory effects where optically detected electron nuclear double resonance (ODENDOR) can, in principle, be observed

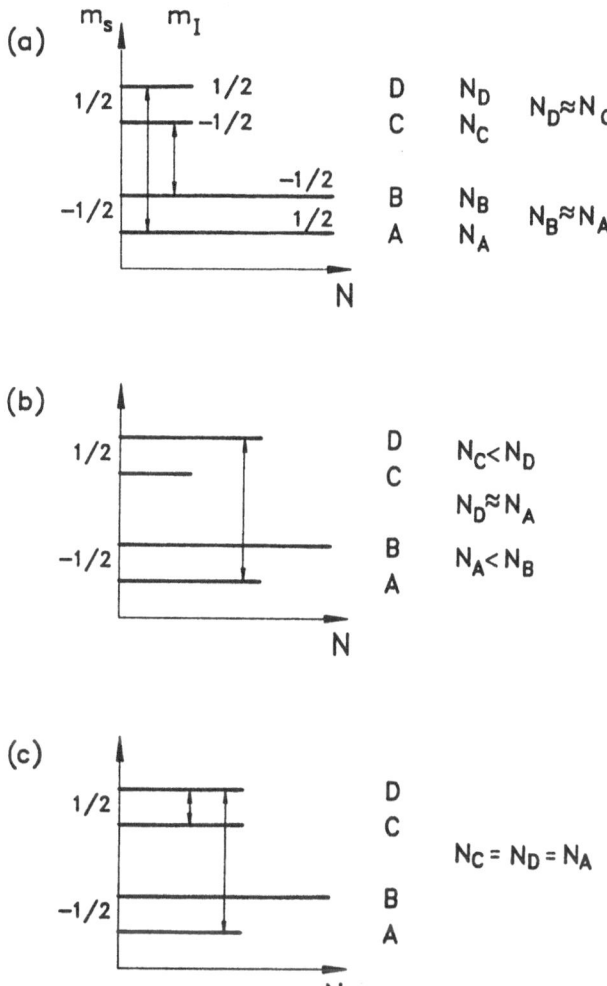

Fig. 5.4a–c. Schematic explanation of optically detected ENDOR. **a** Energy level scheme for $S = 1/2$ and one nucleus with $I = 1/2$. The abscissa schematically indicates the relative occupation N of the levels in thermal equilibrium. **b** Occupation of the levels under a saturating EPR transition between A and D. **c** Occupation of the levels upon an additional strong NMR transition induced between levels C and D

both for the ground and excited states of a defect in both the MCDA and MCPE, similar to EPR.

(ii) In conventional EPR the situation is different. Again, the occupation of EPR levels is driven away from thermal equilibrium by saturating the EPR transition. The NMR transition then (partially) desaturates the EPR transition. Changing the degree of saturation of an EPR transition always has some effect on the experimentally observed intensity of the absorption (χ'') or the dispersion (χ') EPR signal. Both signals can be used to detect the NMR transition indirectly. The effect observed in this way is, however, small for defects in solids. The EPR signal typically changes by not more than 1% if an NMR transition is induced. Despite this small effect, it is possible to measure NMR transitions down to defect concentrations of about 10^{15} cm^{-3}. A signal can also be observed in conventional EPR, if no EPR saturation is possible due to very short spin–lattice relaxation times. An indirect detection of NMR is now no longer possible.

It turns out that saturating the EPR transition is always necessary for the observation of an effect of NMR transitions on the EPR signal intensity. There is, however, a further condition. In order to modify the occupation of the EPR levels, the NMR transitions must be induced at a sufficiently high rate to counterbalance the EPR transition rate, at least to some extent. For short spin–lattice relaxation times, very high microwave power levels are necessary to saturate the EPR transition. Then the experimentally available rf field intensity might not be high enough to affect the occupation of the EPR levels sufficiently.

The method described above of detecting NMR transitions was first invented by *Feher* [5.1,2], and is termed "electron nuclear double resonance" (ENDOR). If NMR transitions are detected by optically detected EPR, then the process is termed "optically detected electron nuclear double resonance" (ODENDOR), which is, in fact, a triple resonance experiment. The conventional detection of ENDOR is sketched in more detail in Fig. 5.5, where the interaction of an electron with $S = 1/2$ with one nuclear spin $I = 1/2$, is assumed. The hf interaction is assumed to be small compared to the electron *Zeeman* energy. In Fig. 5.5a the four resulting levels are denoted by A, B, C and D. The abscissa gives the relative occupation of the levels, and the dashed line indicates the occupation of the levels in thermal equilibrium. The relative occupation differences are given by $\varepsilon_1 = g_e \mu_B B_0 / kT$ and $\varepsilon_2 = g_n \mu_n B_0 / kT$, where $\varepsilon_1 \gg \varepsilon_2$ (Sect. 2.7). Now if a strong EPR transition is induced between the levels A and D, then the situation shown schematically in Fig. 5.5b occurs. The EPR transition $A \rightarrow D$ is saturated. Inducing an NMR transition between the levels C and D now results in a desaturation of the EPR transition; the new level occupation is shown in Fig. 5.5c. The desaturation results in an increase in EPR signal intensity. This signal increase is, how-

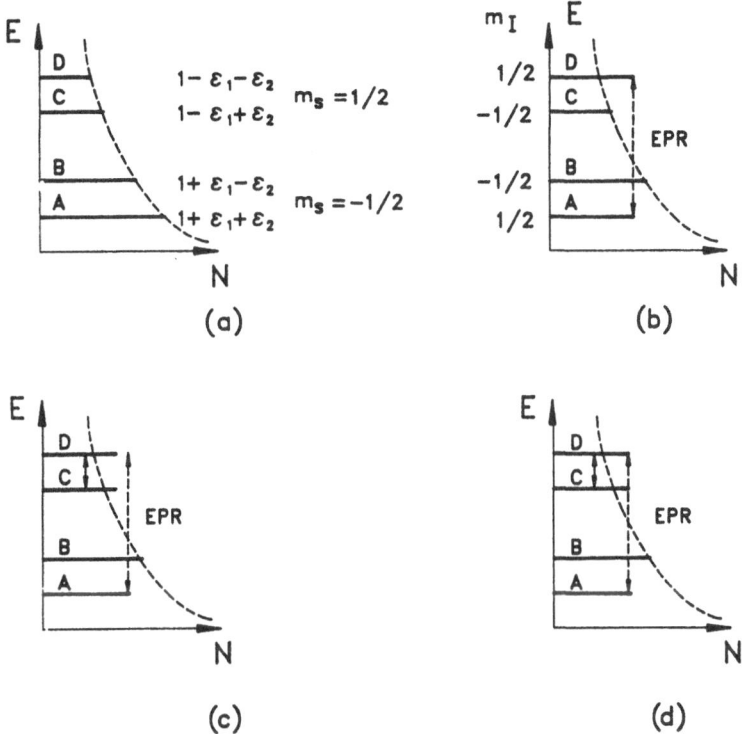

Fig. 5.5a–d. Schematic explanation of ENDOR. **a** Energy level scheme for $S = 1/2$ and one nucleus with $I = 1/2$. The *dashed* line indicates the relative occupation $\varepsilon_1 = g_e \mu_B B_0/kT$, $\varepsilon_2 = g_n \mu_n B_0/kT$ of the levels A–D in thermal equilibrium. **b** Occupation of levels under a saturating EPR transition between A and D. **c** Initial effect of an additionally induced NMR transition between C and D on the occupation of the levels. **d** The EPR transition together with the NMR transition nearly equalizes the occupations of the three levels A, D and C

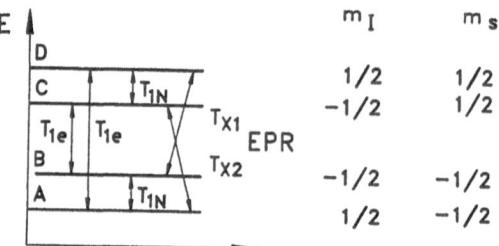

Fig. 5.6. Energy level scheme as in Fig. 5.5 with different relaxation processes. T_{1e} — electron spin lattice relaxation time. T_{1n} — nuclear spin lattice relaxation time. T_{X1}, T_{X2} — cross relaxation times

ever, only transient because the EPR transition rate tends to saturate the EPR again. The situation is then nearly as shown in Fig. 5.5d. However, the saturation of the EPR signal indicated in Fig. 5.5d is no longer as strong as in Fig. 5.5b. The reason for this are different electronic and nuclear spin relaxation processes, which have a different effect in case (d) and in case (b). This is illustrated in Fig. 5.6, where some relaxation processes are indicated for the same level system. All these processes together determine the degree of saturation of, for example, the EPR transition $A \rightarrow D$. There are the electron spin–lattice relaxation time, T_{1e}, the nuclear spin–lattice relaxation time, T_{1n}, and the two cross relaxation times, T_{X1} and T_{X2}. If a strong rf transition takes place between the levels C and D, then the nuclear relaxation path T_{1n} between C and D is by-passed. The cross relaxation path T_{X2} and the relaxation path $C \rightarrow B$ (T_{1e}), in series with the path $B \rightarrow A$ (T_{1n}), now have an increased effect on the total effective relaxation between the levels A and D. This decreased effective electron spin–lattice relaxation time leads to a decreased EPR saturation condition compared to the situation without the rf field. This is a persistent effect, which is present as long as the NMR transition is induced. This technique of detecting an NMR transition is called *stationary ENDOR* and was introduced by *Seidel* [5.3–5]. As can easily be seen from Fig. 5.6, the desaturation effect works as well if the NMR transition is induced between the levels A and B, or if the EPR transition takes place between the levels B and C.

When switching the rf field on and off, the EPR signal exhibits a time dependence as shown schematically in Fig. 5.7. At times below t_1, the EPR signal is saturated and the rf field is off. This steady state condition corresponds to zero ENDOR signal. At t_1 the rf field is switched on instantaneously. The spin system turns from the state indicated in Fig. 5.5b to the state in Fig. 5.5c. When the EPR signal is desaturated, it increases, and the amount of EPR signal increase is termed the ENDOR signal. The time needed for this process is primarily determined by the magnitude of the rf field. A typical value may be 30 μs for a rf field of 0.3 mT rms. When the ENDOR signal has

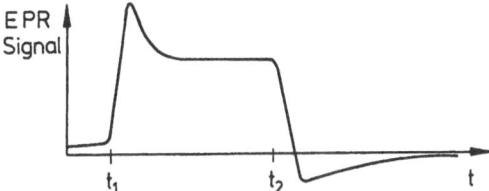

Fig. 5.7. Time dependence of the EPR signal (schematically) when switching on the *rf* field; on at t_1 and off at t_2

passed its maximum value, the so–called transient signal, the saturation of the EPR increases again (compare Fig. 5.5d). The ENDOR signal decreases with a time constant mainly determined by the microwave intensity. It follows the steady state ENDOR signal, which is present as long as the rf field is on. The intensity of this signal is determined by the relative magnitudes of the relaxation times T_{1e}, T_X and T_{1n}. At t_2 the rf is switched off. The system returns slowly to its equilibrium condition after a complicated time behavior. Details of this time behavior are not easily understood using simple arguments; one has to solve a system of rate equations.

The simple model system considered here in order to explain the conventional ENDOR mechanism must not be taken too seriously. It is not able to explain the magnitude of the observed ENDOR effects quantitatively, even when one inserts correct figures for the relaxation times mentioned. One reason for this is the fact that, for typical defects in solids, very many neighbor nuclei take part in very many different relaxation processes. For example, the relaxation between the levels C and D is by–passed by many different relaxation paths involving many other nuclei. Therefore, the by–pass provided by the rf field is no longer as effective. For the same reason, it is almost impossible to predict the absolute magnitude of an ENDOR signal. In particular, it is seldom ever true that ENDOR signals for different defect systems are proportional to their corresponding EPR signals. An ENDOR signal is not a suitable quantity for the measurement of absolute center concentrations. For example, it might well be that the center of interest exhibits a strong EPR signal but almost no ENDOR effect, whereas a small and uninteresting minority of defects shows excellent ENDOR spectra. For these defects some relaxation behavior might favor a large ENDOR effect. Here, one must be aware of the fact that the type of center one is looking at with ENDOR, may be specifically selected by the method applied for its investigation. This is particularly true for multiple resonance methods which are capable of very high sensitivity, like optically detected EPR and ENDOR.

Many other mechanisms for the explanation of ENDOR effects are discussed in the literature; see, for example [5.6]. In order to understand the basic features of the applications of ENDOR spectroscopy to the investigation of defect structures in solids, however, it is not necessary to go too far into these details. It turns out that the ENDOR effects observed experimentally are very different for different classes of materials. Therefore, different authors use different models to explain their observations.

In practical ENDOR spectroscopy it is important to be able to vary the sample temperature over a wide range, in order to select, by trial and error, the best conditions for ENDOR with respect to the different relaxations. There is not much experience, yet, with ODENDOR for defects in solids. Recent experiments using the MCDA method showed very large ODENDOR effects. For example, an investigation of the paramagnetic EL2 defect in GaAs showed that when setting the magnetic field into the flank of the EPR line, the ODENDOR effect was nearly as large as the ODEPR effect (about 15% of the MCDA), while it almost vanished for B_0 in the center of the EPR line [5.7,8]. A different observation was made investigating phosphorous antisite defects in GaP, where the maximum ODENDOR effect was only about 10% of the ODEPR effect, with a maximum for B_0 in the center of the EPR line [5.9]. ODENDOR observed in luminescence was recently reported for oxygen donors in GaP [5.10]. At present, the mechanisms to explain the ODENDOR observations are not well understood. The simple mechanism sketched above is not able to explain the magnitude of the observed ODENDOR effects, which were observed to be very large, despite the fact that many shells of neighbors could be observed. The possible role of cross relaxations and spin memory during the optical cycle have not yet been investigated. If ODENDOR works, the effect is larger than in the conventional detection of ENDOR, and together with the enhanced sensitivity of the optical detection, it provides a very attractive method. It is certainly the only one with the potential for studying defects in thin layers or very dilute systems. The sample temperature should be driven as low as possible in order to have maximum electron spin polarization, and in order to be able to saturate the EPR transition. 1.5 K has proven to be a good choice so far.

5.4 Examples of ENDOR Spectra

In order to study the frequency positions of lines in an ENDOR spectrum, the simple spin *Hamiltonian* for the interaction of an electron spin $S = 1/2$ with one neighbor nucleus with spin $I = 1/2$ is considered:

$$\mathcal{H} = \mu_B \mathbf{S}\tilde{g}\mathbf{B}_0 + \mathbf{I}\tilde{A}\mathbf{S} - g_n\mu_n\mathbf{I}\cdot\mathbf{B}_0. \qquad (5.4.1)$$

The hyperfine tensor $\tilde{A}$ is again written in the form (Sect. 3.5):

$$\tilde{A} = a\tilde{1} + \tilde{B}, \qquad (5.4.2)$$

where the tensor $\tilde{B}$, in its principal axis system, is expressed by:

$$\tilde{B} = \begin{pmatrix} -b + b' & & \\ & -b - b' & \\ & & 2b \end{pmatrix} \qquad (5.4.3)$$

and the isotropic hyperfine interaction constant a is equal to:

$$a = \frac{1}{3}(A_{xx} + A_{yy} + A_{zz}). \qquad (5.4.4)$$

A first order solution of the spin *Hamiltonian* (5.4.1) for nuclear transitions, $\Delta m_S = 0$, $\Delta m_I = \pm 1$, and for $b' = 0$ yields for the ENDOR frequency, f_{ENDOR}:

$$f_{\text{ENDOR}} = \frac{1}{h} |m_S\{a + b[3\cos^2(\theta) - 1]\} - g_n\mu_n B_0| \qquad (5.4.5)$$

The angle θ is the angle between the direction of the static magnetic field $\mathbf{B}_0$ and the z–axis of the ligand hyperfine or super hyperfine (shf) tensor $\tilde{A}$. For distant neighbors the interaction constants a and b tend to become very small. For this case, (5.4.5) predicts just one ENDOR line determined by the nuclear *Zeeman* term $g_n\mu_n B_0$, independent of m_S. This line is termed the *nuclear Zeeman* ENDOR line f_{nuc}. There is one nuclear *Zeeman* line for each isotope. In many ENDOR spectra the nuclear *Zeeman* lines are very prominent. There are also cases where these lines are absent. The mechanism for this behavior is still unclear.

For $S > 1/2$ one observes one ENDOR line for each value of m_S according to (5.4.5). For $m_S = 0$ there is no shf ENDOR line, but only a contribution to the intensity of the nuclear *Zeeman* ENDOR line (e.g., Si:Fe^0, S=1 [5.11]). If the absolute value of the shf interaction term, $|m_S\{a + b[3\cos^2(\theta) - 1]\}|$, is small compared to the absolute value of the nuclear *Zeeman* term, $|g_n\mu_n B_0|$, then one observes two ENDOR lines $f^{(1)}$ and $f^{(2)}$ separated on the frequency scale by $2|m_S\{a + b[3\cos^2(\theta) - 1]\}|$, with the line due to the nuclear *Zeeman* frequency in the middle of these two lines (for $S = 1/2$). This is sketched in Fig. 5.8a. If $|g_n\mu_n B_0|$ is small compared to $|m_S\{a + b[3\cos^2(\theta) - 1]\}|$, one again observes two lines separated by $2|g_n\mu_n B_0|$, but with no lines in between these lines (Fig. 5.8b). If the shf interaction is only slightly larger than the nuclear *Zeeman* term, then one of the two ENDOR lines may be folded back at the origin of the frequency scale, according to the absolute value of the frequency in (5.4.5). Particularly, in this case, the simple first order solution of the spin *Hamiltonian* (5.4.1) must never be taken too seriously. Second order effects may considerably alter the frequency positions, see Sect. 6.8 for details.

If $I > 1/2$ the quadrupole interaction of the nucleus with an electric field gradient must be taken into account. In a similar way as in (5.4.3), the quadrupole tensor $\tilde{Q}$ is described by the two constants q and q' in its principal axis system:

$$\tilde{Q} = \begin{pmatrix} -q + q' & & \\ & -q - q' & \\ & & 2q \end{pmatrix} \qquad (5.4.6)$$

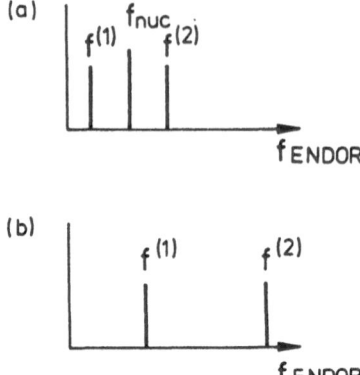

Fig. 5.8. a Position of shf ENDOR lines $f^{(1)}$ and $f^{(2)}$ ($S = 1/2$) relative to the nuclear *Zeeman* line f_{nuc}, if the shf interaction is smaller than the nuclear *Zeeman* term. **b** Position of shf ENDOR lines if the shf interaction is larger than the nuclear *Zeeman* term. There is no longer a line in the middle of $f^{(1)}$ and $f^{(2)}$

To account for the quadrupole interaction, the quadrupole term $\mathbf{I}\tilde{Q}\mathbf{I}$ must be added to the spin *Hamiltonian* (5.4.1). Again, in a simple first order solution, one obtains for the ENDOR frequency f_{ENDOR} for $b' = 0$ and $q' = 0$:

$$f_{\text{ENDOR}} = \frac{1}{h}|m_S\{a + b[3\cos^2(\theta_A) - 1]\} - g_n\mu_n B_0 + m_q 3q[3\cos^2(\theta_Q) - 1]|,$$

(5.4.7)

where m_q is the average value of the two nuclear quantum states m_I, m'_I, between which the nuclear transition takes place. For example, for $I = 3/2$ there are three m_q-values, $m_q = -1$, $m_q = 0$ and $m_q = +1$. The ENDOR lines for $m_q = 0$ are sometimes termed *hf or shf lines* (which, however, has nothing to do with the lines corresponding to interactions with a central nucleus or with ligands). The angles θ_A and θ_Q in (5.4.7) have a meaning analogous to the angle θ in (5.4.5). A quadrupole interaction may also be present for very distant nuclei leading to a corresponding splitting of nuclear *Zeeman* lines. An example of a simple ENDOR spectrum according to (5.4.7) is shown in Fig. 5.9 for interstitial neutral hydrogen centers in KCl [5.12]. The triplet character of the lines comes from quadrupole interactions which are considerably smaller than the shf interactions. All isotopes in KCl, ^{39}K, ^{41}K, ^{35}Cl and ^{37}Cl, have nuclear spin $I = 3/2$.

A splitting of lines for $m_q = 0$ corresponds to higher order effects not covered by (5.4.7). The $+$ and $-$ signs in Fig. 5.9 denote ENDOR lines for $m_S = -1/2$ and $m_S = +1/2$, respectively ($S = 1/2$). The static magnetic field $\mathbf{B}_0$ was in the [110] direction. For each of the lattice sites A, B, C, D of the isotopes, there are ENDOR lines for different values of the two angles θ_A and θ_Q.

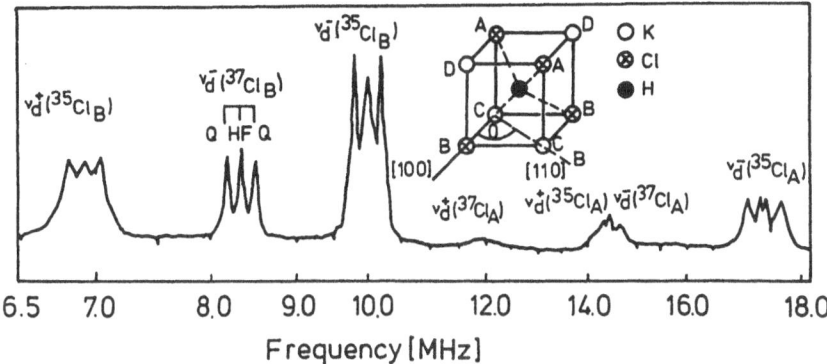

Fig. 5.9. Part of the ENDOR spectrum of interstitial neutral hydrogen centers in KCl. The lines are due to interactions with ^{35}Cl and ^{37}Cl nuclei. The $+$ and $-$ signs belong to $m_S = +1/2$ and $m_S = -1/2$, respectively ($S = 1/2$). The letters A and B indicate the lattice position of the nuclei. $\mathbf{B_0} \parallel [110]$. (After [5.12])

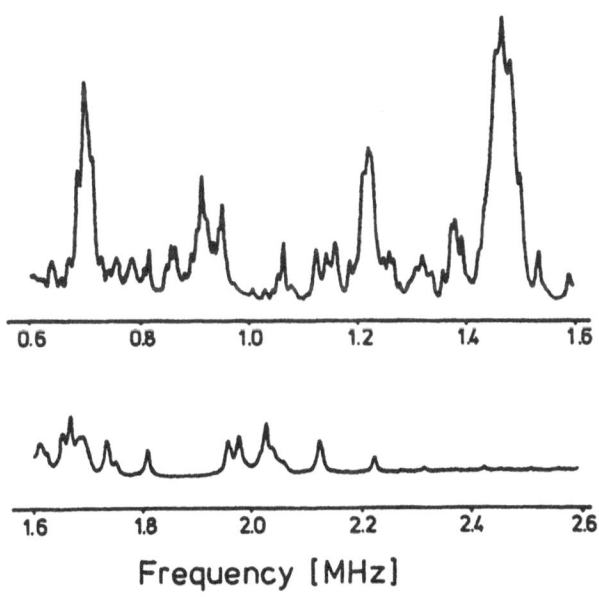

Fig. 5.10. ENDOR spectrum for neutral substitutional hydrogen centers in KCl. $\mathbf{B_0} \parallel ([110] + 5.4°)$ in a $\{100\}$ plane). (After [5.13])

In many cases, ENDOR spectra are not as simple as shown in Fig. 5.9. A different example is shown in Fig. 5.10 for neutral substitutional hydrogen centers on anion sites in KCl [5.13]. There are many lines in a narrow frequency range with multiple overlapping. The tools for a successful analysis of those spectra are described in Chap. 6.

5.5 Relations Between EPR and ENDOR Spectra, ENDOR–Induced EPR

When considering the model systems in Fig. 5.5 or Fig. 5.6 in order to explain how ENDOR works, there is one important point which needs further consideration. It was pointed out that the NMR transition to be detected via EPR must involve exactly the same level which is involved in the EPR transition. Otherwise, it is not clear how a desaturation of the EPR transition could occur with optimum efficiency. The question now arising, is how to adjust the microwave frequency and the magnetic field precisely, so that the EPR transition occurs between the levels A and D in Fig. 5.5. In other words, can the NMR transition $C \rightarrow D$ in Fig. 5.5 still be observed with ENDOR, if, at a fixed magnetic field, the microwave frequency of the spectrometer is not adjusted to the energy difference $A–D$, or, if, at a fixed microwave frequency, different magnetic fields within the EPR spectrum are chosen? It was already mentioned that each of the lines in the calculated resolved EPR spectrum in Fig. 5.2a contains the same information about the unresolved interactions. One might therefore suppose that ENDOR spectra are identical for the magnetic field set to coincide with field positions of the different resolved EPR lines. In order to clarify the question in a more systematic way, it is useful to consider the EPR transitions once again, by going back to the simple model system of (5.1.1). This equation describes the possible EPR transitions for two neighbor nuclei. For the following arguments, it is quite sufficient to again consider a first order solution of the spin *Hamiltonian* in order to describe the spin interactions. However, instead of two neighbor nuclei, there are now N neighbor nuclei ($N > 2$) taken into account. All neighbor nuclei are assumed to have nuclear spin $I = 1/2$. The possible EPR transition energies $h\nu$ are then simply given by:

$$h\nu = \mu_B g_e B + \sum_{i=1}^{N} a_i m_{Ii}. \tag{5.5.1}$$

Since, in an EPR experiment the microwave frequency is kept constant and the magnetic field is varied, (5.5.1) may be expressed in the form:

$$B = \frac{h\nu}{\mu_B g_e} - \sum_{i=1}^{N} \frac{a_i m_{Ii}}{\mu_B g_e}. \tag{5.5.2}$$

All possible EPR lines within the entire spectrum can now be considered as being due to all possible combinations of the values for m_{Ii} under the sum. Since all neighbor nuclei exhibit nuclear spin $I = 1/2$, each m_{Ii} can either be $+1/2$ or $-1/2$. There are 2^N possibilities to assign these two values to the different m_{Ii}. The EPR spectrum, according to (5.5.2), therefore consists of 2^N lines. Since each of these lines has a very small but finite natural line width, the density of these lines around a magnetic field within the field range of the natural line width, is observed as signal intensity of the EPR spectrum at this field (compare Fig. 5.2).

Equation (5.5.2) may now be expressed in a slightly different way with no effect on its physical meaning:

$$B = \frac{h\nu}{\mu_B g_e} - \sum_{i=1,i\neq k}^{N} \frac{a_i m_{Ii}}{\mu_B g_e} - \frac{a_k m_{Ik}}{\mu_B g_e}, \tag{5.5.3}$$

where k is some number between 1 and N. The term with the number k is no longer part of the sum, it is subtracted separately. The spin quantum number m_{Ik} of the neighbor k, again has the two possible values $m_{Ik} = \pm 1/2$. Even if transitions by a suitable radio–frequency field are induced between these two levels during the measurement of conventional EPR, this has no effect on the EPR spectrum. All possible spin states, as discussed in (5.5.2), are still present with the same probability as considered above.

We now use the neighbor nucleus k for an ENDOR experiment. There are four possible levels for the interaction of the electron spin with this neighbor nucleus k. These levels correspond to $m_S = \pm 1/2$ and $m_{Ik} = \pm 1/2$, respectively. At the magnetic field B_0 their energies E are given by:

$$E = \mu_B g_e B_0 m_S + a_k m_{Ik} m_S - g_{nk} \mu_n B_0 m_{Ik}. \tag{5.5.4}$$

These levels form the level scheme sketched in Fig. 5.5 to explain the ENDOR mechanism. There are two possible EPR transition energies for this process:

$$h\nu_{EPR} = \mu_B g_e B_0 + a_k m_{Ik}, \tag{5.5.5}$$

where $m_{Ik} = \pm 1/2$, and there are two possible NMR transition energies for the ENDOR process:

$$h\nu_{NMR} = |a_k m_S - g_{nk} \mu_n B_0|, \tag{5.5.6}$$

where $m_S = \pm 1/2$. The question now is, whether an ENDOR signal of the same neighbor nucleus k can also be observed with the same EPR transition energy (fixed microwave frequency of the spectrometer), but at a different magnetic field, B_0', within the EPR spectrum. The answer is straightforward. Since B_0' is within the EPR spectrum, it must satisfy (5.5.3). Consequently, sets of nuclear quantum numbers m_{Ii}', $i \neq k$ must exist such that:

$$B_0' = \frac{h\nu_{EPR}}{\mu_B g_e} - \sum_{i=1,i\neq k}^{N} \frac{a_i m_{Ii}'}{\mu_B g_e} - \frac{a_k m_{Ik}}{\mu_B g_e} \tag{5.5.7}$$

Comparison of this equation with (5.5.3) immediately yields:

$$B_0 = B_0' + \sum_{i=1, i \neq k}^{N} \frac{a_i(m_{Ii}' - m_{Ii})}{\mu_B g_e}. \tag{5.5.8}$$

This means, that if one uses a different magnetic field within the EPR spectrum, then all the other nuclei with number i, $i \neq k$ are able to add an extra field in a way that the ENDOR process for the neighbor k, according to (5.5.4), is again possible. Consequently, an ENDOR signal of the neighbor k can be measured at any magnetic field within the EPR spectrum. The shape of the EPR spectrum makes no difference. However, due to the nuclear *Zeeman* term in (5.5.6) the ENDOR frequency is shifted by an amount:

$$|\Delta f_{\text{ENDOR}}| = \frac{1}{h}|g_{\text{nk}}\mu_{\text{n}}(B_0' - B_0)|. \tag{5.5.9}$$

This shift can be used to measure the nuclear g value g_{nk} and, thus, to determine the chemical identity of the neighbor k. (For many practical cases this first order expression is, however, not sufficiently accurate, see Sect. 6.8) Scanning the frequency of the rf field results in selecting different neighbors k for the ENDOR measurement. Since k may have any value between 1 and N, the above arguments are valid for the entire ENDOR spectrum containing signals of all neighbor nuclei. One therefore obtains ENDOR spectra with identical information for any value of B_0 within the EPR spectrum. However, there is still an important difference between the ENDOR spectra measured at different magnetic fields. As mentioned above, the signal intensity of the EPR spectrum for a given value of B_0 depends on the number of values according to (5.5.2) which fall into the small interval $B_0 \ldots B_0 \pm \delta B_0$, where δ is of the order of 10^{-5}. Or, in other words, the EPR intensity at the field B_0 is proportional to the probability that a calculated field value according to (5.5.2), for all possible combinations of m_{Ii}, falls into the interval $B_0 \ldots B_0 \pm \delta B_0$. The same arguments hold for the magnitude of the ENDOR process as a function of the magnetic field. The magnitude of the ENDOR signal observed for the neighbor k at the magnetic field B_0', is proportional to the probability that the neighbor nuclei according to (5.5.8) add the extra field necessary to enable the ENDOR process according to (5.5.4–6). The same term which produces the extra field in (5.5.8) determines the EPR signal intensity at the field B_0', according to (5.5.7). Therefore, the ENDOR signal intensity is proportional to the EPR intensity at the magnetic field used for the ENDOR measurement. If one scans the magnetic field over the entire range of the EPR spectrum during an ENDOR measurement, the amplitude of each ENDOR line will reproduce the shape of the EPR spectrum. The ENDOR lines are slightly shifted as a function of the magnetic field (5.5.9), which has nothing to do with their amplitude. When performing such a field scan experiment, the ENDOR frequency must be adjusted accordingly (5.5.9). This kind of experiment is called *ENDOR–induced EPR* (EI-EPR).

(a)

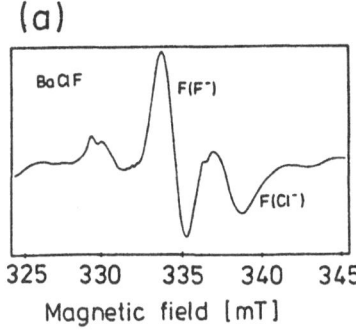

(b)

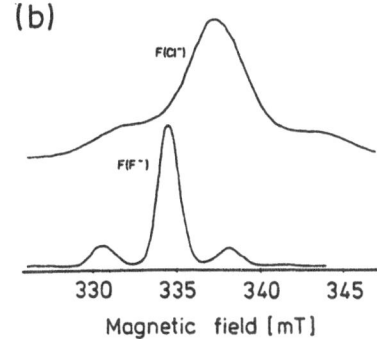

Fig. 5.11. a EPR spectrum of F^- centers in BaFCl. The two types of F^- centers $F(F^-)$ and $F(Cl^-)$ are present simultaneously. (After [5.14]). **b** The EPR absorption spectra of the two types of F^- centers are measured separately by ENDOR–induced EPR. (After [5.14])

So far, only the simplified case has been considered, where all neighbor nuclei have $I = 1/2$. It can, however, easily be seen from the arguments given above, that it is still possible to reproduce the EPR spectrum via an ENDOR signal, even for arbitrary neighbor nuclear spins. The only condition is that the ENDOR signal used to reproduce the EPR spectrum comes from a nucleus with $I = 1/2$ (100% abundant). Otherwise, the EPR spectrum reproduced is different from the original EPR spectrum. This general case is treated in Appendix G.

The possibility of measuring ENDOR–induced EPR spectra provides a very useful experimental tool. It enables one to separate overlapping EPR spectra due to different defects simultaneously present, and to separate the parts of the EPR spectrum of a low symmetry defect which correspond to different defect orientations. If, in an ENDOR spectrum, lines are present from different defects, then each ENDOR line can be assigned to its defect by measuring the corresponding EPR spectrum via the ENDOR line. Applications are summarized in [5.14,15]. As an example, Fig. 5.11a shows the superimposed EPR spectra of the two types of F centers in BaFCl, where electrons can be trapped at Cl^- vacancies ($F(Cl^-)$ centers), or F–vacancies ($F(F^-)$ centers). $F(Cl^-)$ centers can be produced alone, but $F(F^-)$ centers cannot be produced without the simultaneous production of $F(Cl^-)$ centers [5.16–18]. In the EI–EPR spectrum (Fig. 5.11b) using ^{19}F ENDOR lines (nuclear spin $I(^{19}F) = 1/2$), both EPR spectra can be separated, and, for example, both their g_e values can be determined.

The situation is slightly more complicated, if, for the measurement of EI–EPR, an ENDOR line of a nucleus with $I > 1/2$ is used. In this case, the separate term with the number k (5.5.3) contributes more than one EPR

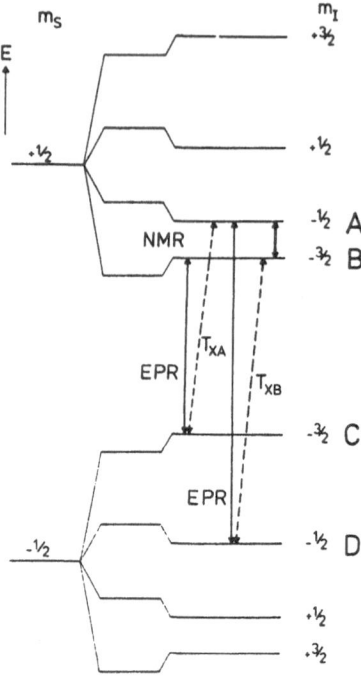

Fig. 5.12. Energy level scheme for the interaction of an electron spin $S = 1/2$ with one nucleus with $I = 3/2$. The nuclear levels are modified by a quadrupole interaction. A NMR transition between A and B can only affect the two EPR transitions shown via simple relaxation processes T_{XA} and T_{XB} in an ENDOR–induced EPR experiment. (After [5.14])

transition to the EPR spectrum. Any NMR transition at the nucleus k, is, however, no longer able to affect all possible EPR transitions for this nucleus at the same time in first order. For the example $I = 3/2$ in Fig. 5.12, the four nuclear levels for each m_S are modified by a quadrupole interaction. The NMR transition is assumed to take place between the levels A and B. Only the two EPR transitions indicated in the figure can be affected by this NMR transition, considering a simple effective relaxation process (compare Fig. 5.6). Consequently, in the EI–EPR spectrum the two additional possible EPR transitions are missing. Therefore, the half width of the EI–EPR spectrum is reduced compared to the original EPR spectrum. This effect is the more pronounced, the higher the hyperfine interaction for the nucleus k. Depending on the NMR transition used for the measurement of the EI–EPR at the nucleus k, the EI–EPR spectrum may also be shifted on the magnetic field scale compared to the original EPR spectrum. In Fig. 5.12, this is the case for NMR transitions between $m_I = -3/2$ and $-1/2$, and between $m_I = 1/2$ and $3/2$. No shift is observed for a transition between $m_I = -1/2$

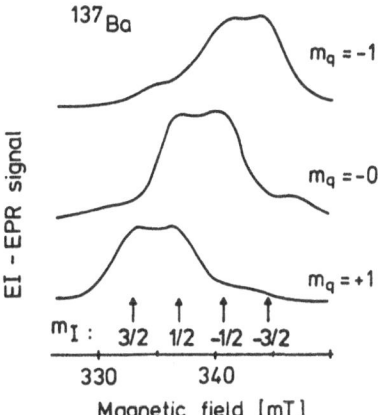

Fig. 5.13. ENDOR–induced EPR spectra for $F(Cl^-)$ centers in BaFCl measured using the three quadrupole lines of ^{137}Ba ($I = 3/2$) for $m_q = -1, m_q = 0$, and $m_q = 1$, respectively. (After [5.14])

and $1/2$. The direction of the shift relative to the position of the original EPR spectrum, can be used to establish the sign of the quadrupole interaction constant relative to that of the hyperfine interaction of the nucleus k, when inducing different NMR–transitions. An example which corresponds to Fig. 5.12 is shown in Fig. 5.13. The two–line EI–EPR spectrum was measured by locking the ENDOR frequency successively to the three ENDOR lines for $m_q = -1$, $m_q = 0$ and $m_q = 1$ of the ^{137}Ba ($I = 3/2$) quadrupole triplet of $F(Cl^-)$ centers in BaFCl. In this case, the signs of the quadrupole and shf interactions are the same. If the signs were different, the low field EPR lines would appear when locking the ENDOR frequency to the $m_q = -1$ line. In addition to the two expected EPR lines, two other lines also appear, weakly indicating that additional relaxations occur [5.14].

The mechanism of EI–EPR works for $S > 1/2$ as well. Again, only a subset of all possible m_S values is contained in the EI–EPR spectrum. When taking different NMR transitions coupled to different m_S states, one can measure different subspectra of the original EPR spectrum, which correspond to different values of m_S. This enables one to determine relative signs of fine–structure, hf or shf interaction constants [5.14,19]. With EI–EPR, it is also possible to investigate a fine–structure splitting which is not resolved in the EPR spectrum, as observed, e.g., for Ni^{3+} defects in GaP [3.4.1]. For further details see [5.14].

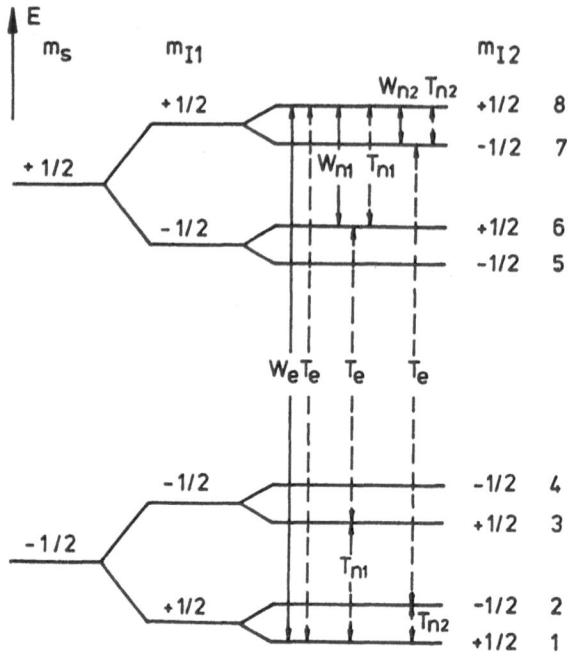

Fig. 5.14. Energy level scheme for the interaction of an electron spin $S = 1/2$ with two neighbor nuclei with $I_1 = I_2 = 1/2$ to explain triple resonance processes. (Details, see text)

5.6 Electron Nuclear Nuclear Triple Resonance (Double ENDOR)

ENDOR–induced EPR provides a useful tool to assign ENDOR lines to their corresponding EPR spectra in the case of overlapping EPR spectra. However, all the many ENDOR lines from these different defects are present, and the large number of lines can make an analysis of the spectra very difficult. ENDOR–induced EPR is only useful as long as the separated EPR spectra are sufficiently different from each other. This is not often the case. It is, therefore, desirable to have some measurement technique which delivers the individual ENDOR spectra for each defect separately, right from the beginning. Such an experiment is indeed possible. As was pointed out in Sect. 5.3, the effectiveness of an ENDOR process depends strongly on the type of relaxation by–passes provided by many other neighbor nuclei not directly involved in the ENDOR process. Such a relaxation by–pass, which affects the magnitude of an ENDOR signal, can also be provided by an additional NMR transition, which is simultaneously induced by a second rf frequency. A simple model system to illustrate this is shown in Fig. 5.14 for the interaction of an electron with two neighbor nuclei with $I_1 = I_2 = 1/2$. It is quite obvious from the figure that the induced NMR transition between the levels 7 and 8, W_{n2}, has less effect on the desaturation of the EPR transition between the levels 1 and 8, W_e, if an additional rf transition, W_{n1}, is induced between the levels 6 and 8. In this case, the relaxation by–pass through the levels 8, 6, 3 and 1 becomes more efficient, diminishing the ENDOR process involving the levels 1, 8, and 7. The same result is present, in principle, if there are more than two neighbor nuclei present, and if the nuclear spin of these nuclei exceeds 1/2. This can be easily verified by rate equation calculations [5.20]. An ENDOR process is always diminished if the two NMR transitions take place at nuclear levels belonging to the same m_S. In the case of Fig. 5.14 the two NMR transitions both take place at $m_S = 1/2$.

It can be also seen from Fig. 5.14 that an ENDOR effect is enhanced if the two NMR transitions take place at different m_S levels. This is particularly obvious if one NMR transition takes place between the levels 7 and 8 for $m_S = 1/2$, and the levels 1 and 2 for $m_S = -1/2$. In this special case, the two transitions take place at the same nucleus, but at different rf frequencies for different m_S levels. This case is called *special triple resonance* in [5.21,22]. Also, in the general case, where two NMR transitions take place at different neighbor nuclei for different m_S values, an enhancement of an ENDOR effect always occurs. In a special measurement technique, the effects described above can be used to separate ENDOR spectra due to different defects. It turns out, experimentally, that an ENDOR signal is changed by not more than about 10% when a second NMR transition is induced. It is therefore necessary to detect the change of the ENDOR signal upon switching on and off the second rf frequency with high sensitivity, using lock–in techniques. Usually, an ENDOR signal is detected by modulating the rf source and observing the corresponding change of the EPR signal using lock–in techniques. In the case of triple resonance, one modulates the second rf frequency with a second frequency considerably lower than the first modulation frequency, and observes the triple resonance signal with a second lock–in amplifier connected in series to the first one. In this measurement configuration no signal is observed if the two rf transitions take place at two nuclei which belong to different defects. No signal is observed, because there is no mutual influence of the two NMR transitions in this case. The detection technique is insensitive to additive effects of two NMR transitions. Thus, it is possible to sensitize one center with the first rf frequency by adjusting the frequency to one of its ENDOR lines for the exclusive measurement of its ENDOR spectrum by the second rf frequency.

An experimental example is shown in Fig. 5.15, again for the two types of F centers present simultaneously in BaFCl [5.16,17]. Figure 5.15a shows an ENDOR spectrum with lines due to both types of defects. For the measurement of the triple resonance spectrum in Fig. 5.15b, the first rf frequency was adjusted to an ENDOR line which belongs to the $F(Cl^-)$ center. The com-

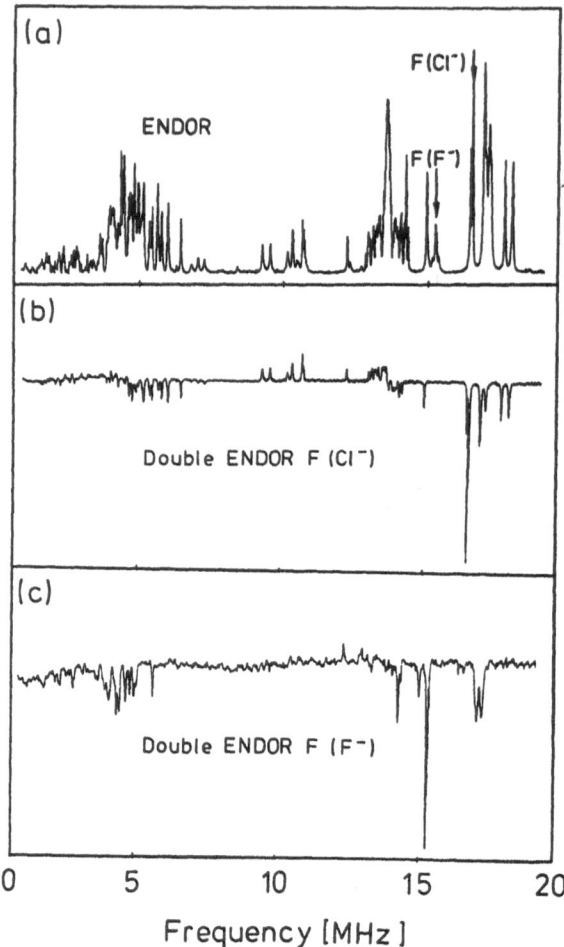

Fig. 5.15a–c. ENDOR and Double ENDOR of F centers in BaFCl **a** ENDOR spectrum with lines of the two types of F centers in BaFCl. **b** Double ENDOR spectrum using an $F(\text{Cl}^-)$ ENDOR line for the measurement. **c** Double ENDOR spectrum using an $F(\text{F}^-)$ ENDOR line for the measurement. All lines in **a** are reflected by the lines in **b** and **c** together. The different signs of the Double ENDOR lines are due to $m_S = 1/2$ and $m_S = -1/2$, respectively. (After [5.21])

plete subset of ENDOR lines due exclusively to the $F(Cl^-)$ ENDOR spectrum appears in triple resonance upon sweeping the second rf frequency. All lines which have negative sign belong to the same m_S value as the ENDOR line of the first rf frequency. The positive lines belong to the opposite m_S. Correspondingly, the $F(F^-)$ spectrum can be obtained by adjusting the first rf frequency to an $F(F^-)$ ENDOR line (Fig. 5.15c). Further important applications of triple resonance are discussed in Sect. 6.6.

5.7 Temperature Dependence and Photo–Excitation of ENDOR Spectra

The high precision with which the frequency position of ENDOR lines can be determined for many defects makes it possible to measure changes in the shf or quadrupole interactions which are caused by variation of the sample temperature. Examples are described in this section; a more detailed discussion of the physical origin of the temperature dependence of the interaction constants is presented in Chap. 7.

In semiconductor physics it is not only important to obtain information about the structure of defects, but also about their energy level position in the gap. Therefore, a way to correlate a determination of the energy levels and that of the defect structure is needed. Such a correlation is provided by photo–EPR/ENDOR and photo–ODEPR/ODENDOR experiments, which are a sort of photo–excitation spectroscopy of EPR/ENDOR spectra.

5.7.1 Temperature Dependence of ENDOR Spectra

ENDOR spectra can often be measured within a wide range of temperatures. It is then possible to measure the frequency shift of ENDOR lines as a function of temperature. The line width of typical "narrow" ENDOR lines in good, single crystals can be as low as about 10 kHz, or even less. Therefore, even small temperature effects on the shf interactions of a few kHz over the measured temperature range are detectable for narrow ENDOR lines. The shf interactions depend in a non linear way on the distance between the neighbor nucleus and the center of the defect. They decrease rapidly with increasing distance (Chap. 7). Therefore, with increasing temperature, one would expect a decrease in shf interactions due to lattice expansion. However, the amplitudes of the lattice vibration with increasing temperature also increase, which results in a larger shf interaction as discussed in detail in Sect. 7.4.6 and Sect. 7.6.5. An example of this is shown in Fig. 5.16 for hydrogen atoms occupying cation vacancies in KCl [5.23]. The ENDOR frequency of a nearest ^{35}Cl neighbor increases approximately exponentially. The hydrogen atom is

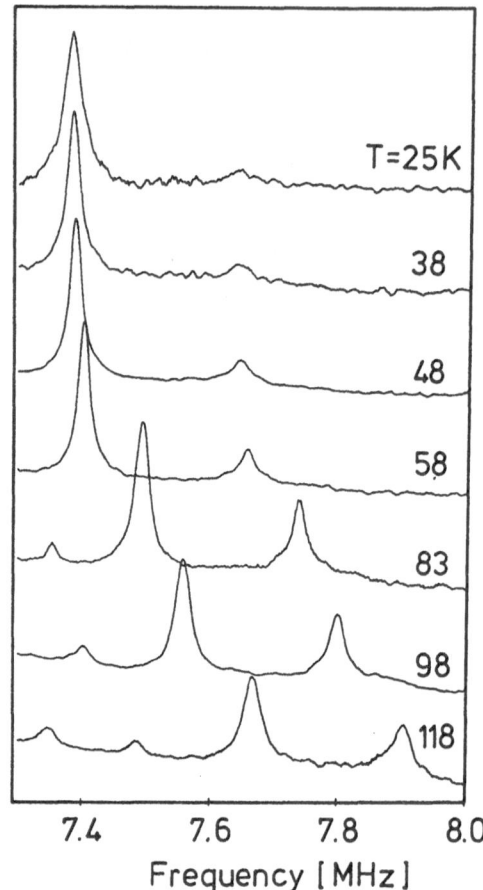

Fig. 5.16. Temperature dependence of ENDOR frequencies of atomic hydrogen centers on cation vacancy sites in KCl doped additionally with Sr^{++} to provide the cation vacancies. The prominent *ENDOR line* is due to nearest ^{35}Cl neighbors. (After [5.21])

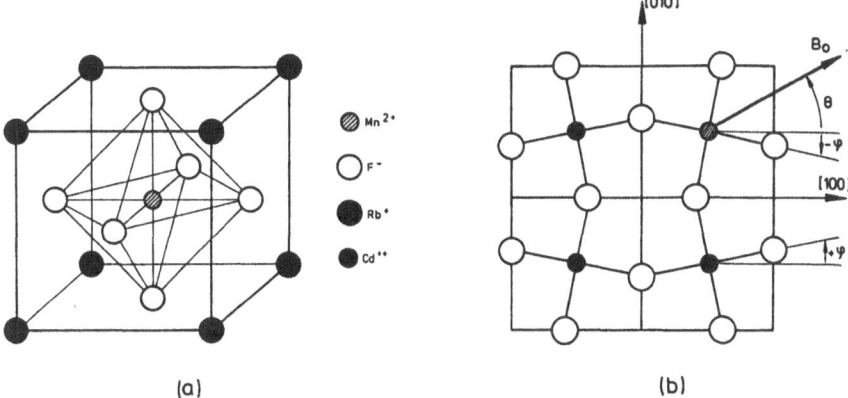

Fig. 5.17. a Local environment of Mn^{2+} doped into $RbCdF_3$ as a probe in the cubic phase of $RbCdF_3$ above the structural phase transition. **b** local environment of Mn^{2+} doped into $RbCdF_3$ for a temperature below the structural phase transition. The orientation of the static magnetic field B_0 used for ENDOR (Fig. 5.18) is also indicated

so much lighter than its neighbors that its local vibrational mode can be approximated by a harmonic motion, and one observes the gradual occupancy of the first excited vibrational state. Note, that there is a dynamical contribution to the shf interaction due to the zero point vibration, even at the lowest temperature (see discussion in Sect. 7.6.5). A very slight increase in shf interaction with temperature was observed for F centers in alkali halides (≈ 1.4 kHz/K for the nearest neighbors in KCl), where the lattice expansion effect is just overcompensated by the lattice vibrations [5.24]. Similar effects were observed for aggregates between F centers and F^- impurities in KCl, where the F^- impurity occupies a next nearest (200) position [5.25], except for the ^{19}F interactions, which strongly decrease with increasing temperature. The reason here is that the small F^- ion replacing the larger Cl^- ion has a lot of space and is driven away from the center by the intermediate K neighbor, which vibrates in a very shallow potential. This vibration shows up in a much larger increase of its shf interaction with temperature compared to the other lattice neighbors [5.26]. The temperature dependence of ENDOR can thus reveal a lot of information on the dynamical properties of defects.

Another effect of temperature is the occurrence of a structural phase transition, which can be investigated with ENDOR if a suitable paramagnetic probe is incorporated into such a crystal. For example, $RbCdF_3$ has a structural phase transition from the cubic phase (Fig. 5.17a) to a tetragonal phase at $T_c = 124$ K, where the CdF_6 octahedra are rotated by $\pm\phi$, (the order parameter) about a cubic axis (see Fig. 5.17b). If Cd^{++} is substituted by a paramagnetic probe such as Mn^{2+} (with a $3d^5$ configuration), then the local

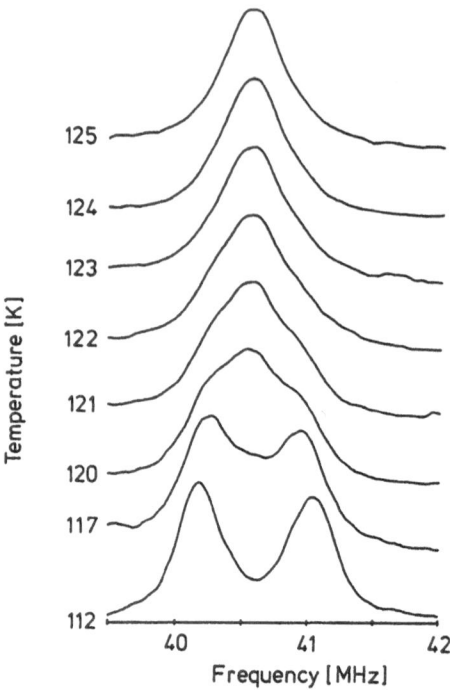

Fig. 5.18. Temperature dependence of an ENDOR line of the nearest ^{19}F neighbors of Mn^{2+} doped into $RbCdF_3$ when varying the temperature through the phase transition temperature at $T_c = 124$ K ($\mathbf{B}_0 \parallel$ [110], $B_0 = 333.1$ mT, $m_S = -\frac{1}{2}$). (After [5.27])

symmetry change of the nearest ^{19}F neighbors can be investigated by ENDOR [5.27]. If, for a certain orientation of the magnetic field B_0 in a {100} plane, the angle between B_0 and the z–axis of the ^{19}F shf tensor is θ for $T < T_c$, it will be $(\theta \pm \phi)$ for $T > T_c$; i.e., instead of one ^{19}F line, one will see two. This ENDOR line splitting is shown in Fig. 5.18, upon lowering the temperature from 125 K to 112 K. In this investigation, a 'local' order parameter could be determined as a function of temperature, which was shown to be different from that of the host crystal due to a lattice relaxation around the Mn^{2+} impurity ion [5.27].

5.7.2 Photo–Excitation of ENDOR Spectra

For a full and technologically relevant characterization of a defect in semiconductor physics, one needs to know its energy level position in the gap, as well as its atom structure. When performing an ENDOR investigation of a defect containing semiconductor crystal, the energy levels of the paramagnetic

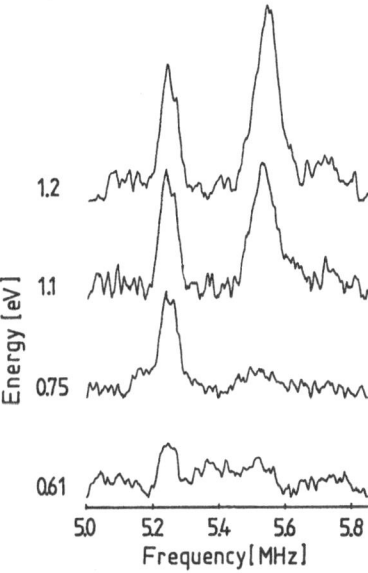

Fig. 5.19. Part of a photo–ENDOR spectrum of a Silicon crystal containing S⁺
and (S–S)⁺ defects. The ENDOR lines belong to ^{29}Si ligands; the line at 5.25 MHz
is due to S⁺ centers, that at 5.5 MHz due to (S–S)⁺ pair defects. (After [5.28])

defects found are generally not known. Defect energy levels are usually determined by optical spectroscopy or deep level transient spectroscopy (DLTS),
and what is needed is a correlation between the magnetic resonance and
energy level determining experiments.

Such a correlation experiment is possible in semiconductors, where the
position of the *Fermi* level can be varied. The defect can be observed by
magnetic resonance if the *Fermi* level is above its energy level, since the level
is then occupied. If the level is emptied due to lowering the *Fermi* level, for
example, by some additional doping, magnetic resonance can no longer be
observed. In this case, illumination of the sample with light of a suitable
photon energy can lift electrons from the valance band into the defect level,
and the magnetic resonance signal appears. This type of photo–excitation
experiment is called photo–EPR or photo–ENDOR, and can, of course, also
be applied to the optical detection EPR and ENDOR (photo–ODEPR and
photo–ODENDOR).

An example is shown in Figs. 5.19,20 for chalcogen defects in B–codoped
Silicon. How two adjacent ENDOR lines of ^{29}Si ligands appear upon sample
illumination with light of increasing photon energy, is shown in Fig. 5.19.
The line at 5.25 MHz appears at lower photon energy than that at 5.5 MHz,
which immediately shows that both ENDOR lines must belong to different
defects, whose energy levels above the valence band can be determined from

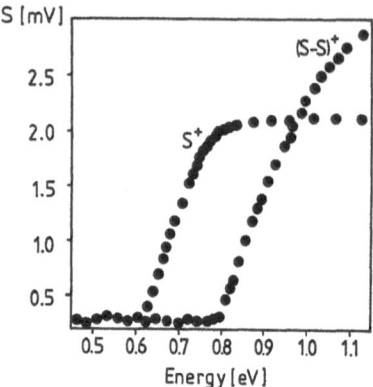

Fig. 5.20. Photo–EPR signal as a function of photon energy for S$^+$ and (S–S)$^+$ defects in Silicon. (After [5.28])

the onset of the ENDOR or EPR signals (Fig. 5.20). The investigation showed that the high frequency ENDOR line belongs to (S–S)$^+$ pair defects, while the other one comes from S$^+$ defects [5.28]. This example shows that the photo–excitation can not only determine the energy levels, but can also discriminate between defects with overlapping ENDOR spectra. (Note, that for a proper level identification, the dynamics of the photo–occupation should be investigated and compared to the corresponding DLTS results).

The photo–ionization of an occupied paramagnetic level could also be used, in principle, for such a correlation spectroscopy. Photo–excitation into an excited optical level should decrease the magnetic resonance signal but not change the conductivity of the sample, while a photo–ionization into the conduction band should also increase the conductivity. Both cases can be distinguished experimentally, and in this way a correlation spectroscopy should be possible. No such experiments have yet been reported.

In ODENDOR an similar excitation spectroscopy can be performed similar to ODEPR, which was described as "tagging" in Sect. 4.4. When using the MCDA technique, the double resonance conditions for a particular ENDOR line are fixed, and the photon energy is varied through the MCDA spectrum while the appearance of the ODENDOR line is monitored. In this way, only that MCDA is measured which belongs to the one defect, or defect orientation, for which the ENDOR line was taken. This can be quite important to separate defects with overlapping MCDA spectra. Analogous experiments can be performed when measuring ODENDOR in emission.

6. Determination of Defect Symmetries from ENDOR Angular Dependences

The analysis of ENDOR angular dependences may sometimes be difficult since there is no systematic procedure leading from the experimental data directly to the structure of the defect. The only way to proceed is by guessing a defect model, and comparing angular dependences calculated for this model with the experimental ones. If no agreement is achievable, the defect model is wrong and must be modified. The interaction parameters, such as the shf and the quadrupole constants, are not known for the defect. They must be "fitted" in order to obtain coincidence between the simulated and the experimental angular dependences. This fit, however, will only work if the structure of the model is correct in all details. Only the angular dependence of the ENDOR spectra provides the essential information about the defect structure. Therefore, a single ENDOR spectrum tells nothing about the validity of a defect model in most cases. The main step in any analysis of an ENDOR angular dependence is the simulation of angular dependences for a given defect model and given interaction parameters. The second step then is the fit of the interaction parameters to the experimental angular dependence.

When considering ENDOR angular dependences, it makes no difference in which way the data were obtained, whether by ENDOR or by ODENDOR, since ultimately, nuclear magnetic resonance (NMR) transitions are detected. In order to derive the structure of a defect, it is, therefore, sufficient to consider NMR transitions in the spin *Hamiltonian* describing all spin interactions in the presence of a magnetic field and of an electric field gradient.

To understand the basic principles which relate NMR angular dependences to the symmetries of the defects, it is not even necessary to start with an exact solution of this *Hamiltonian*. The main features are more easily seen from a simple first order treatment. For practical spectroscopy, however, a very precise calculation of the interactions, and a precise fitting of calculated values to the experimental data is essential. This high precision is essential, in many cases, in order to clarify the structure of a defect unambiguously. For example, it is important to decide whether a small splitting of lines is due to the symmetry of the defect being lower than expected, or whether it is just due to higher order effects from the solution of the spin *Hamiltonian*. Different strategies for the solution of the spin *Hamiltonian* are described in Sect. 6.8.

The NMR data yield information about the symmetry of individual neighbor shells rather than about properties of the entire center. The latter information is usually derived more easily from the EPR spectra. Despite the great power of NMR to unravel details of neighbor shell structures, it is often very difficult to simply obtain the number of neighbors in a shell. The shape of EPR spectra, on the other hand, usually depends strongly on the number of neighbors in the different shells (Chap. 3). A recalculation of the measured EPR spectrum on the basis of all interaction constants obtained by the analysis of NMR transitions, can, therefore, be very helpful to clarify open questions. In cases where this is not successful or feasible, additional information can be obtained by the application of the techniques such as ENDOR–induced EPR and Double ENDOR. This will be discussed later. In general, before starting the tedious business of analyzing NMR angular dependences in detail, some information about the symmetry of the defect should be available.

When considering NMR angular dependences and neighbor shell symmetries in the following, no attempt is made to explain these relations by the application of powerful mathematical tools such as group theory. In order to find a way successfully through the mass of data to be analyzed in a typical multiple resonance experiment, the experimentalist must gain a good 'feeling' for these relations, which enables him to find a first approach to the solution by 'close inspection' of the data. This feeling is probably more easily achieved by using simple arguments to understand the relations, rather than by handling powerful but complex tools.

6.1 Definition of Neighbor Shells

For understanding the symmetry properties of neighbor nuclei, it is very useful to distinguish between different sets of neighbor nuclei by the following consideration.

Let the position of all neighbor nuclei be described in the frame of a three–dimensional rectangular coordinate system, and let the origin of this system be the center of the defect. Center of the defect, in this sense, means the center of gravity of the spin density distribution. Then the position X_i of each nucleus i is defined by a three–dimensional vector $\mathbf{V}_i = (V_{ix}, V_{iy}, V_{iz})$ pointing from the nucleus to the center. Any of these vectors may be transformed into some other vector $\mathbf{W}_i$ by the application of an orthonormal 3×3 matrix $\tilde{M}$ by:

$$\mathbf{W_i} = \begin{pmatrix} M_{xx} & M_{xy} & M_{xz} \\ M_{yx} & M_{yy} & M_{yz} \\ M_{zx} & M_{zy} & M_{zz} \end{pmatrix} \begin{pmatrix} V_{ix} \\ V_{iy} \\ V_{iz} \end{pmatrix}. \tag{6.1.1}$$

Now let each of the vectors $\mathbf{V}_i$ be transformed by the same matrix $\tilde{M}$ into different vectors. This procedure may be repeated for many different matrices $\tilde{M}$. Of special interest is the set $\{\tilde{S}\}$ of all those matrices $\tilde{S}$ which leave the host crystal, together with the defect, unchanged. The matrices of this set are called *symmetry matrices*. This set is a subset of the set $\{\tilde{S}_c\}$ of matrices leaving the host crystal unchanged. $\{\tilde{S}\}$ is identical with $\{\tilde{S}_c\}$ if, and only if, the defect has the same symmetry as the host crystal. If the symmetry of the defect is lower than that of the crystal, then there are matrices of the set $\{\tilde{S}_c\}$ which, again, do not change the structure of the defect, but do change the orientation of the defect within the host crystal. These special matrices are called *orientation matrices* $\tilde{O}$, and they form the set $\{\tilde{O}\}$.

It should be mentioned that it is not necessary to distinguish between symmetry and orientation matrices in order to describe experimental ENDOR angular dependences correctly. The analysis of complex angular dependences can, however, be greatly facilitated by the application of experimental means which are able to distinguish between these two types of transformations (Double ENDOR, ENDOR–induced EPR, see below). It is, therefore, convenient to follow the concept suggested here.

By definition, all symmetry matrices $\tilde{S}$ leave the defect unchanged. Neighbor nuclei may be transformed into different ones which have the same distance from the center. Application of all symmetry matrices of the set $\{\tilde{S}\}$ to one arbitrary neighbor nucleus produces a set of neighbor nuclei which is called a *neighbor shell*. This neighbor shell contains the neighbor nucleus chosen in an arbitrary way. Application of all symmetry matrices to a different neighbor nucleus outside of this neighbor shell yields a different neighbor shell. In special cases, a neighbor shell may consist of just one neighbor nucleus. Since the matrices $\tilde{S}$ are orthonormal, it follows that all members of a neighbor shell have the same distance from the center. On the other hand, neighbor nuclei having the same distance from the center do not necessarily belong to the same neighbor shell. For example, this is the case for some neighbor shells of a substitutional point defect in silicon (T_d defect symmetry).

The neighbor shells of a defect differ by their symmetry properties and by the distance of their nuclei from the center. Although the number of neighbor shells is unlimited for a defect in an unlimited host crystal, there is always a finite number of different neighbor shell symmetries. The T_d defect in silicon, for example, has only four different neighbor shell symmetries.

6.2 Neighbor Shells and Transformation of Interaction Tensors

The nuclei of a neighbor shell are related to each other by symmetry operations. The nucleus i, with a position characterized by the vector $\mathbf{V}_i$, is transformed into the neighbor k, ($k \neq i$) of the same shell, by the application of a suitable symmetry matrix $\tilde{S}$:

$$\mathbf{V}_k = \tilde{S}\,\mathbf{V}_i. \qquad (6.2.1)$$

Now let $\tilde{A}_i$ be the shf tensor of the nucleus i. Then the corresponding shf tensor $\tilde{A}_k$ for the nucleus k is simply obtained by:

$$\tilde{A}_k = \tilde{S}\,\tilde{A}_i\,\tilde{S}^{-1}, \qquad (6.2.2)$$

where $\tilde{S}^{-1}$ is the inverse matrix of $\tilde{S}$. Since $\tilde{S}$ is orthonormal, the matrix $\tilde{S}^{-1}$ in (6.2.2) can be replaced by the transposed matrix $\tilde{S}'$ of $\tilde{S}$. Equation (6.2.2) then simplifies to:

$$\tilde{A}_k = \tilde{S}\,\tilde{A}_i\,\tilde{S}'. \qquad (6.2.3)$$

This relation greatly facilitates the analysis of ENDOR angular dependences, and it is fundamental for the determination of defect structures from ligand hyperfine data (Sects. 6.3,4). It is not necessary, nor even advantageous, to consider interaction tensors of individual neighbor nuclei rather than interaction tensors of different neighbor shells. Each shell is characterized by just one interaction tensor, which is the actual tensor of one arbitrary chosen nucleus in this shell. It is of no importance which of the interaction tensors of the individual nuclei in a shell is considered to be *characteristic* for this shell.

The symmetry relations, as expressed by (6.2.3), are also valid in an analogous way for quadrupole tensors. Again, each neighbor shell is characterized by just one quadrupole tensor. To avoid confusion, the same neighbor nucleus in a shell should be selected to be characteristic of the entire shell, with respect to its hyperfine and quadrupole tensors.

In cases where a neighbor shell consists of just one nucleus, (6.2.3) apparently makes no sense. For example, the central nucleus of a defect forms such a shell. One should expect a very simple angular dependence for this shell, consisting of just the lines for one nucleus. This is generally not the case. If the interaction for the single nucleus is not purely isotropic, one observes an angular dependence which is often indistinguishable from that of other shells containing several nuclei. The reason is the influence of different defect orientations. The distinction between orientation effects and neighbor shell symmetry effects is probably the most difficult task in any analysis of ENDOR angular dependences. In establishing a defect model, one may be severely misled by not observing this point. Details, together with several examples, are discussed later.

The possibility of different isotopes in the same neighbor shell has not yet been considered. They cause different interaction tensors for nuclei in the same shell which are not related to each other by the symmetry relations (6.2.3). This usually causes no significant difficulties for the analysis of ENDOR spectra. In many cases, it is beyond the resolving power of ENDOR spectra to indicate which nucleus in a shell corresponds to a particular isotope. For example, if there are N isotopes with different non–vanishing magnetic moments in a shell, then the data look as if there were N neighbor shells with the same type of symmetry, but with different interaction parameters. Each of these neighbor shells can then formally be considered to consist of just one isotope. The relative intensities of ENDOR lines corresponding to different isotopes usually reflect their natural abundances. The information on the structure of the defect furnished by one of the pseudo–shells is redundant with that of the other $N - 1$ shells. When deriving a defect model from ENDOR, one must be aware of this isotope effect, and not assume additional neighbor shells which are not, in fact, present.

In cases where a small splitting of ENDOR lines due to a mutual interaction of neighbor nuclei is observed, the general behavior of the angular dependences is still the same as discussed above; however, the detailed structure of the small splittings can get rather complicated. Higher order effects in the solution of the spin *Hamiltonian* describing interactions between different neighbor nuclei are treated in Sect. 6.8.

6.3 Interaction Tensor Symmetries and ENDOR Angular Dependence

The knowledge of the interaction tensors for one neighbor in a shell, is sufficient for understanding the angular dependence of the entire shell, provided the symmetry matrices are known. It is, therefore, important to study the symmetry properties of single tensors like the hf, shf, or quadrupole interaction tensors, before considering angular dependences.

The most trivial case for a shf tensor corresponds to $b = 0$ and $b' = 0$. There, one observes for $S = 1/2$ two angularly independent ENDOR lines due to the two values of $m_S = +1/2$ and $m_S = -1/2$, respectively. The symmetry of the interaction tensor, in this case, may be characterized by that of an ideal sphere. Examples are central nuclear hyperfine interactions for high symmetry point defects, such as neutral atomic hydrogen in alkali halide crystals [6.1–4], or T_d–symmetry defects in silicon [6.5–7].

Now we consider the case where the anisotropic part b of the shf tensor $\tilde{A}$ has a finite value, but is smaller than the isotropic part a. Assume the following inequalities: $a > 0$, $b > 0$, $b < a$, $b' = 0$, $m_S < 0$ and $g_n > 0$. If the static magnetic field $\mathbf{B_0}$ is now rotated in the z–x plane starting along $\mathbf{B_0} \parallel z$ ($\theta = 0°$) and ending along $\mathbf{B_0} \parallel x$ ($\theta = 90°$), one then obtains, according to (5.4.5), the angular dependence illustrated in Fig. 6.1. The symmetry of this shf tensor may be visualized by a sphere which has been deformed by stretching it along the z–direction. When rotating the magnetic field in the x–y plane, one obtains a straight line for the ENDOR angular dependence, indicating the cylindrical tensor symmetry. This is no longer obtained if there

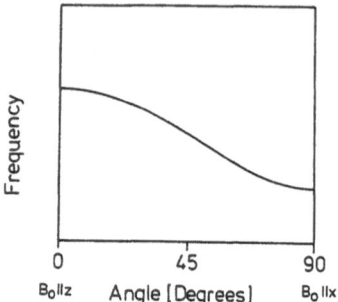

Fig. 6.1. Schematic ENDOR angular dependence obtained for a shf tensor with axial symmetry when rotating the magnetic field in the $z–x$ plane starting along $\mathbf{B_0} \parallel \mathbf{z}(\theta = 90°)$ and ending along $\mathbf{B_0} \parallel \mathbf{x}(\theta = 90°)$

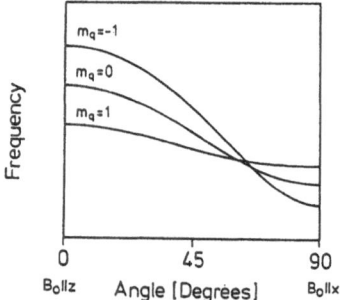

Fig. 6.2. Schematic ENDOR angular dependence obtained for a *shf tensor* with axial symmetry together with a *quadrupole tensor* with axial symmetry. The z–axes of both tensors are parallel. Magnetic field variation as in Fig. 6.1

is a deviation from cylindrical symmetry described by $b' > 0$ (5.4.3). The symmetry of the tensor is now represented by a body obtained by additionally stretching the cylindrically symmetric body by a small amount along the y–direction. By definition, the direction of the largest diameter of the body is identical with the principal z–axis of the tensor, and the direction of the smallest diameter indicates its principal x–axis. It follows that the absolute value of b' can never be larger than the absolute value of b. These symmetry statements for the hf or the shf tensor are also valid for the quadrupole tensor, the only difference being that the quadrupole tensor has no isotropic part.

The effect of the quadrupole interaction on the ENDOR angular dependence is shown in Fig. 6.2. It is assumed there that the principal axis system of the shf tensor is parallel to that of the quadrupole tensor. The magnetic field is varied in the $z–x$ plane, as in Fig. 6.1. The *shf* line ($m_q = 0$) is identical to that in Fig. 6.1. For a magnetic field rotation in the $x–y$ plane one would observe three parallel straight lines for $m_q = -1$, $m_q = 0$ and

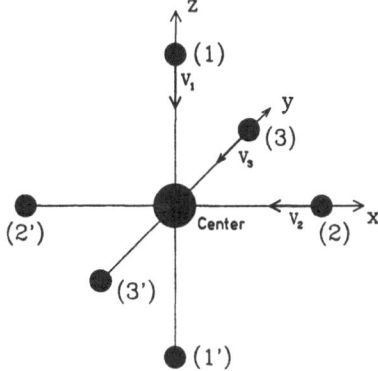

Fig. 6.3. Symmetry of the first neighbor shell of a substitutional center in a cubic host crystal

$m_q = +1$, respectively (for $b' = 0$ and $q' = 0$). These three lines would no longer be straight for $b' \neq 0$ or $q' \neq 0$. According to the values of q' and b', one then observes an angular dependence similar to that of Fig. 6.2.

The Figs. 6.1,2 represent general features based on a first order solution of the spin *Hamiltonian*, which is only a good approximation for $|q'| \ll |q|$, $|q| \ll |b|$, $|b'| \ll |b|$, $|b| \ll |a|$, and where $|a|$ is small compared to the electron *Zeeman* term. For practical cases, an exact treatment of the spin *Hamiltonian* is unavoidable. This is considered in Sect. 6.8.

6.4 Neighbor Shell Symmetries and ENDOR Angular Dependences

6.4.1 Simple Example

So far, the only case considered is that in which the plane of magnetic field variation is always perpendicular to one of the principal axes of the interaction tensor. There are, in fact, defects in cubic crystals where such simple symmetries are found. For example, consider the first 100–neighbor shell of an F–center in KBr [3.4], the symmetry of which is sketched in Fig. 6.3. The center is in the origin of the coordinate system (x, y, z). The first six neighbors (K^+) are indicated by dots and are numbered by (1), (1'), (2), (2'), and (3), (3'), respectively. There are three pairs of neighbors where the two members of each pair are related to each other by an inversion. Since the ENDOR frequencies do not depend on the sign of the magnetic field vector, the ENDOR lines of the two pair members coincide. Neighbors with this property are called *magnetically equivalent*. It is, therefore, sufficient to consider the neighbors (1), (2), and (3). However, the coinciding ENDOR lines may be split due to a

mutual interaction between magnetically equivalent nuclei. Magnetic equiva-
lence of neighbors may thus be detectable in an ENDOR spectrum. Whether
this is the case or not, depends on the size of the interaction constants; see
Sect. 6.8 for details.

Let the neighbor nucleus (1) be chosen as the characteristic nucleus for
the neighbor shell (Fig. 6.3). Due to reasons explained later, the principal axis
system of the interaction tensor of this neighbor is parallel to the coordinate
system (x, y, z), and the parameter b' is equal to zero. The shf interaction
tensor $\tilde{A}_1$ is then of the simple form:

$$\tilde{A}_1 = \begin{pmatrix} A_{xx} & & \\ & A_{yy} & \\ & & A_{zz} \end{pmatrix} = a\,\tilde{1} + \begin{pmatrix} -b & & \\ & -b & \\ & & 2b \end{pmatrix}. \tag{6.4.1}$$

When varying the magnetic field in the z–x plane starting along the z–axis
$(\theta = 0°)$ in order to measure an ENDOR angular dependence, the neighbor
(1) contributes with a line identical to that shown in Fig. 6.1. The neighbor (2)
is evidently related to the neighbor (1) by a 90° rotation around the y–axis.
This rotation is effected by the symmetry matrix $\tilde{S}_2$ which has the form:

$$\tilde{S}_2 = \begin{pmatrix} 0 & 0 & 1 \\ 0 & 1 & 0 \\ -1 & 0 & 0 \end{pmatrix}. \tag{6.4.2}$$

Transformation of the vector $\mathbf{V}_1$ by application of the symmetry matrix $\tilde{S}_2$
yields the vector $\mathbf{V}_2$:

$$\mathbf{V}_2 = \tilde{S}_2\,\mathbf{V}_1 = \begin{pmatrix} 0 & 0 & 1 \\ 0 & 1 & 0 \\ -1 & 0 & 0 \end{pmatrix} \begin{pmatrix} 0 \\ 0 \\ -1 \end{pmatrix} = \begin{pmatrix} -1 \\ 0 \\ 0 \end{pmatrix}. \tag{6.4.3}$$

This is exactly what one expects from an inspection of Fig. 6.3. Accordingly,
the shf tensor $\tilde{A}_2$ of the neighbor nucleus (2) is obtained by:

$$\tilde{A}_2 = \tilde{S}_2\,\tilde{A}_1\,\tilde{S}_2' = \begin{pmatrix} A_{zz} & & \\ & A_{yy} & \\ & & A_{xx} \end{pmatrix}. \tag{6.4.4}$$

The principal z–axis of this tensor is parallel to the x–axis, and the principal x–axis is parallel to the z–axis. Note, that the tensor always has mirror
symmetry with respect to each of its principal axes. It, therefore, makes no
difference whether, for example, the principal x–axis of the tensor is parallel
or antiparallel to the x–axis. The angular dependence for the neighbor (2) is
basically the same as shown in Fig. 6.1, except that the origin of the angular
scale is shifted by 90°. By similar arguments, one finds for the neighbor (3):

$$\tilde{S}_3 = \begin{pmatrix} 1 & 0 & 0 \\ 0 & 0 & 1 \\ 0 & -1 & 0 \end{pmatrix}, \tag{6.4.5}$$

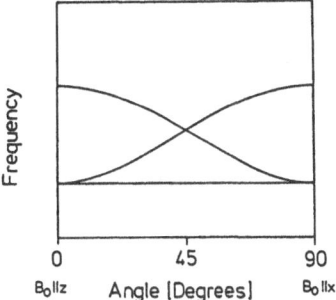

Fig. 6.4. Schematic ENDOR angular dependence for the defect sketched in Fig. 6.3. Magnetic field variation and shf tensor symmetry as in Fig. 6.1

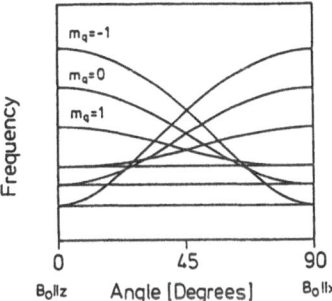

Fig. 6.5. Schematic ENDOR angular dependence for the defect in Fig. 6.3 with shf and quadrupole interactions as in Fig. 6.2

and

$$\tilde{A}_3 = \begin{pmatrix} A_{xx} & & \\ & A_{zz} & \\ & & A_{yy} \end{pmatrix}. \qquad (6.4.6)$$

The magnetic field is always perpendicular to the principal z–axis of the tensor $\tilde{A}_3$ [$\theta = 90°$ in (5.4.5)], and, therefore, the contribution of the neighbor (3) to the angular dependence consists of a straight line. The entire angular dependence is sketched in Fig. 6.4. For a non–vanishing quadrupole interaction, one must add the angular dependence patterns for the different quadrupole states according to (5.4.7); for example, the patterns for $m_q = -1$ and $m_q = +1$ for a nuclear spin of $I = 3/2$. An example is shown in Fig. 6.5.

To sum up, the first neighbor shell for the simple example discussed is characterized by the tensors of the neighbor nucleus (1), and by the symmetry matrices $\tilde{S}_2$ and $\tilde{S}_3$.

6.4.2 General Case

Usually, the situation is not as simple as considered above. For example, in practical cases it is not often possible to align the sample precisely enough so that the magnetic field is exactly varied in the plane which is the most desirable. On the other hand, symmetries are often more complex, and it is no longer possible to understand the experimental angular dependences merely on the basis of simple expressions such as (5.4.5) and (5.4.7).

The first problem is how to describe the influence of the magnetic field direction. A simple approach is to rotate the static magnetic field $\mathbf{B}_0$ in the spin *Hamiltonian* (5.4.1) by the application of a suitable orthonormal 3×3 matrix $\tilde{R}(\mathbf{v}, \theta)$ by an angle θ, in a plane perpendicular to the unit vector $\mathbf{v}$. In the example of Sect. 6.4.1 this vector is $\mathbf{v} = (0, 1, 0)$. The new magnetic field $\mathbf{B}_0(\mathbf{v}, \theta)$ is then obtained by:

$$\mathbf{B}_0(\mathbf{v}, \theta) = \tilde{R}(\mathbf{v}, \theta)\, \mathbf{B}_0, \tag{6.4.7}$$

where

$$\mathbf{B}_0 = \begin{pmatrix} 0 \\ 0 \\ B_0 \end{pmatrix}. \tag{6.4.8}$$

The absolute value of the magnetic field is not changed. The explicit form of the rotation matrix $\tilde{R}(\mathbf{v}, \theta)$ known as *Cayley's* formula [6.8] is given in Appendix B. If the magnetic field in the spin *Hamiltonian* is no longer parallel to the z–direction of the coordinate system, off–diagonal elements in the *Hamiltonian* matrix can become very large. In extreme cases the diagonal elements may even vanish. This causes no problems from a physical point of view, however, all algorithms for a numerical diagonalization of the spin *Hamiltonian* tend to become less efficient. Instead of rotating the magnetic field by the application of $\tilde{R}(\mathbf{v}, \theta)$, it is, therefore, useful to counter–rotate all tensors by the application of $\tilde{R}'(\mathbf{v}, \theta)$. For example, the tensor $\tilde{A}_2$, mentioned in the previous section, is rotated by $\tilde{R}'(\mathbf{v}, \theta)$ in the following way:

$$\tilde{A}_2(\mathbf{v}, \theta) = \tilde{R}'(\mathbf{v}, \theta)\, \tilde{A}_2\, \tilde{R}(\mathbf{v}, \theta) = \tilde{R}'(\mathbf{v}, \theta)\, \tilde{S}_2\, \tilde{A}_1\, \tilde{S}_2'\, \tilde{R}(\mathbf{v}, \theta). \tag{6.4.9}$$

The quadrupole tensor must be treated in a similar way. Also, the electron *Zeeman* term and the fine–structure term must be transformed accordingly.

Transformation of the fine–structure term may, however, be somewhat inconvenient if higher order terms are significant in the case of high values of the electron spin [6.9,10]. The sequence of the transformation matrices in (6.4.9) is essential, and must not be interchanged.

Another problem is the fact that the angle θ is usually not known with sufficient accuracy in an experiment. In the calculations this can be accounted for by replacing the angle θ in (6.4.9) by $\theta - \theta_0$, where the angle θ is set in the experiment, and the angle θ_0 is determined from the measured data, to correct for an offset in the experimental angular scale.

To sum up, there are three independent parameters describing the sample orientation in an angular dependence experiment. These parameters are the angle θ_0 describing an offset of the experimental angular scale, and two additional angles defining the direction of the unit vector $\mathbf{v}$ in (6.4.9).

Another question left open in Sect. 6.4.1 is the orientation of the interaction tensors for the characteristic nucleus of a neighbor shell. (The short expression 'orientation of a tensor' in fact means 'orientation of the principal axis system of a tensor'). It is obvious that the tensor symmetry of a neighbor nucleus must be compatible with the symmetry of the local environment. If there are constraints for the local symmetry, the corresponding tensors must obey these constraints too. This statement provides simple arguments for the orientation and the internal symmetry of tensors.

For example, consider the high symmetry F center in KBr (Sect. 6.4.1). There are three fourfold rotation axes perpendicular to each other going through the center. These axes are the x-, y-, and z-axes of the coordinate system (Fig. 6.3). The only tensor compatible with this local symmetry is characterized by a sphere. Consequently, if there was a central nucleus in the defect, its shf tensor would be isotropic, i.e., $b = 0$ and $b' = 0$. There would be no quadrupole interaction at all for any value of the nuclear spin. The local symmetry would not allow for an electric field gradient. One must, however, be careful when deriving a defect model from the observation of an isotropic hf interaction. Hyperfine anisotropies are sometimes very small, and even undetectable within experimental line widths (particularly in EPR experiments). It is, therefore, necessary to analyze the symmetry properties of many neighbor shells to obtain the correct defect structure.

The symmetry properties and the orientation of neighbor interaction tensors may be obtained as follows.

One considers a straight line going through the neighbor and the defect center, and analyzes the symmetry properties of this interconnection line. There are the following cases to be distinguished:

(i) If the line is a rotation axis with a higher than twofold symmetry, then the principal z-axis of any interaction tensor is parallel to this line and the parameters b' or q' vanish. Since the tensors have cylindrical symmetry, the orientation of the x- and y-axes are of no importance. For example, the first neighbors of the F centers in alkali halides have this symmetry (Fig. 6.3).

(ii) If the line is only identical to the intersection line of two mirror planes (which are perpendicular to each other), then the situation is more complex. Each principal axis of the tensors must lie in a mirror plane. The parameters b' and q' need not necessarily be zero. Which axis lies in which plane does not follow from symmetry arguments. Therefore, the situation may also be different for the ligand hyperfine and the quadrupole tensor. Usually the z-axes of the tensors are parallel to the intersection line of the mirror planes. This is particularly true for the ligand hyperfine tensor, and also in most cases for the quadrupole tensor of near neighbors. However, there might be exceptions for the more distant neighbors. If one has selected the 'wrong' orientation of a tensor z-axis, a correct reproduction of the experimental angular dependence can only be achieved by formally assuming $|b| < |b'|$ or $|q| < |q'|$ in the calculations. An exchange of x- and y-axes only results in a change of the sign of the parameters b' or q'.

(iii) If the interconnection line is only a twofold axis, or if it lies in only one mirror plane, there are no longer distinct orientations of the tensors fixed by symmetry constraints. If the interconnection line is a twofold axis, one of the principal axes of the tensors is always parallel to the interconnection line. In the case of a mirror plane, two principal axes of each tensor are in this plane. In both cases, there is one degree of freedom of tensor orientation, defined by an angle describing the rotation of the tensor around an axis. This axis is either perpendicular to the mirror plane, or it is identical with the interconnection line, in case it is a twofold rotation axis. The angle describing the tensor orientation is often called the *free angle* of tensor orientation. It must be determined from the experimental results. The free angles for the ligand hf and for the quadrupole tensors are usually different. The parameters b' and q' are non-zero in most cases.

(iv) Finally, if there is no longer any symmetry constraint for the interconnection line, there are three free angles of tensor orientation. These angles are, again, different for the shf and for the quadrupole tensor. The parameters b' and q' are non-zero in almost any case. A convenient way to define tensor orientations by one, or by three free angles, δ_1, δ_2, and δ_3, is provided by the *Euler* transformation matrix, $\tilde{E}(\delta_1, \delta_2, \delta_3)$.

The aim of the following considerations is to provide a systematic way to calculate the ENDOR angular dependences. In simulating an angular dependence for a neighbor shell, one assumes certain interaction tensors for the characteristic nucleus. Each tensor is characterized by the interaction constants and by the orientation of its principal axes. In the following, a simple 'recipe' is provided to obtain these tensors for a given defect structure.

The first step is to define a coordinate system (x, y, z) fixed to the host crystal. Then diagonal tensors $\tilde{A}_D$ as in (5.4.2,3), and $\tilde{Q}_D$ as in (5.4.6), are defined in this system. These diagonal tensors are then transformed into the desired interaction tensors as shown in the following example:

Let the neighbor shell considered consist of three neighbor nuclei N_1, N_2 and N_3. The neighbor N_1 has been selected to be characteristic of this shell. Then there are two symmetry matrices, $\tilde{S}_2$ and $\tilde{S}_3$, necessary to transform the tensors of the neighbor N_1, $\tilde{A}_1$, into the tensor of the neighbor N_2, $\tilde{A}_2$, and into the tensor $\tilde{A}_3$ of the neighbor N_3, respectively (6.2.3). In general, the tensor $\tilde{A}_1$ is not diagonal. The necessary transformation of $\tilde{A}_D$ (or $\tilde{Q}_D$) to obtain $\tilde{A}_1$ (or $\tilde{Q}_1$) can be achieved by application of the *Euler* transformation matrix $\tilde{E}$. In many cases, it is, however, useful to introduce an additional transformation $\tilde{T}$, which enables one to achieve a structure of the *Euler* transformation matrix $\tilde{E}(\delta_1, \delta_2, \delta_3)$, where the three *Euler* angles δ_1, δ_2 and δ_3 can be illustrated more easily. The tensor $\tilde{A}_1$ is then expressed by:

$$\tilde{A}_1 = \tilde{T}\ \tilde{E}(\delta_1, \delta_2, \delta_3)\ \tilde{A}_D\ \tilde{E}'(\delta_1, \delta_2, \delta_3)\ \tilde{T}'. \qquad (6.4.10)$$

If $\tilde{A}_1$ has only one free angle, then it is convenient to define the transformation matrix $\tilde{T}$ in a way that the *Euler* angles δ_2 and δ_3 can be chosen to be zero. According to (6.2.3) the other two tensors are obtained by

$$\tilde{A}_2 = \tilde{S}_2\ \tilde{T}\ \tilde{E}\ \tilde{A}_D\ \tilde{E}'\ \tilde{T}'\ \tilde{S}_2' \qquad (6.4.11)$$

and by

$$\tilde{A}_3 = \tilde{S}_3\ \tilde{T}\ \tilde{E}\ \tilde{A}_D\ \tilde{E}'\ \tilde{T}'\ \tilde{S}_3'. \qquad (6.4.12)$$

Now the tensors must be transformed according to (6.4.9), in order to account for the rotation of the magnetic field:

$$\tilde{A}_1(\mathbf{v}, \theta - \theta_0) = \tilde{R}'(\mathbf{v}, \theta - \theta_0)\ \tilde{T}\ \tilde{E}\ \tilde{A}_D\ \tilde{E}'\ \tilde{T}'\ \tilde{R}(\mathbf{v}, \theta - \theta_0), \qquad (6.4.13)$$

$$\tilde{A}_2(\mathbf{v}, \theta - \theta_0) = \tilde{R}'(\mathbf{v}, \theta - \theta_0)\ \tilde{S}_2\ \tilde{T}\ \tilde{E}\ \tilde{A}_D\ \tilde{E}'\ \tilde{T}'\ \tilde{S}_2'\ \tilde{R}(\mathbf{v}, \theta - \theta_0), \qquad (6.4.14)$$

$$\tilde{A}_3(\mathbf{v}, \theta - \theta_0) = \tilde{R}'(\mathbf{v}, \theta - \theta_0)\ \tilde{S}_3\ \tilde{T}\ \tilde{E}\ \tilde{A}_D\ \tilde{E}'\ \tilde{T}'\ \tilde{S}_3'\ \tilde{R}(\mathbf{v}, \theta - \theta_0). \qquad (6.4.15)$$

This is the most general form of transformations which must be applied to the shf or, in an analogous way, to the quadrupole diagonal tensors in order to obtain the ENDOR angular dependence. The *Euler* matrix $\tilde{E}$ and the transformation matrix $\tilde{T}$ are usually different for transformations of the shf and of the quadrupole tensors. The rotation matrix $\tilde{R}$ is identical for all tensors. The interaction constants for the shell are contained in the shf tensor $\tilde{A}_D$ and in the quadrupole tensor $\tilde{Q}_D$. The symmetry of the neighbor shell is contained in the symmetry matrices $\tilde{S}$. An example of how to find these matrices is given below.

6.4.3 Defect Structure and Symmetry Matrices

There is no way to derive the structure of a defect directly by inspection of experimental data. The only possible way is an indirect one, in which one calculates possible experimental results for different defect models, and then compares the theoretical results with the experimental data. Sometimes it can be difficult to establish an unequivocal defect model this way. Therefore, in order not to be misled too easily, it is essential to get a close understanding of the way in which defect structures show up in angular dependence patterns. It is obvious from the previous discussions that the symmetry matrices play a key role in linking defect structures to angular dependence patterns. It is the aim of this chapter to provide a detailed example of the way in which symmetry matrices are determined by the structure of the defect. In the example of Fig. 6.3, it was assumed that the magnetic field is always perpendicular to one principal axis of any interaction tensor, which is, however, no longer true in most other cases. The examples shown below correspond to exact diagonalizations of the spin *Hamiltonian* (Sect. 6.8). The interaction tensors are calculated according to (6.4.13–15).

Still, a *high symmetry defect* is considered where the symmetry of the defect is equal to that of the host crystal. As a consequence, there is only one possible orientation of the defect in the host crystal, or only one position of the defect in the unit cell. The more complicated case of low symmetry defects in environments of higher symmetry is considered later.

Figure 6.6 shows the unit cell of the silicon lattice with a point defect in an interstitial position in the center of the cubic unit cell. Al^{++} [6.11] or Fe^0 in Silicon [6.6] are examples for such a defect. Symmetry matrices may now be found by the following procedure:

First, it is necessary to define a coordinate system (x, y, z) which is fixed to the host crystal, and which will be the unique basis of all calculations. The origin of the coordinate system coincides with the center of the defect. This is indicated in Fig. 6.6. (It is essential that the coordinate system is not fixed to the defect structure. This makes no difference here, but is important for the description of low symmetry defects considered later).

Secondly, one must find the different neighbor shells. There are four nearest neighbors, 1a, 1b, 1c and 1d in <111> directions. By inspection of the crystal structure, it can be verified that each of these neighbors can be transformed into another one by symmetry operations of the crystal. All {110} planes are mirror planes of the crystal structure, and it can be easily seen that the four first neighbors are related to each other by mirror operations provided by these planes. These four neighbors, therefore, form a neighbor shell.

Let the neighbor 1a now be selected to be characteristic of this first neighbor shell. The interconnection line between this neighbor and the defect center is parallel to a $\langle 111 \rangle$ direction. This line has threefold rotation symmetry.

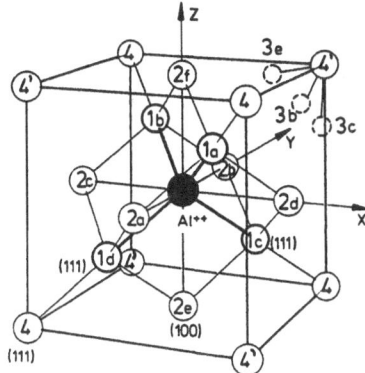

Fig. 6.6. Unit cell of Silicon with a point defect (Al^{++}) in a high symmetry (T_d) interstitial position

Therefore, the principal z–axis of the tensor $\tilde{A}_1$ for the neighbor 1a must coincide with the $[1\ \bar{1}\ 1]$ direction, and the tensor must exhibit cylindrical symmetry with $b' = 0$. $\tilde{A}_1$ has no free angle. According to the recipe, we must find the transformation matrices $\tilde{T}$ and $\tilde{E}$ (6.4.10). Let the *Euler* matrix in (6.4.10) be equal to the unit matrix $\tilde{1}$, which implies that the *Euler* angles δ_1, δ_2 and δ_3 are all zero. Then the transformation matrix $\tilde{T}$ must be defined in a way that the principal z–axis of the tensor $\tilde{A}_1$ is parallel to the $[1\ \bar{1}\ 1]$ direction. This is quite straightforward. Consider a new coordinate system (x', y', z') which is initially parallel to the crystal system (x, y, z). It can be seen from Fig. 6.6 that a rotation of the system (x', y', z') around the $[1\ 1\ 0]$ axis by an angle $\theta = \arctan(\sqrt{2}) = 54.736°$, will transform the z'–axis into a position parallel to $[1\ \bar{1}\ 1]$. This rotation is provided by a matrix $\tilde{R}(\mathbf{v}, \theta)$, where $\mathbf{v} = (1/\sqrt{2}, 1/\sqrt{2}, 0)$ and $\theta = 54.736°$. The desired $\tilde{T}$ matrix for the first neighbor shell $\tilde{T}_1$ is, therefore, equal to this rotation matrix $\tilde{R}(\mathbf{v}, \theta)$. The tensor $\tilde{A}_1$ can now easily be calculated according to (6.4.10), where the tensor $\tilde{A}_D$ has the form of (5.4.2,3) with $b' = 0$. Since $I(^{29}Si) = 1/2$ there are no quadrupole interactions.

The next step is to find the symmetry matrices. It can again be seen from Fig. 6.6, that the neighbor 1b is related to the neighbor 1a by a mirror symmetry provided by the $(1\ \bar{1}\ 0)$ mirror plane. Let the matrix $\tilde{M}(1\ \bar{1}\ 0)$ be the transformation matrix to provide this mirror operation. It can be achieved by a rotation around the $[1\ \bar{1}\ 0]$ direction by $\theta = 180°$ followed by an inversion. According to Sect. 6.2, the symmetry matrix for the nucleus 1b, $\tilde{S}_{1b}$, is now simply equal to $\tilde{M}(1\ \bar{1}\ 0)$. By similar arguments, the symmetry matrix for the nucleus 1c, $\tilde{S}_{1c}$, is equal to $\tilde{M}(0\ \bar{1}\ 1)$, and the symmetry matrix $\tilde{S}_{1d}$ is equal to $\tilde{M}(1\ 0\ 1)$.

The matrices necessary to calculate the angular dependence for the first silicon neighbors, according to (6.4.10-12), are summarized as follows (values rounded to three digits):

$$\tilde{T}_1 = \begin{pmatrix} 0.789 & 0.211 & 0.577 \\ 0.211 & 0.789 & -0.577 \\ -0.577 & 0.577 & 0.577 \end{pmatrix}, \tag{6.4.16}$$

$$\tilde{S}_{1b} = \begin{pmatrix} 0 & 1 & 0 \\ 1 & 0 & 0 \\ 0 & 0 & 1 \end{pmatrix}, \tag{6.4.17}$$

$$\tilde{S}_{1c} = \begin{pmatrix} 1 & 0 & 0 \\ 0 & 0 & 1 \\ 0 & 1 & 0 \end{pmatrix}, \tag{6.4.18}$$

$$\tilde{S}_{1d} = \begin{pmatrix} 0 & 0 & -1 \\ 0 & 1 & 0 \\ -1 & 0 & 0 \end{pmatrix}. \tag{6.4.19}$$

Since all z–axes of the tensors of the first neighbor shell are in $<111>$ directions, the symmetry of this shell is called '111–symmetry'.

The second nearest neighbors are denoted by 2a – 2f. All these neighbors can again be transformed into each other by $\{110\}$ mirror planes. Consequently, they form a neighbor shell containing six neighbors with a symmetry called '100–symmetry', because the interconnection lines from the neighbors to the center are parallel to $<100>$ directions. The interconnection line of each neighbor to the center is identical with the intersection line of two mirror planes perpendicular to each other. The directions of the z–axes of the tensors are, therefore, fixed by symmetry; they point along the $<100>$ directions. However, the parameter b' is not necessarily equal to zero. Let the neighbor 2a be the characteristic neighbor for the 100–shell. The x–axis and the y–axis of its tensor must lie in the mirror planes. Since there are no free angles, the *Euler* transformation matrix is again chosen to be an identity matrix ($\delta_1 = \delta_2 = \delta_3 = 0$). In order to find the T transformation matrix $\tilde{T}_2$, the auxiliary coordinate system (x', y', z') initially parallel to the crystal coordinate system (x, y, z) must be rotated by $\theta = 90°$ around the x–axis, and then, once again, by $\theta = 45°$ around the y–axis. It is important that both rotations are defined with respect to the same coordinate system (x, y, z). The matrix $\tilde{T}_2$ is then:

$$\tilde{T}_2 = \tilde{R}[(0,1,0),45°] \, \tilde{R}[(1,0,0),90°]. \tag{6.4.20}$$

The symmetry matrix $\tilde{S}_{2b}$ for the neighbor nucleus 2b is found in an easier way by first calculating the matrix $\tilde{S}_{2d}$ for the neighbor 2d. It follows immediately that the matrix $\tilde{S}_{2d}$ is equal to $\tilde{M}(1\,1\,0)$. The neighbor 2d, in turn, is related to the neighbor 2b by the $(1\,\bar{1}\,0)$ mirror plane. It follows, therefore, that the symmetry matrix of the neighbor 2b, $\tilde{S}_{2b}$, is given by

$$\tilde{S}_{2b} = \tilde{M}(1\,\bar{1}\,0) \, \tilde{M}(1\,1\,0). \tag{6.4.21}$$

The calculation of the remaining symmetry matrices of the 100–neighbor shell is now straightforward. One obtains for the matrices of this shell:

$$\tilde{T}_2 = \begin{pmatrix} 0.707 & 0.707 & 0 \\ 0 & 0 & -1 \\ -0.707 & 0.707 & 0 \end{pmatrix}, \tag{6.4.22}$$

$$\tilde{S}_{2b} = \begin{pmatrix} -1 & 0 & 0 \\ 0 & -1 & 0 \\ 0 & 0 & 1 \end{pmatrix}, \tag{6.4.23}$$

$$\tilde{S}_{2c} = \begin{pmatrix} 0 & 1 & 0 \\ 1 & 0 & 0 \\ 0 & 0 & 1 \end{pmatrix}, \tag{6.4.24}$$

$$\tilde{S}_{2d} = \begin{pmatrix} 0 & -1 & 0 \\ -1 & 0 & 0 \\ 0 & 0 & 1 \end{pmatrix}, \tag{6.4.25}$$

$$\tilde{S}_{2e} = \begin{pmatrix} 1 & 0 & 0 \\ 0 & 0 & 1 \\ 0 & 1 & 0 \end{pmatrix}, \tag{6.4.26}$$

$$\tilde{S}_{2f} = \begin{pmatrix} 1 & 0 & 0 \\ 0 & 0 & -1 \\ 0 & -1 & 0 \end{pmatrix}. \tag{6.4.27}$$

The shell of the third nearest neighbors is slightly more complicated. Three of these neighbors are indicated in the upper right part of Fig. 6.6. The interconnection lines of each of these neighbors to the center are in $\{110\}$ mirror planes. There are no further symmetry constraints. Let the neighbor 3a be characteristic of the third shell, which is called the 110–shell. Two principal axes of the tensor $\tilde{A}_{3a}$ of this neighbor must be in the $(1\,\bar{1}\,0)$ mirror plane, and the parameter b' may not be equal to zero. It is assumed that the z–axis and the x–axis are in the plane. If the experimental results formally require $|b'| > |b|$, then this assumption is apparently wrong, and the z–axis must be exchanged with the x–axis. If the results then require $b' < 0$, then the x–axis may be exchanged with the y–axis to allow for a positive parameter b' (if desired). There is one free angle defining the tensor orientation in the mirror plane. Let the T matrix $\tilde{T}_3$ now be defined in such a way that the first $Euler$ angle δ_1 is a free parameter, and the other $Euler$ angles are equal to zero, and that $\delta_1 = 0$ corresponds to an orientation of the principal z–axis of the tensor $\tilde{A}_{3a}$ parallel to the $[1\,1\,0]$ axis in the $\{1\,\bar{1}\,0\}$ mirror plane. This means that the $Euler$ matrix is an identity matrix if the principal z–axis of the tensor is parallel to $[110]$. In order to find the $\tilde{T}$ matrix as for the previous shells, the auxiliary system (x', y', z') must be rotated by $90°$ around the y–axis, and then by 45° around the z–axis. The matrix $\tilde{T}_3$ is then:

$$\tilde{T}_3 = \tilde{R}[(0,0,1),45°] \; \tilde{R}[(0,1,0),90°]. \tag{6.4.28}$$

Analogous to (6.4.10) the tensor $\tilde{A}_{3a}$ is now obtained by:

$$\tilde{A}_{3a} = \tilde{T} \; \tilde{E}(\delta_1,0,0) \; \tilde{A}_D \; \tilde{E}'(\delta_1,0,0) \; \tilde{T}'. \tag{6.4.29}$$

The <111> axes in the silicon lattice have threefold rotation symmetry. The neighbor 3b is therefore related to the neighbor 3a by a 120° rotation around the [111] axis, and the neighbor 3c is related to the neighbor 3a by a rotation of −120°, respectively. Thus, the corresponding symmetry matrices are easily shown to be:

$$\tilde{S}_{3b} = \tilde{R}[(1,1,1),-120°], \tag{6.4.30}$$

$$\tilde{S}_{3c} = \tilde{R}[(1,1,1),+120°]. \tag{6.4.31}$$

The third neighbor shell has twelve neighbors since there are three other <111> directions, $[1\ \bar{1}\ \bar{1}]$, $[\bar{1}\ 1\ \bar{1}]$ and $[\bar{1}\ \bar{1}\ 1]$, which are related to the $[1\ 1\ 1]$ direction in the same way, by three mirror matrices, as the neighbors 1b, 1c and 1d are related to the neighbor 1a in the first 111–shell. In order to obtain the symmetry matrices for the remaining nine neighbors in the third shell, one must multiply each of the three matrices $\tilde{T}_3$, $\tilde{S}_{3b}$ and $\tilde{S}_{3c}$ by these mirror matrices from the left.

It is useful to consider angular dependence patterns characteristic of the above three neighbor shell symmetries. In order to study their main features, it is not necessary to select specific interaction constants. In the following examples, these constants are therefore chosen in a somewhat arbitrary way. Let B_0 be 350 mT, $S = 1/2$, $g_e = 2.0023$ and $g_n = -1.1095$ (for ^{29}Si, 4.7% abundant). With these values, f_{nuc} is given by $f_{nuc} = g_n \mu_n B_0/h = 2.9601$ MHz. The magnetic field is rotated in the $(1\ 1\ 0)$ plane by an angle θ_B starting along the $[0\ 0\ 1]$ direction ($\theta_B = 0°$) and ending along $[1\ \bar{1}\ 0]$ ($\theta_B = 90°$). If one now selects the isotropic shf constant to be $a/h = 10$ MHz, and the anisotropic shf constant to be $b/h = 2$ MHz, one obtains the pattern shown in Fig. 6.7 for the first neighbor shell. As qualitatively predicted by (5.4.5), there are two patterns spaced roughly by $2f_{nuc}$ corresponding to $m_S = \pm 1/2$ (compare Fig. 5.8b). The slight differences in the shapes of the two patterns are due to higher order effects not contained in (5.4.5), according to which they should have identical shapes and be shifted from each other by precisely $2f_{nuc}$. The different lines in each pattern can be assigned directly to their corresponding neighbor nuclei as indicated in Fig. 6.7. Maxima and minima of the lines occur at field positions where (5.4.5) also has these maxima and minima. When applying this equation, one must bear in mind that the angle θ in (5.4.5) has a meaning quite different from that of the angle θ_B. In (5.4.5) θ is the angle between the magnetic field and the principal z–axis of the shf tensor. For a

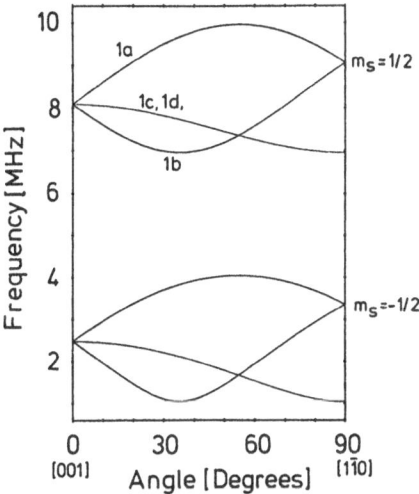

Fig. 6.7. ENDOR angular dependence for the first neighbors of the defect shown in Fig. 6.6. The magnetic field was varied in the (110) plane starting along [001] ($\theta_B = 0°$) and ending along [1$\bar{1}$0] ($\theta_B = 90°$). For the interaction constants and definition of the angle θ_B see text

given magnetic field, θ usually has a different value for each nucleus in a shell. The angle θ_B, on the other hand, describes the magnetic field position with respect to the host crystal. Since the z–axes of the tensors for the neighbors 1a and 1b are accidentally in the plane of magnetic field variation, the angles θ for these neighbors differ just by constant offsets from the angle θ_B. The angles θ for the neighbors 1c and 1d depend on θ_B in a more complex way. These neighbors have identical values of θ for all positions of the magnetic field in the (1 1 0) plane. Their ENDOR lines coincide. They are, therefore, said to be 'equivalent' with respect to a magnetic field variation in this plane. (However, they are not magnetically equivalent, since they are not related to each other by inversion symmetry). Whether or not neighbors are equivalent with respect to a specific plane of magnetic field variation, cannot, however, be conclusively seen from their θ angles. An example of this is shown below. If the shf interactions are considerably smaller than the nuclear *Zeeman* term, the angular dependence patterns in first order have mirror symmetry on the frequency scale with respect to f_{nuc}. An example is shown in Fig. 6.8 for $a/h = 1$ MHz and $b/h = 0.2$ MHz, again, for the first shell considered in Fig. 6.7.

For the calculations shown in Figs. 6.7,8 it was assumed that the parameters a and b are both positive. If both are negative, then one obtains identical patterns; however, their assignment to $m_S = 1/2$ and $m_S = -1/2$ is reversed. If the constants a and b have opposite signs, then the angular dependence

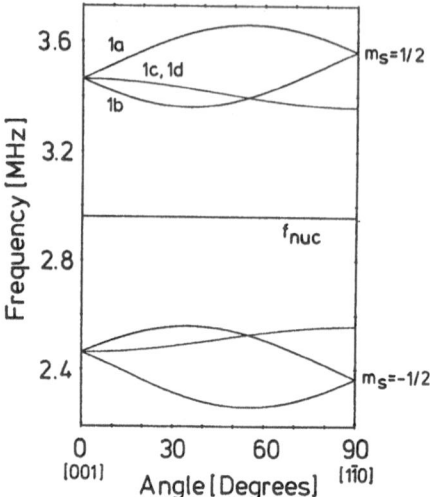

Fig. 6.8. ENDOR angular dependence for the same symmetry as in Fig. 6.7 with considerably reduced interaction constants compared to the calculation in Fig. 6.7. The two patterns have now mirror symmetry with respect to the line f_{nuc} of the nuclear *Zeeman* term

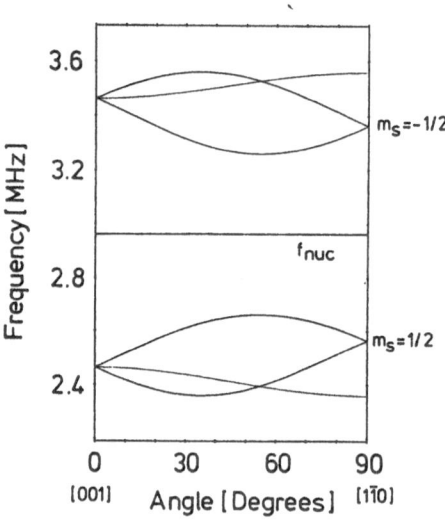

Fig. 6.9. ENDOR angular dependence as in Fig. 6.8, but with $a < 0$ and $b > 0$

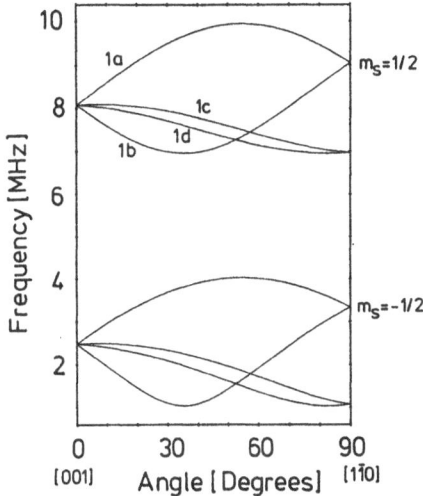

Fig. 6.10. ENDOR angular dependence as in Fig. 6.7, but the plane of magnetic field variation is rotated around the z-axis by 5° with respect to its (110) position. The neighbors *1c* and *1d* are no longer equivalent

patterns look like those in Fig. 6.9 for $a/h = -1$ MHz and $b/h = 0.2$ MHz, which provide clear experimental evidence as to whether the constants a and b have opposite signs or not. In order to obtain their absolute signs, however, one must know m_S for each pattern.

So far, only a magnetic field variation precisely in the (1 1 0) plane has been considered. In actual experiments it is sometimes very difficult to align the sample precisely enough. Figure 6.10 shows one example of possible consequences. In the calculations it was assumed that the plane of magnetic field variation is rotated around the z–axis by 5°, with respect to its [1 1 0] position. It follows that the neighbors 1c and 1d are no longer equivalent. The individual ENDOR lines of each neighbor nucleus in the shell are now visible. Measurements on purposely 'misadjusted' samples may sometimes be used to facilitate the analysis of complex angular dependences.

The second nearest neighbors with 100–symmetry have the pattern shown in Fig. 6.11 ($a/h = 10$ MHz, $b/h = 2$ MHz, $b' = 0$, magnetic field in the (1 1 0) plane). There are only two lines for each m_S state, since the neighbors 2a – 2d are equivalent, and the neighbors 2e and 2f are equivalent, as can be seen from Fig. 6.6, considering the behavior of the angle θ_A in (5.4.7). However, these arguments hold only as long as the parameter b' is accidentally equal to zero. Indeed, the symmetry properties of the 100–shell do allow b' to be non–zero. In this case, the pattern looks like that in Fig. 6.12 ($a/h = 10$ MHz, $b/h = 2$ MHz, $b'/h = 0.2$ MHz). Now, only the neighbors 2a and 2d, and the

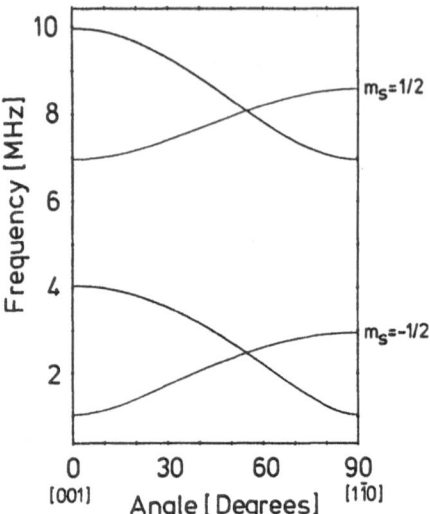

Fig. 6.11. ENDOR angular dependence for the second shell (100–symmetry) of the defect shown in Fig. 6.6. Magnetic field variation and interaction constants as in Fig. 6.7

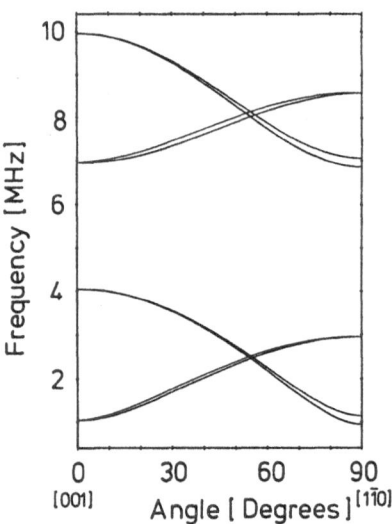

Fig. 6.12. ENDOR angular dependence as in Fig. 6.11, but with $b' = 0.2$ MHz

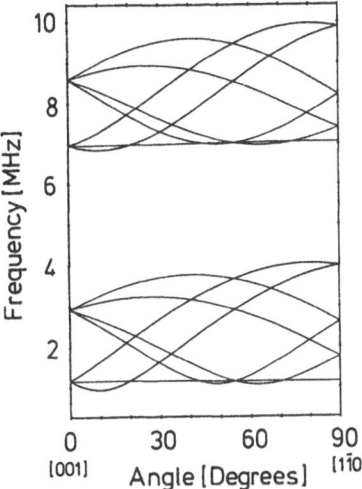

Fig. 6.13. ENDOR angular dependence for the third shell (110–symmetry) of the defect shown in Fig. 6.6. Magnetic field variation as in Fig. 6.7. For the interaction constants and the free angle see text

neighbors 2b and 2c are still equivalent, but only for the magnetic field in the (1 1 0) plane.

For the third nearest neighbors the angular dependence is more complicated. There is one free angle δ_1 not fixed by symmetry constraints. Figure 6.13 shows an example calculated for $a/h = 10$ MHz, $b/h = 2$ MHz, $b'/h = 0.2$ MHz and $\delta_1 = 10°$. There are twelve neighbors in this 110–shell; five neighbors are equivalent in pairs for $\mathbf{B}_0$ in the {110} plane. For the sample misalignment considered in Fig. 6.10, the lines for all neighbors are visible; see Fig. 6.14. The shape of the pattern for the 110–shell depends strongly on δ_1. Particular problems can arise when the z–axes of the interaction tensors are accidentally nearly parallel to <111> or <100> directions. In these cases, the 110–shell patterns may be easily mistaken for 111–shell symmetries or 100–symmetries, respectively. This may even lead to the assumption of a wrong defect model; see [4.47] for a comment on a wrong analysis published in the literature. An example of this kind of problem is shown in Fig. 6.15 for $a/h = 10$ MHz, $b/h = 2$ MHz, $b'/h = 0.2$ MHz and $\delta_1 = 35.264°$. The pattern is very similar to the 111–pattern in Fig. 6.10. Only the small splittings due to the finite value of b' make the difference. This small splitting may be obscured by a large ENDOR line width. Particularly in ODENDOR, one usually observes larger line widths than in conventional ENDOR.

Very careful and precise experiments may be necessary to distinguish between the different shell symmetries. Another way to distinguish between the different shell symmetries is provided by the fact that the shells contain dif-

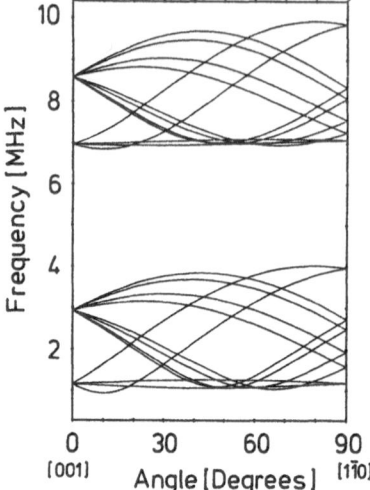

Fig. 6.14. ENDOR angular dependence as in Fig. 6.13, but plane of magnetic field variation tilted as in Fig. 6.10

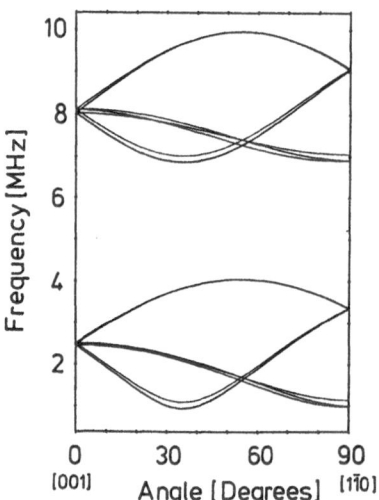

Fig. 6.15. ENDOR angular dependence for the third shell as in Fig. 6.13, but the z-axes of the shf tensors are parallel to $<111>$ directions ($\delta_1 = 35.26°$). The pattern is now very similar to that for the first shell in Fig. 6.7

ferent numbers of nuclei. For example, the 111–shell has four nuclei, whereas there are twelve in the 110–shell. Since the shape of the EPR spectrum is very sensitive to the number of interacting nuclei (if the interaction energy is not too small), an additional careful analysis of the EPR spectrum may be very advantageous and is recommended as a check on the ENDOR analysis. For details see Sect. 6.6.

Beside the first three neighbor shells, there are further shells to be considered. In Fig. 6.6 there are eight neighbors denoted by 4 and 4'. However, although these eight neighbors all have the same distance from the center, they do not belong to the same shell. It can be seen from the figure that there are no symmetry operations of the host crystal, including the defect, which transform one of the neighbors denoted by 4 into one of the set denoted by 4'. In fact, the neighbors 4 form a 111–symmetry neighbor shell and the neighbors 4' form a different 111–neighbor shell. Both these shells may have quite different interaction constants despite the fact that all eight neighbors have the same distance from the center. To describe the angular dependence of these shells, the same transformation matrices may be used as considered for the first 111–shell. Since all ENDOR lines are insensitive to an inversion of the magnetic field, it makes no difference whether a neighbor is in a [1 1 1] or in a [$\bar{1}$ $\bar{1}$ $\bar{1}$] position.

There is one additional neighbor shell symmetry not yet considered for the high symmetry defect in silicon. This shell has 24 nuclei of very low symmetry with three free angles for the tensor orientations. Neighbors of this kind are not included in Fig. 6.6, in order not to overload the figure. The way to find the T matrix and the symmetry matrices for these neighbors is basically not different from that of the other shells. One needs a larger defect model than sketched in Fig. 6.6 to see the 24 neighbors. A construction kit for three–dimensional models may be helpful. Exactly the same neighbor shell symmetries, as discussed above for the high symmetry interstitial defect, are also present for a substitutional point defect in silicon. The only difference is that the sequence of the neighbor types is different for both systems. It is, therefore, basically impossible to decide solely on the basis of symmetry properties whether, for example, this high symmetry defect in silicon is substitutional or interstitial. There is an extended discussion of this problem for defects in silicon in the literature, see for example [6.12]. Therefore, a conclusive decision on the defect model can often be made only if one knows the distance of a neighbor shell from the center, as well as its symmetry properties. In almost all cases it is not even necessary to know the distance of each neighbor shell, as long as it is at least known how shells of the different symmetry types follow each other with increasing distance from the center. It is perhaps one of the biggest deficiencies of the magnetic resonance methods, that they provide no direct experimental information on neighbor shell distances. This information can only be indirectly derived by linking experimentally obtained interaction constants, via theoretical models, to neighbor shell distances (see Chap. 7 for a further discussion). Sufficient theoretical information, in that sense, is already available if one knows that the center wave function falls off monotonically with distance.

The basic way to find transformation matrices for neighbor shell nuclei, as discussed above, is generally applicable provided the symmetry of the defect is equal to the symmetry of the host crystal. In particular, there are no additional problems if the unit cell of the host crystal is no longer cubic, such as for F–centers in BaFCl already discussed. In systems of low symmetry the neighbor shells tend to contain fewer neighbor nuclei, and free angles are likely to appear more often. There might even be neighbor shells with only one neighbor nucleus.

6.5 Low Symmetry Defects in Higher Symmetry Environments

So far, it has been assumed that the crystal containing the defect has the same symmetry as the original crystal without the defect. However, the introduction of a defect usually causes some relaxation of the lattice around the defect. The relaxation often takes place under conservation of symmetry, which is then not detectable in the ENDOR angular dependence.

A different situation arises if the symmetry of the defect is lower than that of the local environment of the defect site. An example is a defect with an orbital degeneracy in the ground state, where the static *Jahn–Teller* effect [6.13] causes a symmetry–lowering relaxation of the lattice around the defect (for example, Ag^{++} centers in NaCl [6.14]). Another example of defects with a symmetry lower than that of the host crystal are pair defects or aggregate centers, where additional impurities or intrinsic lattice defects, such as interstitials or vacancies, are associated with the paramagnetic center. In all these cases, the new feature of these defects is that they exist in different orientations in the host lattice. The differently oriented defects of a species behave as completely different defects. They are well separated in space in the crystal, and interaction processes between differently oriented defects usually do not occur (except for very high concentrations). They can, therefore, be separated with triple resonance (Double ENDOR) techniques in the same way as different defect types. Formally, the only common feature of the differently oriented defects of the same species, is that their orientations and, thus, their angular dependences, are related to each other by symmetry operations.

A simple example of a low symmetry defect is shown in Fig. 6.16 [6.11]. It is, again, the interstitial Al^{++} defect in silicon, however, the silicon neighbor 1c (Fig. 6.6) is replaced by a diamagnetic Al^- ion. The defect has a threefold symmetry axis in [1 1 $\bar{1}$] orientation. It is obvious that there are four different <111> orientations of this defect, where the Al^- replaces the neighbors 1a, 1b, 1c or 1d, respectively.

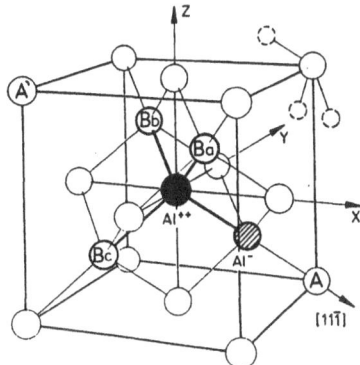

Fig. 6.16. Unit cell of silicon with the Al^{++}–Al^- pair defect

In order to find all the transformation matrices for this defect one goes through the following procedure:

As a first step, one selects one defect orientation and defines this orientation to be the 'basic' one. Let the orientation of the Al^{++}–Al^- defect shown in Fig. 6.16 be the basic one.

To establish the symmetry matrices, one then goes through all the steps discussed for the high symmetry defect in Sect. 6.4. For the low symmetry defect this is even easier, since there are only two different neighbor shell symmetries. One type of neighbor shell consists of only one atom exactly on the $[1\,1\,\bar{1}]$ defect axis. Two of those shells are indicated in Fig. 6.16 by their corresponding atoms A and A', respectively. The Al^- ion itself also forms a shell of this symmetry. The corresponding interaction tensors have cylindrical symmetry with b' and q' equal to zero. The z–axes of these tensors are parallel to the threefold defect axis, and there are no free angles for these tensors. The T transformation matrix is obtained by rotating the auxiliary system $(x'$, y', $z')$ [initially parallel to the coordinate system (x, y, z), Fig. 6.16] around the $[\bar{1}\,1\,0]$ axis by $\theta = 180° - \arctan\sqrt{2} = 125.26°$. The T matrix for this shell, T_A, is then equal to $R[(-1,1,0), 125.26°]$. One obtains:

$$\tilde{T}_A = \begin{pmatrix} 0.211 & -0.789 & 0.577 \\ -0.789 & 0.211 & 0.577 \\ -0.577 & -0.577 & -0.577 \end{pmatrix}. \tag{6.5.1}$$

The second type of neighbor shell consists of three neighbors, related to each other by a threefold rotation around the defect axis; for example, the neighbors denoted by B_a, B_b and B_c, respectively, in Fig. 6.16. Let the neighbor B_a be characteristic of this shell. Each of these three neighbors are located in $\{110\}$ mirror planes, but there are no further symmetry constraints. This situation is analogous to the 110–shell in the high symmetry case of Sect. 6.4. By similar arguments, as for (6.4.28) one obtains for the T matrix

$\tilde{T}_B$ of this shell:

$$\tilde{T}_B = \tilde{R}[(0,0,1), -45°]\tilde{R}[(0,1,0), 90°], \tag{6.5.2}$$

$$\tilde{T}_B = \begin{pmatrix} 0 & 0.707 & 0.707 \\ 0 & 0.707 & -0.707 \\ -1 & 0 & 0 \end{pmatrix}. \tag{6.5.3}$$

The orientation of the z–axis of the neighbor B_a is then defined by the Euler angle δ_1 with respect to the $[1\,\bar{1}\,0]$ direction. The symmetry matrices, $\tilde{S}_{Bb}$ and $\tilde{S}_{Bc}$, for the neighbors B_b and B_c, are obtained according to (6.4.30) and (6.4.31), respectively:

$$\tilde{S}_{Bb} = \begin{pmatrix} 0 & 0 & -1 \\ 1 & 0 & 0 \\ 0 & -1 & 0 \end{pmatrix}, \tag{6.5.4}$$

$$\tilde{S}_{Bc} = \begin{pmatrix} 0 & 1 & 0 \\ 0 & 0 & -1 \\ -1 & 0 & 0 \end{pmatrix}. \tag{6.5.5}$$

The equations (6.5.1–5) give all the transformation matrices necessary to calculate the angular dependences for the basic orientation of the defect shown in Fig. 6.16. Once again taking the same interaction constants as for the calculation of the angular dependence in Fig. 6.13 ($S = 1/2$, $g_e = 2.0023$, $B_0 = 350$ mT, $g_I = -1.1095(^{29}\text{Si})$, $I = 1/2$, $a/h = 10$ MHz, $b/h = 2$ MHz, $b'/h = 0.2$ MHz and $\delta_1 = 10°$), one obtains for the shell B the result shown in Fig. 6.17. Only the ENDOR lines for $m_S = +1/2$ are shown. As expected, only a few lines out of the lines in Fig. 6.13 are seen. An analogous result is obtained for the shell A. Just one out of the three lines for each m_S state in Fig. 6.7 is reproduced. If the shell of type A consists of the aluminum neighbor with nuclear spin $I = 5/2$, there may be five lines for each m_S instead of one, due to a quadrupole interaction.

The next step is to find all the transformation matrices which rotate the entire defect in such a way that the Al^-–ion replaces, one after the other, the silicon neighbors B_a, B_b and B_c, respectively, in addition to its position shown in Fig. 6.16. Those transformation matrices were called "orientation matrices" $\tilde{O}$ in Sect. 6.1. For the Al^{++}–Al^- pair center in Fig. 6.16, the orientation matrices are easily obtained. It is quite obvious from Fig. 6.16, together with Fig. 6.6, that in this special case the orientation matrices $\tilde{O}_k$, $k = 2, 3, 4$ can be obtained by similar arguments as for the symmetry matrices for the first neighbor shell of the high symmetry defect (Fig. 6.6).

In order to obtain the complete angular dependence, including defect orientations, all interaction tensors must be additionally transformed by the orientation matrices. To describe these transformations in a more systematic way, it is useful to make the following definitions:

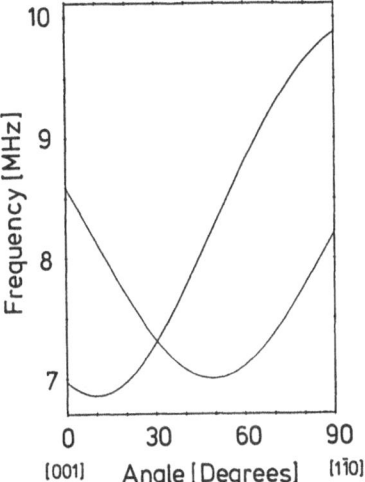

Fig. 6.17. ENDOR angular dependence of the shell B (Fig. 6.16) for the orientation of the $Al^{++}-Al^{-}$ pair center shown in Fig. 6.16 (basic orientation). Only the pattern for $m_S = 1/2$ is shown. Interaction constants and magnetic field variation as for the calculation in Fig. 6.13

An orientation matrix $\tilde{O}_1$ is introduced which is always identical to the identity matrix $\tilde{1}$:

$$\tilde{O}_1 = \tilde{1}. \tag{6.5.6}$$

Additionally, "neighbor matrices", $\tilde{N}_i$, $i = 2, \ldots, n$ are defined, where n is the number of neighbors in a shell (for shells with more than one neighbor), and $\tilde{N}_i$ is the product of the T matrix of the shell with the corresponding symmetry matrix $\tilde{S}_i$, $i = 2, .., n$ of the neighbor i in the shell:

$$\tilde{N}_i = \tilde{T}\tilde{S}_i, \quad i = 2, \ldots, n. \tag{6.5.7}$$

The neighbor matrix $\tilde{N}_1$ for the neighbor $i = 1$ is always defined as being identical to the T matrix of the shell:

$$\tilde{N}_1 = \tilde{T}. \tag{6.5.8}$$

Using these definitions, the complete transformations for the shf and the quadrupole tensors can now be written in the form [compare (6.4.10–12)]:

$$\tilde{A}_{k,s,i}(\mathbf{v}, \theta - \theta_0) = \tilde{R}'(\mathbf{v}, \theta - \theta_0)\tilde{O}_k\tilde{N}_{s,i}\tilde{E}_s\tilde{A}_{D,s}\tilde{E}'_s\tilde{N}'_{s,i}\tilde{O}'_k\tilde{R}(\mathbf{v}, \theta - \theta_0). \tag{6.5.9}$$

All matrices with index s are different for each neighbor shell. For high symmetry defects with just one defect orientation, the index k is equal to one, and, accordingly, for neighbor shells with just one neighbor, such as, for example, the central nucleus of a defect (hyperfine interaction), the index i is equal to one. In the spin *Hamiltonian*, there are no neighbor matrices for the electron *Zeeman* term or for the fine–structure term. These two terms must, however, be transformed by the orientation matrices. For the electron *Zeeman* term $EZ_k(\mathbf{v}, \theta - \theta_0)$ one obtains:

$$EZ_k(\mathbf{v}, \theta - \theta_0) = \mu_B \mathbf{S} \tilde{R}'(\mathbf{v}, \theta - \theta_0) \tilde{O}_k \tilde{T}_g \tilde{E}_g \tilde{g}_{eD} \tilde{E}'_g \tilde{T}'_g \tilde{O}'_k \tilde{R}(\mathbf{v}, \theta - \theta_0) \mathbf{B}_0,$$

(6.5.10)

where $\tilde{g}_{eD}$ is the diagonal electronic g matrix. Its orientation is not necessarily fixed by symmetry constraints in low symmetry host crystals. The matrices $\tilde{T}_g$ and $\tilde{E}_g$ provide the necessary transformations analogous to (6.4.10). Also, the fine–structure term FS depends on $\mathbf{v}$, $\theta - \theta_0$, $\tilde{O}_k$, and on the transformations $\tilde{T}_{FS}$ and $\tilde{E}_{FS}$, analogous to $\tilde{T}_g$ and $\tilde{E}_g$ in (6.5.10). The fine–structure term may have a rather complicated transformation behavior in cases where the electron spin S exceeds the value of 3/2. For $S \le 3/2$ the FS tensor is a 3×3 matrix which can be transformed in a way quite analogous to the g_e matrix. Details are found in the literature [6.9,10]. If the fine–structure term is of the simple form $FS = \mathbf{S}\tilde{D}\mathbf{S}$ it can also be expressed by:

$$FS_k(\mathbf{v}, \theta - \theta_0) = \mathbf{S} \tilde{R}'(\mathbf{v}, \theta - \theta_0) \tilde{O}_k \tilde{T}_{FS} \tilde{E}_{FS} \tilde{D}_D \tilde{E}'_{FS} \tilde{T}'_{FS} \tilde{O}'_k \tilde{R}(\mathbf{v}, \theta - \theta_0) \mathbf{S},$$

(6.5.11)

where $\tilde{D}_D$ is the diagonal fine–structure tensor.

If one again assumes that mutual indirect interactions between different neighbors can be neglected, then the angular dependences can be calculated using spin *Hamiltonians* of the form:

$$\mathcal{H}_{k,s,i}(\mathbf{v}, \theta - \theta_0) = EZ_k(\mathbf{v}, \theta - \theta_0) + FS_k(\mathbf{v}, \theta - \theta_0)$$

$$+ \mathbf{I}_{s,i} \tilde{A}_{k,s,i}(\mathbf{v}, \theta - \theta_0) \mathbf{S} - g_{ns} \mu_n \mathbf{I}_{s,i} \cdot \mathbf{B}_0$$

$$+ \mathbf{I}_{s,i} \tilde{Q}_{k,s,i}(\mathbf{v}, \theta - \theta_0) \mathbf{I}_{s,i}.$$

(6.5.12)

In many cases it is suitable to select the crystal coordinate system (x, y, z) in such a way that the g matrix $\tilde{g}_e$ is diagonal in this system. The electron *Zeeman* term is then of a simpler form, where $\tilde{T}_g$ and $\tilde{E}_g$ are both equal to the identity matrix. In the example of Fig. 6.16, the coordinate system has not been chosen this way. However, this is of no importance in this case, since no anisotropy of the g value is considered here. It should be stressed, once again, that the coordinate system is fixed to the crystal, rather than to the defect in its basic orientation. All neighbor transformations and all defect orientations are described with respect to this unique system. In order to obtain the complete angular dependence for the low symmetry defect, one must calculate and plot the angular dependence according to the *Hamiltonian* (6.5.12) for

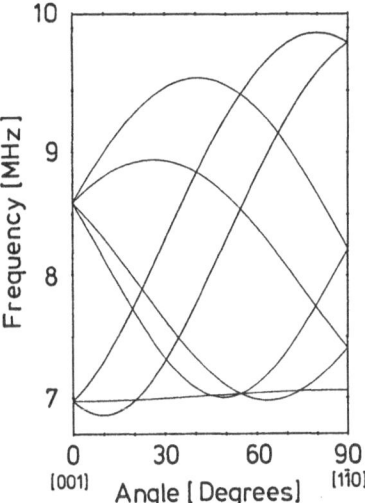

Fig. 6.18. ENDOR angular dependence of the shell B for all orientations of the defect in Fig. 6.16. Magnetic field and interaction constants as in Fig. 6.17

all values of k, s, i and m_S. Usually, the calculations for some values of k and i lead to redundant results; some lines coincide. This is reflected experimentally by the relative intensity of the observed ENDOR lines, and this fact can sometimes be used to facilitate analyses. In general, however, one must be very careful when drawing conclusions from ENDOR line intensities. There are many other mechanisms which affect line intensities beside the multiplicity of transitions due to symmetry constraints. If there are symmetry related lines for identical values of m_S close together in frequency, then the intensity information is usually more reliable, provided the intensity of the ENDOR lines one wants to compare is not additionally influenced by the structure of the EPR spectrum.

The complete angular dependence for the first silicon neighbors for a defect with the symmetry of the Al^{++}–Al^- pair center (shell type B, Fig. 6.16) is again shown in Fig. 6.18 for the interaction constants already used in the previous calculation. The pattern in no way differs from that already shown in Fig. 6.13 for the 110–neighbors of the high symmetry defect. This is, indeed, not surprising. The product of the four orientation matrices $\tilde{O}_k$ with the three neighbor matrices $\tilde{N}_i$, as contained in (6.5.9), yields exactly the same 12 different matrices as already discussed in Sect. 6.4.3 for the 110–shell of the high symmetry T_d defect. In a similar way, the angular dependence for the type A shell in Fig. 6.16 cannot be distinguished from the 111–shell of the T_d defect shown in Fig. 6.7. This means that there is no way to distinguish the defect model shown in Fig. 6.6 from that in Fig. 6.16, just on the basis of the ENDOR angular dependences discussed so far, and there is danger that one may be completely misled. It is exactly this point which sometimes presents a severe problem for the analysis of ENDOR data. The problem is not restricted to the special symmetry situation used as an example in the above discussions. Therefore, it is often essential to gain additional information about whether observed ENDOR lines belong to different orientations of a defect, or to different neighbors of the same shell.

6.6 Ways to Distinguish Between High and Low Symmetry Defects

It is often impossible to decide, on the basis of angular dependence patterns, whether one is dealing with a low symmetry defect present in different orientations, or with a high symmetry defect with more nuclei in the neighbor shells. For example, it follows from Sect. 6.5 that, for the 110–shell of a T_d defect, several different additional defect structures are theoretically possible which yield the same angular dependence. The same pattern can be produced as well by two neighbors and six different orientations, or by three neighbors and four orientations. This is formally expressed in (6.5.9) by the fact that the interaction tensors are transformed by a product of neighbor matrices $\tilde{N}_{s,i}$ and orientation matrices $\tilde{O}_k$.

The only definite determination of which of these possible structures is present, is achieved by Double ENDOR, if it works. Since differently oriented defects are completely independent of each other in the crystal, a Double ENDOR signal can only occur if the two ENDOR lines used for the measurement belong to the same orientation. Thus, formally, the orientation matrices $\tilde{O}_k$ in (6.5.9) can be measured independently of the neighbor matrices $\tilde{N}_{s,i}$, and the defect structure can be definitely determined. For example, in a Double ENDOR experiment, if one adjusts one rf frequency to an ENDOR line which belongs to exclusively one orientation of the defect, then with the second rf frequency, only the subset of those lines is measured which belongs to the same orientation. It makes no difference there whether the ENDOR lines belong to different shells or not. In this example, exactly one specific orientation matrix $\tilde{O}_{k'}$ is selected for the measurement of the entire angular dependence, and for the simulation of this angular dependence, only this $\tilde{O}_{k'}$ must be taken into account. An example is shown in Fig. 6.19 for the trapped hole center in α–Al_2O_3:Mg^{++} [6.15]. In Fig. 6.19a ENDOR lines are shown for the interaction of the hole center (O^-) with Al neighbors ($I = 5/2$). Five ENDOR lines are expected for each Al neighbor due to quadrupole interactions. In order to decide which of the measured lines belong to different orientations of the center, in a Double ENDOR experiment one rf frequency is adjusted for each angle to the line indicated by the arrow. The result is shown in Fig. 6.19b. Only five lines out of all the lines in Fig. 6.19a appear. Detailed Double ENDOR experiments reveal that there are six different po-

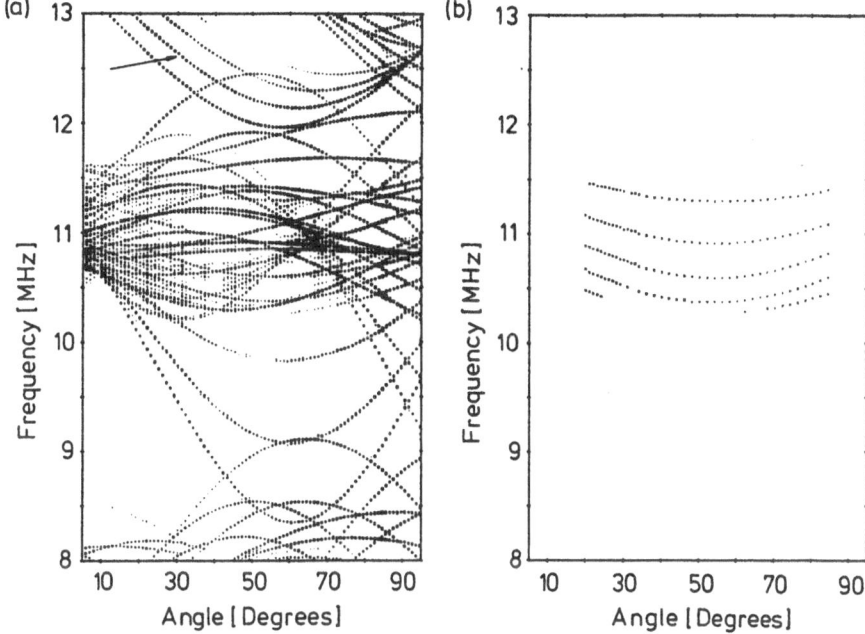

Fig. 6.19. a Part of the ENDOR angular dependence of O$^-$ defects in α-Al$_2$O$_3$:Mg^{++}. The lines around 11 MHz belong to the shf and quadrupole interactions with one Al–neighbor (I(^{27}Al) = 5/2)) for six different positions of the defect in the unit cell. **b** Double ENDOR spectrum for the interactions of **a** for one selected defect position. For the measurement the first rf frequency for the different angles was adjusted to the values indicated by the *arrow*. (After [6.15])

sitions of the center in the unit cell, and that each Al shell consists of just one neighbor.

It may happen that ENDOR lines belonging to different center orientations, for example, $\tilde{O}_{k'}$ and $\tilde{O}_{k''}$, coincide. If, in a Double ENDOR experiment the first rf frequency is adjusted to this ENDOR line, then the resulting angular dependence exclusively contains lines which are due to the orientations $\tilde{O}_{k'}$ or $\tilde{O}_{k''}$. In cases of complex structures, it is necessary to take all possible defect structures into account which are formally compatible with the experimental angular dependence pattern. For each of these possible structures it must then be determined, theoretically, which of the possible orientations $\tilde{O}_k$ contribute to each of the experimental ENDOR lines. It can then be predicted which of the $\tilde{O}_k$ are contained in the Double ENDOR spectra measured with the first rf frequency adjusted to the different ENDOR lines. At the same time, those ENDOR lines can be theoretically selected which will give the most suitable information about a defect model in a possible Double ENDOR experiment. This procedure was performed in detail, for example, to clarify the structure of Ag^0-Ag^+ centers in NaCl [6.16].

The information power of Double ENDOR is excellent, but in many cases this technique does not work satisfactorily, due to experimental difficulties. Most of these difficulties arise from low signal intensities. This is a particular problem in hosts where the magnetic nuclei have only a low abundance, as in Si. So far, this excludes successful Double ENDOR experiments for almost all centers in Si. In order to clarify whether or not one is dealing with a low symmetry defect in these cases, one must therefore apply the second best method. As can be seen from (6.5.12), only the hf or the shf and the quadrupole tensors are transformed by the product of $\tilde{O}_k$ and $\tilde{N}_{s,i}$ matrices. The g_e matrix and the fine–structure term are transformed by the same $\tilde{O}_k$ matrices, but not by matrices such as the $\tilde{N}_{s,i}$ matrices. It should, therefore, be possible to decide between the contributions of $\tilde{O}_k$ and $\tilde{N}_{s,i}$ to an ENDOR angular dependence, by taking information from the EPR spectrum into account. This may work as long as the $\tilde{g}_e$ matrix anisotropy, or the anisotropy of the fine–structure term is sufficiently large. If the EPR spectrum is split due to such anisotropies, ENDOR spectra can be measured separately for different orientations, just by suitable selection of the magnetic field within the EPR spectrum during the ENDOR experiment. One example where this was successfully applied is the structure analysis of thermal donors in silicon [6.17,18]. High magnetic fields, and, consequently, high microwave frequencies are helpful to enlarge EPR line splittings from g_e matrix anisotropies. If these anisotropies are too small to split the EPR spectrum, ENDOR–induced EPR measurements on all ENDOR lines may still be able to provide the necessary information. A small g_e matrix or fine–structure anisotropy then shows up in slightly different positions of the different ENDOR–induced EPR spectra which can be determined with high accuracy. However, if the nuclear spin of the ENDOR line used is higher than $I = 1/2$, one must be careful not to be misled by the shifting effect of an ENDOR-induced EPR spectrum due to the selection of different m_I sublevels.

If the g_e matrix or fine–structure anisotropies are too small, the methods described above are no longer applicable. There are low symmetry defects with no measurable anisotropy of the EPR spectrum such as chalcogenide pairs in silicon [6.19,20]. Then it is helpful to recalculate the shape of the EPR spectrum using all shf data determined by ENDOR. All the interaction constants a, b and b' can be formally determined from the ENDOR angular dependences independent of the specific defect model giving rise to an angular dependence pattern. It is clear from (6.5.9), that there is only one tensor $\tilde{A}_D$ which is compatible with a given angular dependence pattern, but there are several possibilities for the matrices $\tilde{O}_k$ and $\tilde{N}_{s,i}$ giving rise to the same pattern. For the shape of the EPR spectrum it can make a big difference whether the neighbor shells of the defect contain only a few neighbors with many different orientations of the entire defect, or whether there is only one orientation of the defect with high symmetry and, accordingly, more neighbors in the neighbor shells. In the EPR spectrum, all the contributions from the differently oriented individual defects are simply superimposed. If there is no $\tilde{g}_e$ matrix or fine–structure anisotropy, the number of the different orientations has no effect on the width of the EPR spectrum. The number of interacting neighbor nuclei, on the other hand, can influence the width of the EPR spectrum markedly. This may be illustrated for the two different Al centers in silicon discussed in Sect. 6.5. Due to the four 111–orientations of the Al^{++}–Al^- defect (Fig. 6.16), an angular dependence pattern is observed for the single Al^- neighbor, which is indistinguishable from that of four Al neighbors in <111> positions. Also, a configuration of two or three Al neighbors in <111> positions would yield the same ENDOR angular dependence, as long as their interaction tensors are axial ($b' = 0$, $q' = 0$) and aligned precisely along <111> directions. In the case of one or of four Al neighbors, the tensors are axial and parallel to <111> directions by symmetry, if there are no further defects or impurities attached to the center. Since, for the Al^{++}–Al^- pair center the shf interaction with the Al^- predominantly determines the width of the EPR spectrum, the number of aluminum neighbors can definitely be seen from the shape of the EPR spectrum. There is just one Al neighbor [6.11].

There are cases where, in addition to all the adversities mentioned above, the interaction with the neighbor shell of interest is so small, that the number of neighbors in this shell no longer has a noticeable effect on the EPR spectrum. This is the worst situation, where a complete and definite determination of the defect structure is usually no longer possible.

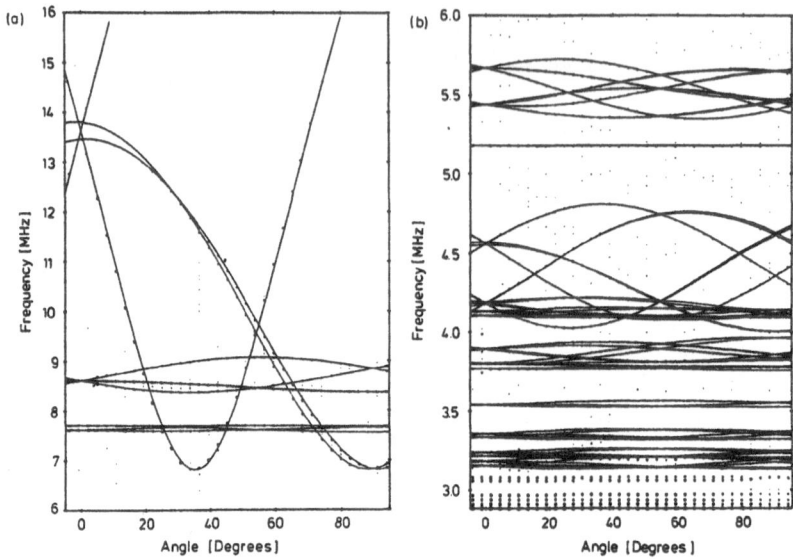

Fig. 6.20a,b. ENDOR angular dependence for the Te$^+$–defect in Si. The *dots* are experimentally determined peak positions, the *solid lines* are calculated values after fitting of free parameters in the spin *Hamiltonian*. The dots above 3.2 MHz not covered by solid lines are due to noise in the original ENDOR spectra. (After [6.5])

6.7 Role of EPR Spectrum for an ENDOR Analysis

Apart from the reasons given in the previous section, there are additional reasons for the necessity of a careful analysis of the EPR spectrum as an integral part of an ENDOR investigation.

ENDOR experiments are sometimes difficult due to bad signal–to–noise–ratios, and it may, therefore, happen that ENDOR lines of neighbors are not measurable. In other cases, the anisotropy of the shf interactions for a neighbor shell may be so small that it is impossible to determine the symmetry type of this shell. Therefore, it is also always advisable to consider the EPR spectrum carefully. For example, Fig. 6.20 shows the angular dependence of Te$^+$ centers in Si [6.5]. The dots are the experimentally determined peak positions, and the solid lines are calculated with the interaction constants of 12 neighbor shells. The straight line around 5.2 MHz corresponds to a neighbor shell with an extremely small anisotropy of the shf interaction. Therefore, it is not possible to assign this neighbor shell exclusively on the basis of the ENDOR angular dependence, to one of the three possible symmetry types, as discussed in Sect. 6.4, which, however, correspond to different numbers of neighbor nuclei. There are four neighbors in a 111–shell, six neighbors in a 100–shell, and twelve neighbors in a the 110–shell.

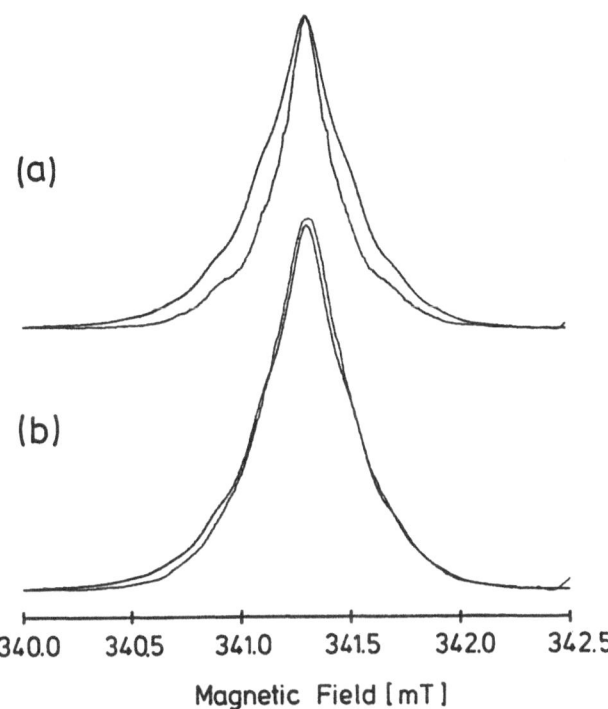

(a)

(b)

340.0   340.5   341.0   341.5   342.0   342.5

Magnetic Field [mT]

Fig. 6.21. a *Outer curve*: Measured integrated EPR spectrum of the Te^+ defect in Si. *Inner curve*: Recalculated EPR absorption spectrum taking into account all analyzed ENDOR shf data with the assumption that the shell around 7.6 MHz (Fig. 6.20a) consists of only six 100–neighbors. **b** *Outer curve*: Measured EPR spectrum as in Fig. 6.21a. *Inner curve*: Calculated EPR spectrum with the assumption that the *structure* at 7.6 MHz (Fig. 6.20a) belongs to twelve 110–neighbors and that the *line* at 5.2 MHz (Fig. 6.20b) belongs to six 100–neighbors. (After [6.5])

Another problem arises for the pattern around 7.6 MHz (0°). From the shape of this angular dependence pattern it is not clear whether it corresponds to a 100–symmetry or to a 110–symmetry. The outer curve in Fig. 6.21a corresponds to the measured integrated Te$^+$ EPR spectrum. The inner curve in Fig. 6.21a is calculated, taking into account all analyzed ENDOR shf interactions, with the assumption that the neighbor shell around 7.6 MHz consists of only six 100–neighbors. The width of the calculated EPR spectrum is definitely too small. In this calculation, it was assumed that the width of an individual EPR line is smaller, by orders of magnitude, than the half width of the EPR spectrum. The twelve neighbor shells analyzed correspond to a total number of 108 ^{29}Si nuclei. Because $S = 1/2$, these 108 neighbors yield a maximum number of 216 lines in the ENDOR spectrum. The corresponding EPR spectrum, however, consists of $2^{108} = 3.25 \times 10^{32}$ lines. Despite the fact that many of these lines are extremely small due to the low abundance of ^{29}Si (4.7%), it is obvious that the many EPR lines all overlap, giving rise to the broad EPR structure. Assuming that the structure at 7.6 MHz in the ENDOR spectrum belongs to twelve neighbors with 110–symmetry, and that of the angular independent line at 5.2 MHz is due to a neighbor shell with six neighbors with 100–symmetry, one obtains the correct EPR line shape (Fig. 6.21b). At the same time, it is clear that no ENDOR lines were overlooked. It can be checked by suitable model calculations that the assumption of any additional ENDOR interaction in the frequency range above about 4 MHz in Fig. 6.21, results in a misfit of the calculated EPR spectrum. Moreover, it is definitely clear from the good fit in Fig. 6.21b, that the electron spin is, indeed, $S = 1/2$. If a higher electron spin is assumed, then, according to (5.4.5), correspondingly lower values of the interaction constants a and b result. These lower values of a and b would yield a calculated EPR spectrum with a half width considerably smaller than found experimentally. Generally, a calculation of the EPR spectrum using all analyzed ENDOR data provides quite a powerful tool to make a defect structure analysis unequivocal. The only condition is, that really all hf and shf interactions, down to lowest detectable ones, are carefully analyzed and included in the calculation of the EPR spectrum. If it is not possible to sufficiently analyze small ENDOR interactions, then one must be very careful not to overinterpret the evidence of the calculated EPR spectrum.

Considering the extremely large number of EPR transitions for a defect where very many neighbor nuclei contribute with reasonable interaction energies, it seems difficult to calculate the entire EPR spectrum without using excessive computer time, in particular, when several isotopes of neighbor nuclei are present in the different neighbor shells. However, as long as a first order calculation of the EPR spectrum is sufficiently adequate, a way exists to obtain theoretical EPR spectra, such as the one in Fig. 6.21, within a few seconds or minutes, using a standard personal computer. The algorithm is explained in Appendix G.

6.8 Solution of the Spin Hamiltonian

The solution of the spin *Hamiltonian* is the central problem in any practical ENDOR analysis. In most practical cases, the first order solution in perturbation theory, used so far to show the basic features of spectra, is not sufficient. The spin *Hamiltonian* must be diagonalized. In its general form the *Hamiltonian* matrix to be diagonalized can have a very high dimension. Equation (6.5.12) covers only the contribution of one neighbor nucleus. The full *Hamiltonian* contains the sum over all partial *Hamiltonians* $\mathbf{H}_{k,s,i}(\mathbf{v}, \theta - \theta_0)$ for all neighbor nuclei i in a neighbor shell, and for all neighbor shells s. For the simple example with $S = 1/2$, and with a total number of 20 nuclei with $I = 1/2$, the dimension of the full *Hamiltonian* matrix becomes $(2S+1) \cdot (2I+1)^{20} = 2097152$ which can not be handled by any means. Therefore, one has to find suitable approximations for the solution of the general problem. However, once one is dealing with approximations, one must always take into account the errors produced by these approximations. It may even be rather difficult to distinguish between genuine defect symmetry properties and pseudo–effects.

6.8.1 Concept of Effective Spin

The main reason why the dimension of the *Hamiltonian* matrix becomes that large is the simultaneous interaction of the unpaired electron spin with so many neighbor nuclei. It is, indeed, only necessary to include nuclei simultaneously if there are significant mutual interactions between different nuclei. This is considered in Sect. 6.8.4.

   If only one neighbor nucleus is considered, the spin *Hamiltonian* can still be significantly simplified by the "rigid electron" approximation. It is applicable as long as the hf or shf interaction energies are sufficiently small compared to the electron *Zeeman* energy. As a rule of thumb, the interaction energies should be smaller than about $h \times 100$ MHz (for experiments in the X–band). The interaction of the electron spin with the different neighbor nuclei is, then, too small to influence the electronic spin function very much. In this case, all nuclei "see" a "rigid electron". It is evident then that, the interaction with each nuclear spin can be considered individually, although interactions might take place with many neighbor nuclei at a time. Then the spin *Hamiltonian* just describes the interactions of one nuclear spin with the magnetic field, with an electric field gradient, and with the rigid electron. The latter interaction is accounted for in the electron nuclear interaction term by simply replacing the electron spin operator $\mathbf{S}$ by its expectation value $<\mathbf{S}>$. One then obtains the following "nuclear spin *Hamiltonian*":

$$\mathcal{H}^N = \mathbf{I}\tilde{A} <\mathbf{S}> -g_n\mu_n\mathbf{B}_0 \cdot \mathbf{I} + \mathbf{I}\tilde{Q}\mathbf{I}. \qquad (6.8.1)$$

The dimension of $\mathcal{H}_N$ is only $2I + 1$, which is a great advantage. Before using (6.8.1), one must calculate all values of $<S>$, which is a vector with the three components $<S>_x, <S>_y$ and $<S>_z$, and where each of the components depends on all interactions which determine the electron wave function. The components also depend on the direction of the magnetic field. The simplest case is an isotropic g_e value and only the electron *Zeeman* term determining the electron wave function. If the static magnetic field $\mathbf{B_0}$ is always parallel to the z-axis of the crystal coordinate system, one then obtains for $<S>$:

$$<S>= (0,0,m_S). \tag{6.8.2}$$

An additional strong isotropic hyperfine interaction leads to a similar expression of $<S>$:

$$<S>= (0,0,\delta m_S), \tag{6.8.3}$$

where δ is a factor of the order of 1. Its small deviation from 1 is approximately proportional to $(ah)^2/W_{EZ}$ [6.21].

The expectation value $<S>$ is often called "effective spin". In (6.8.3) the product δm_s is called "effective m_S", m_S^{eff}. Using the "effective spin approximation" for the calculation of ENDOR angular dependences then means to follow the concept described above.

6.8.2 Nuclear Spin Hamiltonian

For the analysis of ENDOR angular dependences it is very important to have a good intuition about which shape of an angular dependence pattern corresponds to which type of neighbor shell symmetry. It was the aim of Sects. 6.1–5 to give some guidelines on how to obtain angular dependence patterns for different types of shell symmetry, and how to assign observed patterns to their corresponding type of symmetry. It was pointed out in which way the symmetries and the orientations of interaction tensors affect the shape of angular dependence patterns. But the size of the interaction constants a, b, b', q and q' can also affect the shape of the patterns in a way that makes it very difficult to assign the patterns to their type of symmetry. These effects have not been considered so far, and are now discussed employing some characteristic examples.

For nuclear spin $I = 1/2$ there are no quadrupole interactions. A first order solution of $\mathcal{H}_N$ is a good approximation only if the following conditions hold:

(i) $<S>= (0,0,m_S^{\text{eff}})$

(ii) $|a| \ll W_{EZ}$, where W_{EZ} is the electron *Zeeman* energy,

(iii) $|b| \ll |a|$,

(iv) $|b'| \ll |b|$.

Then for f_{ENDOR}, one obtains the known expression [for $b' = 0$, compare (5.4.5)]:

$$f_{ENDOR} = \frac{1}{h}|m_S^{eff}\{a + b[3\cos^2(\theta) - 1]\} - g_n\mu_n B_0|. \qquad (6.8.4)$$

The restrictions needed for the validity of (6.8.4) are severe for most applications. Also, a higher order perturbation treatment of $\mathcal{H}_N$ is only sometimes feasible; therefore, an exact diagonalization is generally the only applicable solution. For $I = 1/2$ this is rather easy. For each nucleus i in a shell s for a defect orientation k and for a magnetic field $\mathbf{B}_0$ with an orientation described by $\tilde{R}(\mathbf{v}, \theta - \theta_0)$ a *Hamiltonian* of the form

$$\mathcal{H} = \mathbf{I}\tilde{A} <\mathbf{S}> -g_n\mu_n\mathbf{I}\cdot\mathbf{B}_0, \qquad (6.8.5)$$

must be solved, where $<\mathbf{S}>= (<S>_x, <S>_y, <S>_z)$, $\tilde{A} = \tilde{A}_{s,k,i}(\mathbf{v}, \theta - \theta_0)$, $g_n = g_{ns}$, $\mathbf{I} = \mathbf{I}_{s,i}$, and $\mathbf{B}_0 = (0, 0, B_0)$; compare (6.5.12). The nuclear spin operator $\mathbf{I}$ is written in the form $\mathbf{I} = (I_x, I_y, I_z)$, where:

$$I_x = \frac{1}{2}(I^+ + I^-), \qquad (6.8.6)$$

$$I_y = -\frac{i}{2}(I^+ - I^-), \qquad (6.8.7)$$

and i is the imaginary unit. The operators I^+, I^-, and I_z are described in Sect. 2.3. By applying these operators to the nuclear spin functions $|I, m_I\rangle$ with $I = 1/2$, and $m_I = 1/2$ or $m_I = -1/2$, it is quite straightforward to obtain the 2×2 *Hamilton* matrix for the two wave functions, $|1/2, 1/2\rangle$, and $|1/2, -1/2\rangle$, respectively. The matrix is of the form:

$$\mathcal{H} = \begin{bmatrix} x & k + il \\ k - il & y \end{bmatrix}, \qquad (6.8.8)$$

where the quantities x, y, k and l are real numbers. This matrix can easily be analytically diagonalized, and one obtains the ENDOR frequencies f_{ENDOR} as the absolute value of the difference of the eigenvalues divided by h:

$$f_{ENDOR} = \frac{1}{h}|[(x^2 - 2xy + y^2) + 4(k^2 + l^2)]^{1/2}|. \qquad (6.8.9)$$

Generally, (6.8.9) yields a rather complicated expression which must be handled by a computer. Rather simple solutions are obtained for special cases, as e.g., for a completely isotropic hyperfine interaction. For the ENDOR frequency f_{ENDOR} one obtains:

$$f_{ENDOR} = \frac{1}{h}|(a^2(<S>_x^2 + <S>_y^2 + <S>_z^2) + B_0^2 g_n^2 \mu_n^2$$

$$-2aB_0 g_n\mu_n <S>_z)^{1/2}|. \qquad (6.8.10)$$

If, in addition, $<\mathbf{S}>$ has only a z-component, $<S>_z = m_S^{\text{eff}}$, then one simply obtains:

$$f_{\text{ENDOR}} = \frac{1}{h}|a\, m_S^{\text{eff}} - g_n \mu_n B_0|. \tag{6.8.11}$$

The exact solution of the nuclear spin *Hamiltonian* is then equal to the first order solution.

In another special case of practical relevance, the effective spin has only the z-component m_S^{eff}, and the hf or shf interaction has axial symmetry ($b' = 0$). If θ is the angle between the direction of $\mathbf{B}_0$ and the principal axis z of the hf or shf tensor, then all the transformation matrices in (6.5.12) can be expressed by just one matrix $\tilde{T}$ which is of the simple form:

$$\tilde{T} = \begin{pmatrix} \cos(\theta) & 0 & \sin(\theta) \\ 0 & 1 & 0 \\ -\sin(\theta) & 0 & \cos(\theta) \end{pmatrix}. \tag{6.8.12}$$

Then, expressing (6.8.9) by a *Taylor* series in powers of b, one obtains:

$$f_{\text{ENDOR}} = \frac{1}{h} \, \Bigg| \, m_S^{\text{eff}}\{a + b[3\cos^2(\theta) - 1]\} - g_n\mu_n B_0 +$$

$$+ \frac{9b^2(m_S^{\text{eff}})^2 \sin^2(\theta) \cos^2(\theta)}{2(a\, m_S^{\text{eff}} - g_n\mu_n B_0)}$$

$$+ \text{(higher order terms)} \, \Bigg|. \tag{6.8.13}$$

In the limit of $|b| \ll |a|$ the first order expression (6.8.4) is obtained. However, strong deviations from the simple relation are predicted for increasing anisotropy of the hf or shf tensor. It is also obvious from the third term in (6.8.13) that, in this case, there is no longer a linear relation between the nuclear g value and the ENDOR frequency. For small deviations of the magnetic field, a linear dependence of f_{ENDOR} on the magnetic field is still observed experimentally, but the value for g_n obtained in this way may be completely wrong. Another effect not expected from the first order solution arises from the denominator of the third term in (6.8.13). This term can become very large, even for relatively small values of b, if the denominator tends to zero. In most cases a and g_n are both positive. One then observes strong deviations of the ENDOR angular dependence patterns from the first order solution for $m_S^{\text{eff}} > 0$, whereas, these deviations are less pronounced for $m_S^{\text{eff}} < 0$. Figure 6.22 shows a model calculation, according to (6.8.9), for a variation of the magnetic field in a (110) plane for a defect with T_d-symmetry in Si (compare Fig. 6.8). The nuclear *Zeeman* frequency $(g_n\mu_n B_0)/h = 2.9601$ MHz is indicated by the straight line. The upper angular dependence pattern corresponds to a 111-symmetry neighbor shell with shf constants $a = 5.3$ MHz and $b = 1.06$ MHz, $m_S^{\text{eff}} = -1/2$. The lower pattern corresponds to $m_S^{\text{eff}} = 1/2$. A

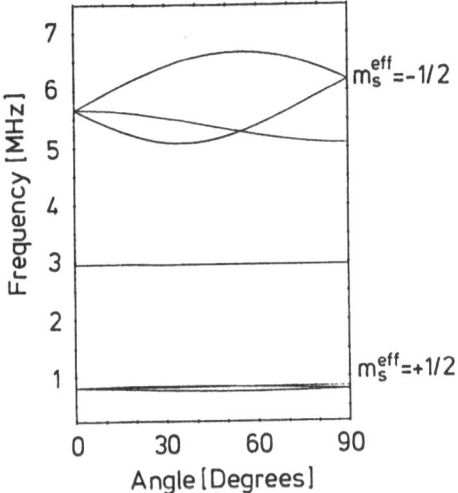

Fig. 6.22. ENDOR angular dependence for the same symmetry and magnetic field variation as in Fig. 6.8, but with the shf constants $a = 5.3$ MHz and $b = 1.06$ MHz, and the nuclear g value $g_n = +1.1095$. The patterns for the two values of m_S^{eff} have a very different shape. (Compare Fig. 6.8)

very strong deviation from the first order solution is observed for this pattern. One would expect both patterns to exhibit mirror symmetry with respect to $f_{\text{nuc}}(W_{\text{shf}}/h < f_{\text{nuc}})$. If the hf or shf interaction energy is slightly higher than the nuclear *Zeeman* energy, one then obtains the patterns of Fig. 6.23 with $a = 6$ MHz and $b = 1.2$ MHz. The lower pattern is now nearly the mirror image to what one expects from (6.8.4) (compare Fig. 6.8). When analyzing complex experimental angular dependences one may easily be misled by this effect. For example, one might suppose that the lower pattern in Fig. 6.23 belongs to a different neighbor shell with a smaller value of b, than the neighbor shell corresponding to the upper pattern.

Equation (6.8.4) predicts that any line in an ENDOR angular dependence can be described by a sine function with suitable amplitude, periodicity and phase. As long as $|b| \ll |a|$, this is also the case if the magnetic field is no longer varied in a plane containing the hf tensor axis **z** (see the neighbors $1c, 1d$ in Fig. 6.6). For higher anisotropies, however, this is no longer true. Shapes of angular dependence lines may then deviate very much from the shape of sine functions. An example is shown by the solid lines in Fig. 6.24 for the nucleus $1b$ in the upper pattern of the 111–neighbor shell (compare Fig. 6.7 for the 111–symmetry). These lines are calculated values after fitting of free parameters in the spin *Hamiltonian*. They belong to shf and quadrupole interactions of the first As–neighbors ($I(^{75}\text{As}) = 3/2$) of the V^{3+}–center in GaAs ($S = 1$) [6.22,23].

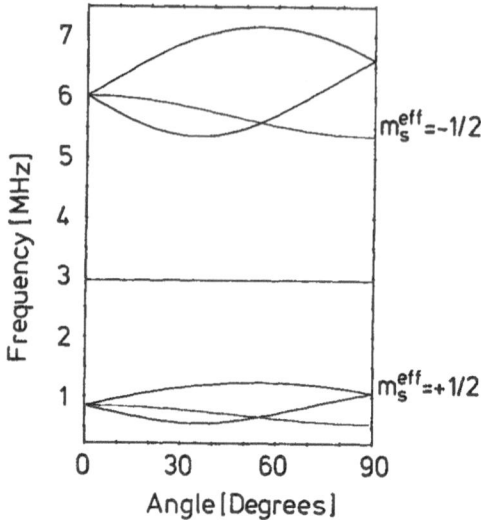

Fig. 6.23. ENDOR angular dependence as in Fig. 6.22, but with the shf constants $a = 6$ MHz and $b = 1.2$ MHz

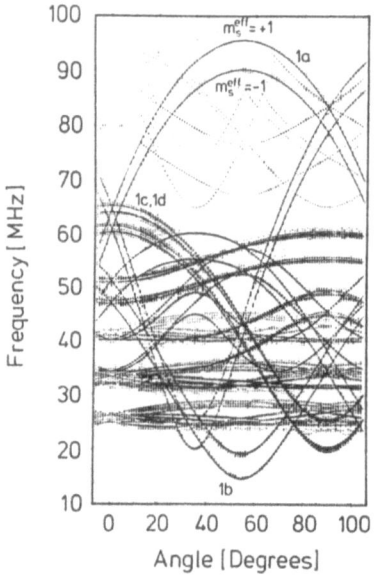

Fig. 6.24. *Dots*: Experimental ENDOR angular dependence for As and Ga ligands of the V^{3+} center in GaAs ($S = 1$). Symmetry of neighbor shells and magnetic field variation as in Fig. 6.10. *Solid lines*: Calculated values of the spin *Hamiltonian* after fitting of free parameters. (After [6.22])

For $I > 1/2$ a quadrupole interaction must be taken into account:

$$\mathcal{H} = \mathbf{I}\tilde{A} <\mathbf{S}> -g_n\mu_n\mathbf{I} \cdot \mathbf{B_0} + \mathbf{I}\tilde{Q}\mathbf{I}. \tag{6.8.14}$$

A first order solution of this equation is adequate as long as, in addition to all conditions for the validity of (6.8.4), the following is fulfilled:

$$|q| \ll |a|, \text{ and } |q'| \ll |q|. \tag{6.8.15}$$

For the ENDOR frequency one obtains:

$$f_{\text{ENDOR}} = \frac{1}{h} \left| m_S^{\text{eff}}\{a + b[3\cos^2(\theta_A) - 1]\} - g_n\mu_n B_0 + \right.$$

$$\left. +m_q 3q[3\cos^2(\theta_Q) - 1]\right|, \tag{6.8.16}$$

where θ_A is the angle between the direction of the magnetic field and the principal axis z of the hf tensor, and θ_Q is the corresponding angle for the quadrupole tensor. Instead of one line according to (6.8.4), one now obtains $2I$ lines. For example, for $I = 3/2$, one obtains the three lines corresponding to $m_q = -1$, $m_q = 0$ and $m_q = 1$. The quadrupole tensor contains no isotropic part, therefore, parallel or nearly parallel quadrupole lines are only observed in an ENDOR angular dependence if the quadrupole principal axis z, is perpendicular to the plane of magnetic field variation, and if q' is zero (in a first order approximation). One such example is given in Fig. 6.19b for the five nearly parallel quadrupole lines of the Al nucleus ($I_{\text{Al}} = 5/2$).

The quadrupole term in (6.8.16) predicts that the quadrupole splitting should be zero, if θ_q is equal to 54.7°. However, this is no longer observed if the anisotropic shf term in (6.8.16) is sufficiently large. In this case, the shf interaction has some influence on the direction of quantization of the nuclear spin, too. One example of this is shown in Fig. 6.25 for interstitial hydrogen centers in KCl [6.1]. Due to the symmetry of the first ^{39}K neighbor shell, the angle θ_q is equal to 54.7° for the magnetic field along the [100] direction in Fig. 6.25. However, the cross–over point of the three quadrupole lines ($I(^{39}$K$) = 3/2$) is not observed for $\theta = 0°$, as expected, but for about $\theta = 17°$, due to this effect. If the quadrupole interaction is of the same order of magnitude as the shf interaction, then the angular dependence pattern can be rather confusing. Any relevance of (6.8.16) is lost. A still relatively simple example is that of Fig. 6.24. All the solid lines correspond to a quadrupole triplet of ^{33}As with $I = 3/2$ for the two values of $m_S^{\text{eff}} = 1/2$, and $m_S^{\text{eff}} = -1/2$, respectively. The calculations can only be performed by an exact diagonalization of the nuclear spin *Hamiltonian*. However, since the dimension of the Hamilton matrix $2I + 1$ is higher than 2, only a numerical diagonalization is possible. An additional effect of a large quadrupole interaction is visible in Fig. 6.24, where nuclear transitions with $\Delta m_I = \pm 2$ are no longer undetectably weak in intensity. Such forbidden lines are the dots in the upper part of the figure above about 70 MHz. These lines appear roughly at twice the frequency of the corresponding allowed lines.

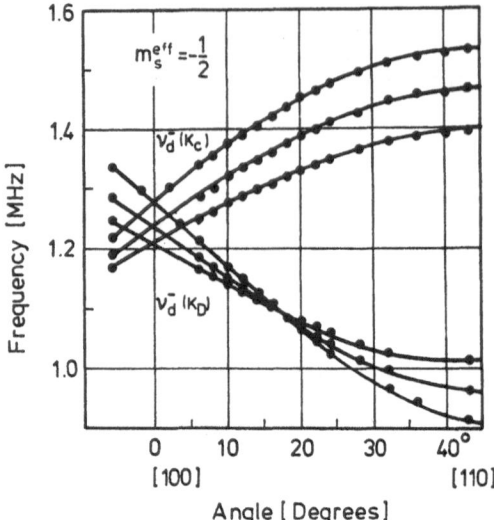

Fig. 6.25. *Dots*: Experimental ENDOR angular dependence ($m_S^{\text{eff}} = -1/2$) for the first K neighbors of the interstitial neutral hydrogen center in KCl. The magnetic field was varied in a {100} plane starting along $B_0 \parallel [100]$ (0°) and ending along $B_0 \parallel [110]$ (45°). *Solid lines*: Calculated values of the spin *Hamiltonian* after fitting of free parameters. (After [6.1])

6.8.3 Calculation of Effective Spin

For the calculation of the effective spin all "large" electron interactions must be taken into account. These are the electron *Zeeman* term and the fine–structure terms. It may also be necessary to include "large" hf, or even shf interactions. Whether this is necessary or not depends on the desired accuracy of ENDOR frequency calculations. The number of lines in an ENDOR angular dependence is directly proportional to the number of different values of the effective spin. If two different ENDOR lines corresponding to different effective spin values are no longer separated due to their line width, then it makes no sense to consider these effective spin values as separate values. This somehow defines the lowest electron–nuclear interaction to be included in the calculation of the effective spin. The situation is more difficult if one wants to calculate an electron–nuclear interaction using the effective spin concept, where, on the other hand, this interaction is large enough as to be included in the calculation of the effective spin. This problem cannot be solved. Either one accepts a limited accuracy of the calculation and does not include the interaction of interest in the calculation of the effective spin, or one must diagonalize the entire *Hamiltonian* without the effective spin formalism. Any mutual interaction between different neighbor nuclei provided by the electron spin cannot be handled by the effective spin concept either.

Usually, the dimension of the *Hamilton* matrix is larger than two. Perturbation theory approaches may be helpful in special cases, but the only generally applicable way is a numerical solution of the *Hamiltonian* using suitable computer algorithms. A large selection of computer programs is commercially available. Computer algorithms which diagonalize the *Hamilton* matrix also yield the corresponding wave functions, $|m_S, m_{I1}, ..., m_{IN}\rangle$. With these functions the effective spin is then calculated as:

$$<\mathbf{S}> = \langle m_{IN}, ..., m_{I1}, m_S | \mathbf{S} | m_S, m_{I1}, ..., m_{IN} \rangle. \qquad (6.8.17)$$

For this calculation, all the interactions determining the wavefunctions in (6.8.17) must be known. They may be known from preceding EPR investigations, or from preliminary ENDOR analyses. Each of the three components of the effective spin is characterized by different indices. There are the indices $m_S, m_{I1}, ..., m_{IN}$, characterizing the electron quantum state and the quantum states of the nuclei $1, ..., N$ with strong interaction. In addition, the indices describing the direction of the magnetic field and the orientation of the defect must be considered. Therefore, in addition to the indices describing the neighbor shell, the nucleus in a shell, and the orientation of the defect, each line in an ENDOR angular dependence is labeled by the values $m_S, m_{I1}, ..., m_{IN}$ of the effective spin.

6.8.4 Mutual Interactions Between Neighbor Nuclei

There are two mechanisms for those interactions:

(i) There is a direct spin–spin interaction between the nuclear spins $\mathbf{I}_1$ and $\mathbf{I}_2$ of different nuclei which is of the form:

$$\mathcal{H}_{NN} = \mathbf{I}_1 \tilde{A}_{NN} \mathbf{I}_2, \qquad (6.8.18)$$

where $\tilde{A}_{NN}$ is a nuclear dipole–dipole interaction tensor. This effect, however, plays no significant role in ENDOR spectra. It just contributes to the ENDOR line widths.

(ii) An indirect coupling between different nuclear spins can be provided by the electron spin. In this coupling mechanism the interaction of the electron spin with one neighbor nucleus modifies the electron spin states, and these modified spin states, in turn, can have a noticeable effect on the interaction with a different neighbor nucleus. This indirect coupling between different neighbor nuclei can lead to a splitting of ENDOR lines. Each ENDOR line of a neighbor nucleus, 1, then splits into a total number of lines N_{tot} given by:

$$N_{tot} = (2I_2 + 1)(2I_3 + 1)...(2I_N + 1), \qquad (6.8.19)$$

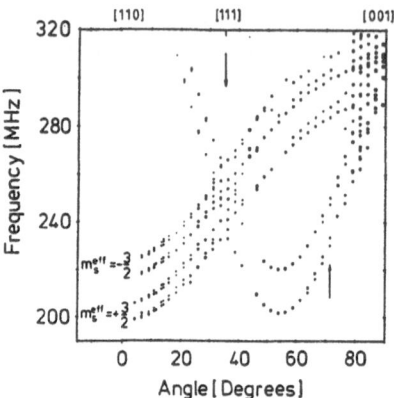

Fig. 6.26. Experimental ENDOR angular dependence of the first P 111–neighbor shell of the paramagnetic Ga vacancy in GaP with a second order hf structure. The magnetic field was varied in a {110} plane starting along $\mathbf{B}_0 \parallel$ [110] (0°) and ending along $\mathbf{B}_0 \parallel$ [001] (90°). (After [6.26])

if there are $N - 1$ neighbor nuclei present with nuclear spins I_i, $i = 2, \ldots, N$. The size of the splitting W_{split} is, by rough estimation, proportional to the square of the shf energy W_{shf} divided by the electron *Zeeman* energy W_{EZ}:

$$W_{split} \propto \frac{W_{shf}^2}{W_{EZ}}. \tag{6.8.20}$$

This interaction is sometimes called "second order hf structure" or "pseudo-dipolar coupling" [6.24,25]. An experimental example is shown in Fig. 6.26. The dots show a part of the first 111–neighbor shell of P neighbors of the paramagnetic Ga vacancy in GaP ($S = 3/2$) [6.26]. There are four ENDOR lines instead of two for $m_S = 3/2$ and $m_S = -3/2$, respectively. (See arrow at 70° in Fig. 6.26). Each of the two lines expected from symmetry is split by about 5 MHz, due to indirect coupling between the first four P neighbors. Their shf interaction energy W_{shf} of about $h \times 200$ MHz is sufficiently large, compared to the electron *Zeeman* energy of about 10 GHz, to give rise to this effect. For the determination of the defect model, these line splittings can be very helpful because one can determine the number of nuclei in the neighbor shell by an analysis of such a splitting. On the other hand, it may sometimes be difficult to tell whether an observed line splitting is due to this effect or to low symmetry effects of the defect, or both. In this case, Double ENDOR experiments can give a definite answer. For the Ga–vacancy in GaP (Fig. 6.26), this is demonstrated in Fig. 6.27. Figure 6.27a shows the ENDOR spectrum for $\mathbf{B}_0 \parallel$ [111], as indicated by the arrow at about 35° in Fig. 6.26. Figure 6.27b shows the corresponding Double ENDOR spectrum, where the first rf frequency was adjusted to the most prominent line in the

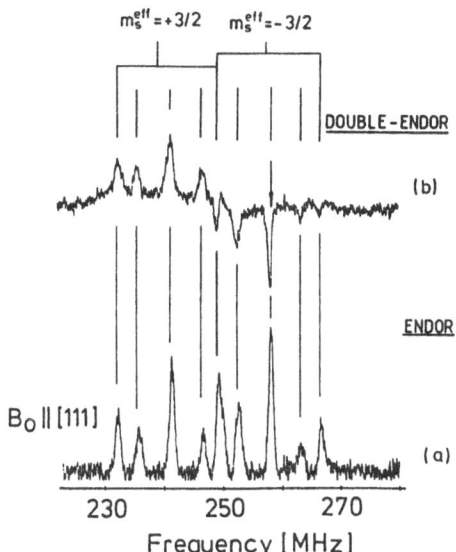

Fig. 6.27. a ENDOR spectrum out of the angular dependence in Fig. 6.26 for $B_0 \parallel [111]$ (indicated by the arrow at about 35° in Fig. 6.26). **b** Corresponding Double ENDOR spectrum with the first rf frequency adjusted to the most prominent line of the spectrum in Fig. 6.26a. (After [6.26])

ENDOR spectrum (arrow in Fig. 6.27). Each ENDOR line is reflected by a corresponding Double ENDOR line, demonstrating that all lines definitely belong to the same defect. It can, therefore, be excluded that some of the lines belong to different orientations of a low symmetry defect. The defect must, therefore, have the same high symmetry as the host crystal, and the observed lines are due to pseudo–dipolar couplings.

The effect of indirect coupling is particularly large if two or even more neighbor nuclei have identical shf tensors. This can be the case for all directions of the magnetic field (magnetically equivalent neighbors), or just for some special magnetic field directions. An example is shown in Fig. 6.24 for the V^{3+} center in GaAs [6.22]. The solid lines are calculated without taking indirect couplings into account. The 111–pattern appears twice, according to $m_S = 1$ and $m_S = -1$, respectively (configuration $3d^2$, $S = 1$). The solid lines corresponding to the neighbors $1c$ and $1d$ are split, due to a small deviation of the plane of magnetic field variation from a (110) plane (also Fig. 6.10). However, as can be seen from the dots around 0° and 65 MHz, there is an additional splitting of the lines for the neighbors $1c$ and $1d$ due to indirect couplings. The neighbors $1c$ and $1d$, are almost equivalent for a magnetic field nearly in a (110) plane, and, therefore, this splitting is large. The pseudo–dipolar splitting for the neighbors $1a$ and $1b$ is buried in the ENDOR line width because the effect is much smaller for these nonequivalent neigh-

bors. If there is no fine–structure interaction, the size of line splittings due
to indirect spin coupling of two equivalent neighbor nuclei can be calculated
analytically, using perturbation theory approximations [6.21,24,25] if b and q
are sufficiently small (also [6.27]). In all other cases, the *Hamiltonian* containing all electron interactions, and the interactions with the neighbor nuclei
between which a pseudo–dipolar coupling takes place, must be diagonalized
numerically.

6.8.5 Large hf or shf Interaction for One Nucleus

If the hf or shf interaction is no longer negligibly small compared to the
electron *Zeeman* term, then the electron spin system is influenced by this
interaction, and the electron spin quantum number m_S, defined with respect
to $\mathbf{B}_0$, is no longer a good quantum number. In this case, for a nuclear spin
I one always observes $2I$ ENDOR lines for each value of m_S, even if there
is no quadrupole interaction. An example is the large Vanadium hf interaction for GaAs:V^{3+} centers [6.22]. For an isotropic g_e value, no fine–structure
interaction, and isotropic hf interaction with no quadrupole interaction, the
ENDOR frequencies can be calculated in second order perturbation theory
[6.27]:

$$f^{\pm}_{\text{ENDOR}} = \frac{1}{h}||m_S|a \mp g_n\mu_n B_0 - W(2m_q \pm 1)|, \qquad (6.8.21)$$

where

$$W = \frac{a^2}{2g_e\mu_B B_0}, \qquad (6.8.22)$$

and

$$m_q = \frac{m_i + m_i'}{2}. \qquad (6.8.23)$$

The + and − signs correspond to $m_S > 0$ and $m_S < 0$, respectively. In practical cases, it might be difficult to distinguish between the effect described by
(6.8.21) and a quadrupole interaction. For the V^{3+} center in GaAs, this was
straightforward, since the V^{3+} ENDOR lines turned out to be completely isotropic, which is not possible for quadrupole interactions. In general, however,
an exact diagonalization of the spin *Hamiltonian* is necessary.

6.8.6 Numerical Calculation of EPR Angular Dependences

For the calculation of <S> it is sometimes necessary to analyze EPR spectra
in detail, in order to obtain all interaction parameters which influence the
effective spin. For a detailed analysis of experimental EPR angular dependences, it is always necessary to simulate measured EPR data precisely by

calculating them for a given set of interaction parameters. The only generally applicable tool for these calculations is, again, an exact numerical diagonalization of a suitable *Hamiltonian* matrix. This is no problem, if the magnetic field is known, and one just wants to calculate the energy eigenvalues. All necessary symmetry transformations are discussed in Chap. 6 and in order to find the *Hamiltonian* matrix, again, only the relations (6.8.6,7), together with those in Sect. 2.3, must be applied. Almost any EPR spectrum is, however, measured at a fixed microwave frequency with the magnetic field as a varying quantity. To simulate this in a calculation is much more difficult if a first order solution of the EPR spin *Hamiltonian* is not adequate. There are different approaches to the problem:

(i) One diagonalizes the EPR spin *Hamiltonian* for many different fixed values of the magnetic field, and afterwards looks for all those magnetic field values which correspond to at least one EPR transition frequency similar to the microwave frequency during the experiment. One obtains a set of different magnetic fields which are approximately equal to the experimentally obtained field values. The accuracy is directly related to the spacing between the field values used for the diagonalizations. It can be further increased by linear or quadratic interpolation. Finally, any precision can be achieved by by a systematic iteration procedure. For each magnetic field, and for each step of such an iteration, a complete new diagonalization of the entire EPR *Hamiltonian* is necessary, which requires much computer time.

(ii) A very comfortable way to obtain all magnetic field values simultaneoulsy for one given microwave energy was published by *Belford* et al. [6.28]. The only disadvantage is that one must diagonalize a matrix whose dimension is N^2, where N is the dimension of the EPR *Hamiltonian* matrix. The computer time needed for the diagonalization of a matrix with dimension N is roughly proportional to N^3. It is therefore evident, that this method may be advantageous only for problems with a rather low dimension, for example $S = 5/2$ or lower, with no nuclear interactions included.

(iii) The smallest amount of computer time, for most practical cases (dimension of *Hamiltonian* matrix higher than about four), is needed by a strategy described by *Scullane* et al. [6.29]. As in case (i), the *Hamiltonian* matrix is diagonalized for some fixed magnetic field values. The spacing between these fields, however, can be much larger than in case (i). For example, each field value can be higher than the preceding one by about 10% (the field values are not equally spaced). For each eigenvalue of each *Hamiltonian* matrix, the corresponding magnetic field value is then corrected individually using fourth order perturbation theory. The mathematics to be handled in the computer program is more complicated than for ways (i) and (ii), but a computer program for even a fourth order perturbation algorithm runs considerably faster than the many diagonalizations needed otherwise. Any accuracy for the final magnetic field values is possible, depending on the spacing of the initial field values.

6.8.7 Fitting of Free Parameters in a Simulated ENDOR Angular Dependence

All procedures necessary to numerically simulate ENDOR angular dependences for a given defect model have new been discussed. Once the experimentalist has assumed a defect model, the symmetry of each neighbor shell, and the distance of each shell from the defect center are fixed. However, the actual values of the interaction constants are not known. In order to decide whether the defect model is really correct, it is necessary to check whether the calculated angular dependences coincide precisely with the experimental ones. This coincidence can only be achieved if:

(i) The defect model is correct in all details.

(ii) All "free" interaction parameters are properly determined to achieve this coincidence. These parameters are determined by "fitting" the calculated angular dependence to the experimental one by systematic variation of the interaction parameters.

If an attempted fit is not successful, then either the defect model is wrong and must be modified, or the fitting calculations did not work properly despite the correct defect model. There are many computer programs commercially available which perform an automatic least–squares parameter fitting of free parameters of a function which is not known analytically. This is exactly the problem to be solved. However, there is another severe difficulty which has nothing to do with these computer algorithms.

It was mentioned that each line in an ENDOR angular dependence is characterized by quite a number of different parameters (for example, m_S^{eff}, $m_{I1}, \ldots, m_{IN}, m_q$, number of symmetry matrix of the corresponding neighbor nucleus, orientation of the center). However, in the experimental angular dependence, it is not normally clear which parameters belong to a given experimental line. So far, no computer algorithm exists to automatically find the proper assignment of parameters to the different lines in the experimental angular dependence. Even if the defect model is correct, the fitting algorithm will not work if there is a mistake in this assignment. So far, it is the task of the experimentalist to find this assignment, somehow, and to communicate it to the computer. This task can be very difficult since the correct assignment can only finally be seen from a successful fit. Depending on the experience

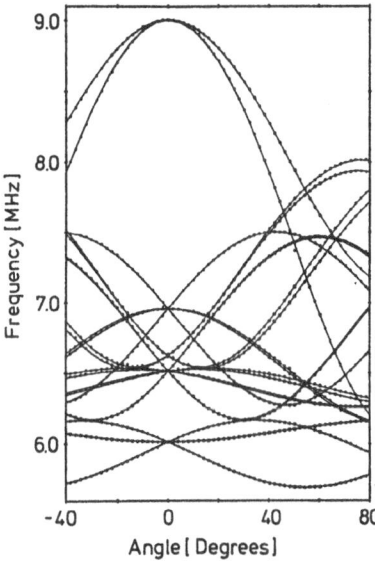

Fig. 6.28. Part of the ENDOR angular dependence for some neighbor shells of the Fe^0 defect in Si demonstrating the good agreement between experimental line positions (*dots*) and calculated values after fitting of free parameters in the spin *Hamiltonian* (*solid lines*). Magnetic field variation as in Fig. 6.7. (After [6.6])

of the experimentalist, the analysis of an ENDOR angular dependence may be a rather long process of trial and error. To solve the problem in the most efficient way it is, therefore, very essential for the experimentalist to have a good knowledge about the possible shapes of angular dependence patterns for the different symmetries, particularly if strong interactions cause significant deviations from the typical patterns of first order solutions. This knowledge can only be achieved by making numerous model calculations for different sets of interaction parameters and all feasible symmetries.

If the assignment of all lines in an experimental angular dependence to the parameters describing the type of the corresponding lines is clear, then the fit of the free parameters should work perfectly, leading to a perfect agreement between calculation and experiment. If this is not the case for some lines, then either an approximation used in the calculations was not adequate, or the symmetry of the defect deviates slightly from that assumed in the model. An example for a good agreement between experiment and calculation is seen in Fig. 6.28 for some neighbor shells of the Fe^0 defect in Si [6.6]. The good agreement shows that the symmetry of the defect model assumed is precisely correct. Such a good agreement can only be achieved if the orientation of the sample during the measurement of the angular dependence is known precisely. This precision is not obtainable for an example with X-ray techniques. The determination of the precise crystal orientation is, therefore, part of the fit procedure. It is necessary to fit the parameters describing the plane **v** of magnetic field variation and the angle offset θ_0 (6.5.12), as well.

An obvious question is whether the information provided by the experimental angular dependence is sufficient to determine all free parameters unequivocally. There is a maximum of eleven free parameters for each neighbor nucleus:

(i) hyperfine parameters a, b, b',

(ii) the angles $\delta_1, \delta_2, \delta_3$ describing the orientation of the hf or shf tensor (if not given by symmetry constraints),

(iii) quadrupole parameters q, q',

(iv) three additional angles for the quadrupole tensor orientation (if not given by symmetry constraints).

In addition, there are three parameters (angles) describing the plane **v** of magnetic field variation and the angle offset θ_0. These parameters are, however, the same for all neighbor shells.

Only three parameters can be determined from one line in the angular dependence. In first order, this line is a sine function in the frequency scale with amplitude and phase. Its periodicity is determined by the symmetry of the host crystal and the plane of magnetic field variation. In order to determine the remaining free parameters, one needs the information for the same neighbor nucleus provided by measurements in a maximum of three different planes of magnetic field variation. For high symmetry host crystals, these different planes are formally provided simultaneously by tensor transformations described by neighbor matrices and orientation matrices (Sects. 6.1–5). For low symmetry host crystals, however, it might be necessary to select different planes by rotating the sample about different axes in different angular dependence experiments. The orientation of these axes, with respect to the crystal lattice, must be known. They can be determined by X–ray techniques. Only in systems with very high symmetry (for example high symmetry centers in cubic host crystals) can the orientation of the plane of magnetic field variation also be seen from the ENDOR angular dependence.

6.8.8 Examples of Results Obtained from Analysis of ENDOR Angular Dependences

A successful analysis of an ENDOR angular dependence, as shown in Fig. 6.20 or in Fig. 6.28, means that all experimental line positions indicated by dots are precisely covered by solid lines of the calculated values, and that one has solved the problem of distinguishing between low and high symmetry defects (Sect. 6.6). One then obtains the following results:

(1) The symmetry of the defect model used for the calculations is correct. Just from the analysis of the angular dependence, however, there is still the possibility that different defect models with the same symmetry properties of neighbor shells are indistinguishable.

(2) The chemical identity of all neighbor nuclei is known. For example, additional impurities attached to the defect, which are not visible in EPR, can be analyzed.

(3) The electron spin of the defect can be unambiguously determined (Sect. 6.7).

(4) For each analyzed neighbor shell one obtains the hf or shf constants a, b, and b', and for $I > 1/2$ the quadrupole constants q and q'. This is a very detailed collection of experimental data to check the validity of theoretical models of the electronic structure of the defect. With the help of sometimes rather simple theoretical arguments, it is possible to link obtained values of interaction constants to neighbor shell distances. It is then possible to decide which of the defect models with identical symmetry properties is the only correct one.

From a theoretical point of view, the most detailed understanding of ENDOR shf data is achieved for ionic host systems. As an example, Table 6.1 summarizes results for the neutral atomic hydrogen center at anion sites in KCl [6.4]. In this case, there is no ambiguity in assigning the shf constants to definite neighbor shells. Using simple theoretical models (Chap. 7), the spin density of the unpaired electron spin of the center can be determined for each neighbor shell of Table 6.1. Figure 6.29 shows these results. The numbers give the spin density in 10^{-4} at. units at the different neighbor sites, according to Table 6.1. A detailed theoretical interpretation of these spin density values is given in [6.30]. For a review on ENDOR investigations of defects in ionic crystals see [6.31].

The situation is much more difficult for defects in semiconductors. One example is Se^+ defects in Si [6.7]. The shf results for eight neighbor shells are given in Table 6.2. So far, there are no sufficiently detailed theoretical arguments available to determine the distance of the analyzed neighbor shells from the center on the basis of the shf data. Therefore, in Table 6.2, the shells can only be assigned to their type of symmetry. Recently, a good theoretical explanation for the interaction constants of the first two shells was achieved [6.32]. The symmetry of the defect is T_d, $S = 1/2$. Due to this symmetry, the defect is either substitutional or interstitial in the center of the unit cell. Total energy calculations clearly show that the defect must be substitutional [6.12].

Table 6.1. Shf and quadrupole constants of the atomic neutral hydrogen center at anion sites in KCL ($T = 40°$ K). The shf and quadrupole constants are uncertain to ± 1 kHz. The free angles δ_{1A} of the shf tensor and δ_{1Q} of the quadrupole tensor are the angles between the [100]–direction and the z–axes of the corresponding tensors. For the lattice positions of the nuclei see Fig. 6.29. (After [3.28])

shell	constant	value [kHz]
$^{39}\text{K}_I$	a	253
	b	219
	q	198
$^{35}\text{Cl}_{II}$	a	257
	b	312
	b'	-3
	q	-88
	q'	-94
$^{35}\text{Cl}_{IV}$	a	37
	b	54
	q	± 45
$^{39}\text{K}_V$	a	4
	b	11
	b'	≈ 0
	δ_{1A}	$26° \pm 0.2°$
	q	± 39
	q'	± 17
	δ_{1Q}	$13.5° \pm 0.2°$

Table 6.2. Shf constants of the Se^+ defect in Si. The symmetry for the neighbor shell in parentheses is not clearly established. The free angles δ_{1A} for the 110-shells are the angles between a $< 110 >$–direction and the tensor z–axes. The Experimental uncertainty of the shf constants is ± 10 kHz and for the angles it is $\pm 2°$. (After [6.7])

shell	a[MHz]	b[MHz]	b'[MHz]	$\delta_{1A}[0]$
111	28.89	12.52		
	9.63	0.57		
	1.24	0.09		
100	(3.13)	(<0.003)	(<0.003)	
110	7.54	0.52	0.13	50
	3.87	0.45	-0.11	58
	4.27	-0.11	-0.11	34
	3.17	-0.07	-0.01	53

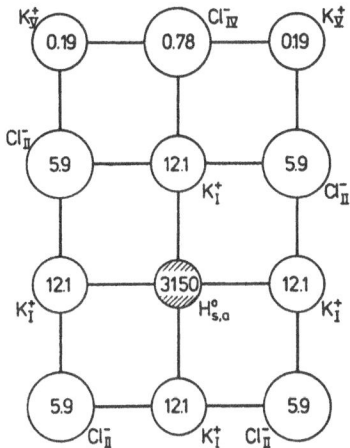

Fig. 6.29. Schematic representation of the spin density distribution at the sites of the neighbor nuclei of the atomic neutral hydrogen center at anion sites in KCl (in 10^{-4} at. units). (After [6.4])

Finally, Table 6.3 summarizes shf and quadrupole results for the interstitial (T_d) Al^{++}–defect (Fig. 6.6), and the Al^{++}–Al^- pair defect (Fig. 6.16) in Si [6.11]. So far, there are also no good theoretical arguments to assign the shells mentioned in Table 6.3 to their distance from the center. It turns out, however, that the data for some neighbor shells are rather similar for the point defect and the pair center. One might, therefore, suppose that the spin density distribution of the Al^{++} defect is not very much disturbed by the addition of a neighboring Al^-. Observing this point, there is much less freedom for the assignment of shf data to their corresponding neighbor shells, and the models for the two defects can be established. This is one example of the way additional information about a defect model can be obtained by lowering the symmetry by adding further impurities.

6.9 Software Treatment of ENDOR Spectra

Very many data have to be collected in an ENDOR experiment in order to obtain the defect model from the analysis of shf interaction anisotropies. One megabyte of data is stored during a typical ENDOR investigation. The angular dependences of the shf interactions can be very complex, particularly for low defect symmetries. In many cases, a decision for a defect model can only be made by the precise analysis of very fine details of the angular dependence patterns. In order to obtain these patterns, peak positions in the measured ENDOR spectra have to be determined with the highest possible accuracy.

Tables 6.3. Shf- and quadrupole constants of the Al^{++} and the $Al^{++}-Al^-$ defects in Si. For the structure of the Al^{++} see Fig. 6.6; for that of the $Al^{++}-Al^-$ pair defect see Fig. 6.16. All tensor z–axes are in $\{110\}$–planes. The experimental uncertainties are $\pm$ 20 kHz for the shf and quadrupole constants and $\pm 2°$ for the positions of tensor z–axes. (After [6.11])

Al^{++}			
TENSOR Z–AXIS	a[MHz]	b[MHz]	b'[MHz]
<111>	52.31	3.91	
<111>	50.09	$--.05$	
<100>	22.6	1.44	0
<111>	11.31	.16	
<110> $-15°$	3.51	.89	$--.04$
<111>	2.03	.21	
<110> $-58°$	1.24	.16	.02

$Al^{++}-Al^-$					
TENSOR Z–AXIS	a[MHz]	b[MHz]	b'[MHz]	q[MHz]	SHELL
<111> $-0.6°$	60.76	6.08	0		1
<111>	37.75	1.1		.39	(Al–neighb.)
<111> $+12°$	56.76	$--.41$	$--.18$		4
<111>	31.32	3.98			4*
<100> $+7°$	24.41	1.42	$--.23$		
<100> $+5°$	25.46	1.02	.18		2
<111> $\pm0°$	13.53	.16			4'
<110> $-13°$	5.32	1.14	0		
<110> $-12°$	4.8	.93	$--.03$		3
<110> $\pm0°$	2.73	.22			

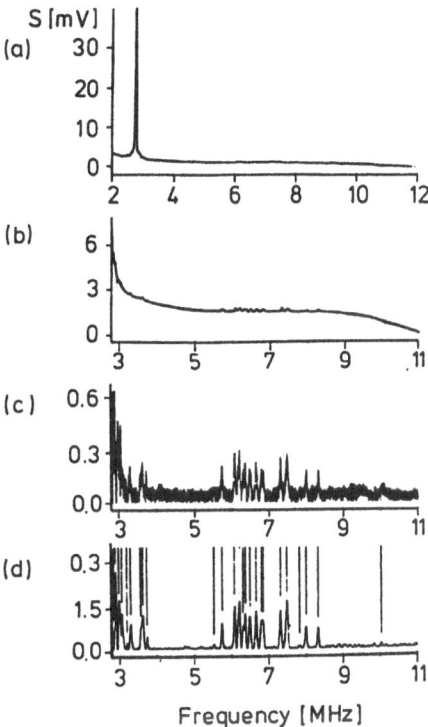

Fig. 6.30. a ENDOR spectrum for Fe^0 centres in silicon, $\mathbf{B}_0$ in (110) plane, the angle between $\mathbf{B}_0$ and the [100] direction is 40°, $T = 10$ K, after [6.6]. **b** The same ENDOR spectrum as in (a), lower part omitted. **c** The spectrum of Fig. 6.30b after subtraction of the background signal. **d** The same spectrum after application of a second order polynomial digital filter. The position of peaks is indicated by the *vertical lines*

This is not quite straightforward as can easily be seen from the spectrum in Fig. 5.10. There are many superimposed lines in a narrow frequency range. It is almost impossible to find the correct positions of the lines just by "close inspection" of the spectrum. Special computer algorithms are necessary to find the true position of the superimposed lines on the frequency scale. However, due to the tremendous amount of data, interactive computer strategies widely used for the exact lineshape analysis of optical spectra are not acceptable. This would be far too time consuming for the experimentalist in the case of ENDOR data. The computer algorithms used have to be designed for completely unattended operation, and they have to deal with difficulties like background signals, noise and superimposed lines.

This is demonstrated for an ENDOR spectrum obtained for Fe^0 centers in silicon; see Fig. 6.30 [6.6]. The very prominent line around 3 MHz corre-

sponds to the nuclear *Zeeman* frequency of ^{29}Si, and its high amplitude is due
to distant ENDOR effects. This line is not very important for defect structure analysis since it just verifies that the sample does contain ^{29}Si. The lines
characteristic of the interaction of the Fe^0 with the first neighbor shells are
the tiny ones in the frequency range from about 5.5 – 8.5 MHz. They show up
more clearly in Fig. 6.30b, where the lower frequency part of the spectrum is
omitted. A large contribution of a background signal is now visible. This background has to be eliminated in order to ensure proper operation of succeeding
algorithms for peak position determination. The shape of the background may
be different for each ENDOR spectrum; sometimes it might even look much
more complicated than the example in Fig. 6.30b. However, it is possible to
get rid of the background signal without making assumptions about functions
which might describe the background shape. Various ways are known in the
literature [6.33]. The basic principle of a successful strategy is described in
the Appendix C. The result after application of this algorithm is shown in
Fig. 6.30c. A large amount of noise is now clearly visible. However, this noise
gives no significant information about the signal–to–noise–ratio (SNR) of the
ENDOR signal during the experiment. A considerably lower level of noise
could have been achieved in the same experiment, and with the same total
time of experiment, just by selecting larger frequency increments during the
measurement, with a correspondingly larger signal integration time per single
step. The only consequence would be a loss of resolution. In principle, it is
possible to achieve the lowest amount of noise for the resolution desired, and
for a fixed total time of experiment, by optimizing the step size, and, correspondingly, the integration time per step. However, this is not a good way to
proceed. An extra bonus of an improved SNR be achieved for the same resolution and the same total time for the experiment by selecting a very small
frequency step, with consequently higher noise level, but with the application
of suitable digital filters [6.34–38] to the measured data (Appendix D). For
the example in Fig. 6.30c, the result after application of an optimized digital
filter is shown in Fig. 6.30d. The spectrum now looks considerably nicer than
that in Fig. 6.30a. A big advantage of the digital filter is that it never shifts
line positions, which is essential for its application to ENDOR spectra. Even
the spectrum in Fig. 6.30d is not yet ready for the automatic determination
of line positions. There are still overlapping lines. However, the line width can
be greatly reduced by the application of a successive deconvolution algorithm
[6.39].

An example is shown in Fig. 6.31. The spectrum in Fig. 6.31a is an original ENDOR spectrum (this example, obtained for neutral hydrogen centres in KCl [6.4] is typical). Multiple overlapping lines are clearly visible.
Figure 6.31b shows the same spectrum after application of the deconvolution
algorithm sketched in Appendix E. All lines are now clearly resolved. Their
precise position can now be determined easily by a further simple algorithm.
It should be mentioned that there is no physical meaning behind the applica-

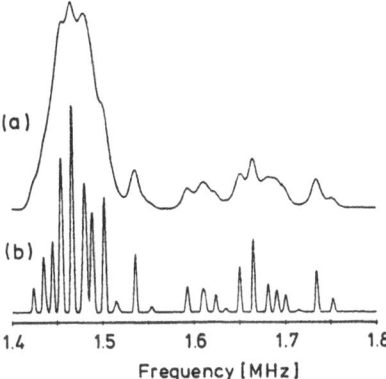

Fig. 6.31a,b. Demonstration of deconvolution algorithm. **a** ENDOR spectrum as measured (part of the spectrum shown in Fig. 5.10). **b** the same spectrum after the application of deconvolution

tion of this deconvolution. The line width yields no valuable information for the purposes discussed here. This is quite different from the situation for the line intensity. In the angular dependence plot, dots and dot chains belonging together are more easily recognized if the corresponding line intensity is taken into account. The line intensity may be characterized by the dot size or by a suitable color scale on a computer terminal. To define the function used for the deconvolution, two parameters have to be optimized experimentally. These are the line width and the line shape. The line shape parameter describes the amount of mixing of a *Gaussian* and a *Lorentzian* line. The same line shape parameters have to be used for the entire ENDOR spectrum. Usually, this causes no major problems, since the widths of ENDOR lines over a relatively narrow frequency range are quite similar. The deconvolution algorithm mentioned is surprisingly forgiving with respect to line shape misfits. Artificial lines are generated only if the line width taken for deconvolution is much smaller than the ENDOR line to be deconvoluted. It goes without saying that deconvolution increases the noise level. However, since the deconvolution algorithm discussed is a successive one, the amount of line narrowing, and thus, the increase in the noise level, can be easily selected.

7. Theoretical Interpretation of Superhyperfine and Quadrupole Interactions

The purpose of the present chapter is to provide the minimum theoretical background required for the interpretation of multiple resonance experiments.

Conventional EPR is a relatively discriminating technique for the investigation of point defects. It provides valuable information about site symmetry and charge state, and, in favorable circumstances, resolved hyperfine structure permits the identification of one or more nuclei. The much greater efficiency of multiple magnetic resonance methods for the detailed analysis of defect structures depends ultimately on the information contained in the many additional resolved superhyperfine interactions. Theory is employed in conjunction with multiple magnetic resonance methods, both as a guide in the interpretation of data for the purpose of structure determination and, at a deeper level, in order to achieve a quantitative understanding of the electronic structure as inferred from spin *Hamiltonian* parameters. With some theoretical guidance, superhyperfine structure can be associated, on the basis of the symmetry of its angular variation, with specific shells of ions surrounding the defect center. It can then provide a virtual map of the electronic wave function for unpaired spins, as well as information about the positions of the ions in their relaxed configuration.

7.1 Structures of Point Defects

The symmetries of point defects range from the highest point symmetry compatible with a site in the host crystal to much lower symmetries reflecting either complex defects or spontaneous symmetry breaking. The latter arises either from the static *Jahn–Teller* effect, associated with an orbitally degenerate ground state, or from off–center displacements of ions. The dynamic *Jahn–Teller* effect, which preserves the point symmetry, has a profound but subtle effect on EPR spectra [7.1]; however, since it is inimical to the detection of hyperfine interactions, it will not be emphasized in the present discussion.

An impurity ion in a solid can be described to a first approximation by a model, non–relativistic *Hamiltonian* which incorporates the interactions of its open shell electrons with the central potentials of its nucleus and core electrons, their spin–orbit interactions, and their interactions with the external crystal field and with one another,

$$\mathcal{H} = \sum_i \left[-\frac{\hbar^2}{2m}\nabla_i^2 - \frac{Ze^2}{r_i} + U(\mathbf{r}_i) + \zeta \mathbf{l}_i \cdot \mathbf{s}_i + V_c(\mathbf{r}_i) \right] + \frac{1}{2}\sum_{i \neq j} \frac{e^2}{r_{ij}}, \quad (7.1.1)$$

where the sums are over electrons in the open shell. The operators $\mathbf{l}_i$ and $\mathbf{s}_i$ are, respectively, the orbital and spin angular momentum of the ith electron. In the simplest picture, the crystal potential $V_c(\mathbf{r})$ in (7.1.1) arises from the charge distribution of neighboring ions (ligands) and reflects the symmetry of that distribution. (In a more rigorous formulation, effects of covalent bonding are included). Its strength, relative to that of the remaining terms in the *Hamiltonian*, depends on the physical system under consideration, and governs the nature of the consequent crystal field splittings of energy levels. Appropriate variants of this *Hamiltonian* are applicable to centers associated with point structural defects as well. The much weaker *Zeeman*, hyperfine and quadrupole interactions, which are not included in (7.1.1), will be considered at length in subsequent sections, since they provide the basis for the experimental techniques under consideration. Note that the cgs system of units is employed in this chapter from Sects. 7.1–5.

Point imperfections with a wide variety of structures have been identified in insulators and semiconductors. Their electronic wave functions range from extremely compact to extremely diffuse. This variation of scale is reflected in the strength of electron–lattice coupling, e.g., as measured by the crystal field splitting and by the *Huang–Rhys* factor, S, both of which have maximum values when the extension of the wave function is comparable with the lattice parameter. The *Huang–Rhys* factor, which is a measure of the difference in the equilibrium lattice configuration between the ground state and an optically excited state, is introduced in Sect. 4.2 and is illustrated in Fig. 4.2 for the idealized case of a two–level system linearly coupled to a single configuration coordinate representing a symmetry–adapted combination of ion displacements. Examples of point imperfections are presented below in approximate order of increasing wave function extension.

7.1.1 Impurities in Insulators

Substitutional metal impurities in insulators can exist in a variety of charge states, many with unpaired spins. Consequently, they are prominent among paramagnetic centers and have long been the focus of investigations in this field.

Lanthanide and actinide impurity ions, with open $4f$ and $5f$ shells, respectively, provide examples of extremely compact wave functions. Since the valence charge densitiy is concentrated at a larger radius than is the density of the open shell electrons, it fails to contribute to their shielding from the nuclear charge, but shields them instead from the influence of neighboring ions. As a consequence of both compactness and shielding, crystal field splittings of energy levels are small compared with the relatively large separations of fine–structure levels induced by spin–orbit interaction, and the *Huang–Rhys* factor is typically of the order $S \cong .01$.

Transition metals of the iron, platinum and palladium groups, with open 3d, 4d and 5d shells, respectively, have significantly less compact wave functions than the lanthanides and actinides. The enhanced crystal field splitting exceeds the diminished spin–orbit splitting for these elements, and approaches the mutual electrostatic interaction of the open shell electrons. The *Huang–Rhys* factor is typically of order $S \cong 1$ for these elements.

Heavy metal impurities are typified by substitutional thallium in alkali halides. In some charge states they have an open p–shell ($6p$ in thallium). In these complexes, the crystal field, spin–orbit and mutual electrostatic interactions are all comparable, and the *Huang–Rhys* factor is of order $S \cong 10$.

7.1.2 Color Centers

Color centers, associated with point lattice defects, can also exist in several charge states, some of which are paramagnetic. Hole centers tend to have more compact wave functions than electron–excess centers. The prototype of the hole center is the V_K center in alkali halides, which consists of an X_2^- molecule ion occupying an adjacent pair of anion (X^-) sites, or, equivalently, a hole shared by two adjacent anions. A *Huang–Rhys* factor $S \cong 1$ is typical.

The F center in alkali halides is the prototype of electron–excess color centers and, indeed, of color centers in general. It consists of a single electron trapped at an anion vacancy, as does the F^+ center in alkaline–earth oxides. Since there is no central nucleus, crystal field splitting is primarily responsible for the entire electronic structure. The largest *Huang–Rhys* factors are encountered for these defects, ranging up to $S \cong 100$.

Impurity related color centers are a common feature, including, for example, the laser–active $Tl^0(1)$ center in alkali halides, which consists of a neutral thallium atom adjacent to a halogen vacancy. The *Huang–Rhys* factor is typically $S \cong 10$.

7.1.3 Defects in Semiconductors

The wave functions of deep level impurities are typically more diffuse than those of color centers, but are largely confined to a small number of shells of neighbors about the defect site, and share some of the character of wave functions derived from several energy bands of the host crystal. Intrinsic defects such as antisite defects in compound semiconductors (e.g., Ga on an As site in GaAs) are similar in that respect. *Huang–Rhys* factors are typically of order $S \cong 1$.

So–called effective–mass centers, including shallow donor and acceptor states in semiconductors and *Wannier* excitons in ionic crystals, have exceedingly diffuse wave functions extending over many unit cells. Consequently, they sample bulk properties of the solid and are relatively insensitive to displacements of near neighbors. *Huang–Rhys* factors are typically of order $S \cong .01$, comparable with lanthanide and actinide impurities whose wave function extension is at the opposite extreme.

7.2 Origin of Zeeman, Hyperfine and Quadrupole Interactions

The energy eigenvalues of the *Hamiltonian* of (7.1.1) may have symmetry–induced degeneracies, and, in any event, they always retain a residual (*Kramers*) degeneracy when the number of electrons is odd; they then split into *Zeeman* levels in the presence of an external magnetic field. The technique of electron paramagnetic resonance (EPR) relies on magnetic dipole transitions between these *Zeeman* levels induced by microwave radiation, as described in Chap. 2. The *Zeeman* levels may be further split into hyperfine levels by virtue of the interaction of the electrons with nuclear magnetic dipole and electric quadrupole moments. These moments arise from circulating currents in the nucleus, and from its finite size and aspherical shape, respectively. Transitions between hyperfine levels induced by radio–frequency fields are employed in both nuclear magnetic resonance (NMR) and electron nuclear double resonance (ENDOR) spectroscopy, as described in Chap. 5. In this section, we explore the origins of these interactions and derive tractable expressions for their contributions to the *Hamiltonian*. A single nucleus at the origin of coordinates is assumed, apart from crystal field effects; extension to a system of electrons which have hyperfine and quadrupole interactions with several nuclei is deferred to subsequent sections.

This rather formal development is included for completeness, but the reader who wishes to avoid it may safely omit Sect. 7.2 without serious loss of continuity.

7.2.1 Origin of the Hamiltonian

The ultimate source of the *Hamiltonian* employed in this work is the *Dirac* equation for a single electron in external electric and magnetic fields,

$$i\hbar\frac{\partial\psi}{\partial t} = [-\boldsymbol{\alpha}\cdot(c\mathbf{p}+e\mathbf{A}) - \beta mc^2 - e\phi]\psi, \qquad (7.2.1)$$

where ϕ and $\mathbf{A}$ are electric scalar and magnetic vector potentials, respectively, $\mathbf{p} = -i\hbar\nabla$ is the momentum operator, $\psi(\mathbf{r}, t)$ is a four–component wave function, and α_x, α_y, α_z and β are 4×4 matrices which anti–commute in pairs. The physical constants e, m, c and $\hbar$ have their usual meanings.

Two components of ψ become vanishingly small for positive energy solutions in the non–relativistic limit. One proceeds by eliminating the two small components in favor of the two large ones; by expanding in powers of v/c, retaining terms out to $(v/c)^2$; and by effecting a *Foldy–Wouthuysen* transformation which insures normalization of the remaining two–component wave function to the same order of approximation [7.2,3]. The approximate *Hamiltonian* obtained by this procedure has the form

$$\mathcal{H} = \frac{\mathbf{p}^2}{2m} - e\phi + \frac{e\hbar}{2m^2c^2}\mathbf{s} \cdot \mathbf{E} \times \mathbf{p} + \frac{e}{mc}\mathbf{A} \cdot \mathbf{p} + 2\mu_{\mathrm{B}}\mathbf{H} \cdot \mathbf{s}$$

$$+ \frac{1}{[1 + (e\phi/2mc^2)]^2}\frac{\hbar e^2}{2m^2e^4}\mathbf{s} \cdot \mathbf{E} \times \mathbf{A} - \frac{\mathbf{p}^4}{8m^3c^2} + \frac{e\hbar^2}{8m^2c^2}\nabla \cdot \mathbf{E},$$

$$(7.2.2)$$

where $\mathbf{s}$ is the spin angular momentum operator, which can be represented by a 2×2 matrix operating on a two–component wave function, and $\mathbf{E} = -\nabla\phi$ is the electric field intensity. Generalization of this *Hamiltonian* for a many–electron system is accomplished by summing over electrons and incorporating internal interactions.

Equation (7.1.1) corresponds to just the first three terms on the right–hand side of (7.2.2), adapted to a many–electron system. The second term accomodates electrostatic interactions with a central nucleus, with other electrons and with a crystal field. The third term is the spin–orbit interaction which, for a many–electron system with a central potential, may be written in the form

$$\mathcal{H}_{SO} = -\frac{e\hbar^2}{2m^2c^2}\sum_i \frac{1}{r_i}\frac{d\phi}{dr_i}\mathbf{l}_i \cdot \mathbf{s}_i. \qquad (7.2.3)$$

When applied within a single open shell of an impurity ion as in (7.1.1), it can be further simplified by replacing the coefficient of $\mathbf{l}_i \cdot \mathbf{s}_i$ by its expectation value ζ, calculated with the appropriate radial wave function.

The fourth and fifth terms on the right–hand side of (7.2.2) accomodate both *Zeeman* and hyperfine interactions. The parameter $\mu_{\mathrm{B}} \equiv e\hbar/2mc$ is the *Bohr* magneton. The sixth term, which is bilinear in electric and magnetic fields, is negligible everywhere except in the immediate vicinity of the nucleus, but it makes a finite integrated contribution to the hyperfine interaction.

The seventh and eighth terms are relativistic corrections which preserve the symmetry and degeneracies of the first two terms. Although they are important for precise atomic structure calculations on heavy atoms, they have only an indirect influence on magnetic resonance spectra via the wave functions, which affect the numerical values of hyperfine constants; the appropriate correction factors are discussed in Sect. 7.4.2. These terms will not be considered further in the present context.

7.2.2 Wigner–Eckart Theorem

In order to facilitate the derivation of tractable expressions for hyperfine and quadrupole interactions, we digress to consider a simple, elegant and powerful theorem which finds ubiquitous application. The theorem concerns spherical tensor operators T_{LM} which transform as bases for irreducible representations of the rotation group. The *Wigner–Eckart* theorem states that matrix elements of spherical tensor operators between angular momentum eigenkets $|jm\rangle$ are expressible in the form [7.4]

$$\langle j'm'|T_{LM}|jm\rangle = \langle j' \| T_L \| j\rangle \langle jLmM|j'm'\rangle, \qquad (7.2.4)$$

where the reduced matrix element $\langle j' \| T_L \| j\rangle$ is independent of the quantum numbers m, M and m', and the factor $\langle jLmM|j'm'\rangle$ is a *Clebsch–Gordan* coefficient.

The utility of the *Wigner–Eckart* theorem in the present context derives from an obvious corollary: The matrix elements of two spherical tensor operators of the same rank, T_{LM} and U_{LM}, are proportional within a manifold of constant j, L and j'. This corollary permits the simplification of an interaction *Hamiltonian* by substitution of one spherical tensor operator for another, within a constant of proportionality equal to the ratio of reduced matrix elements, as will be apparent from subsequent developments.

7.2.3 Zeeman Interaction

The magnetic vector potential $\mathbf{A}(\mathbf{r})$ corresponding to a uniform external magnetic field $\mathbf{H}$ can be written in the form $\mathbf{A}(\mathbf{r}) = (1/2)\mathbf{H} \times \mathbf{r}$. It follows from the identity

$$\mathbf{H} \times \mathbf{r} \cdot \mathbf{p} = \mathbf{H} \cdot \mathbf{r} \times \mathbf{p} = \hbar \mathbf{H} \cdot \mathbf{l} \qquad (7.2.5)$$

that the fourth and fifth terms on the right–hand side of (7.2.2) can be combined to obtain the *Zeeman Hamiltonian* for the many–electron system in the form

$$\mathcal{H}_Z = \mu_B \mathbf{H} \cdot \sum_i (\mathbf{l}_i + g_e \mathbf{s}_i). \qquad (7.2.6)$$

The factor 2 in the fifth term of (7.2.2) has been replaced in (7.2.6) by the factor $g_e = 2.002319$, which incorporates radiative corrections arising from quantum electrodynamics.

7.2.4 Hyperfine Interaction

The magnetic moments associated with both orbital and spin angular momentum of the electrons interact with magnetic fields which arise from circulating currents within the nucleus. It follows from Ampere's law that the magnetic vector potential $\mathbf{A}(\mathbf{r})$ at a point $\mathbf{r}$ outside the nucleus is given in terms of the current density $\mathbf{J}(\mathbf{r}')$ at interior points $\mathbf{r}'$ by

$$\mathbf{A}(\mathbf{r}) = \int \frac{\mathbf{J}(\mathbf{r}')}{c|\mathbf{r} - \mathbf{r}'|} d\tau'. \tag{7.2.7}$$

The integrand in (7.2.7) can be expanded in powers of r'/r and transformed to yield the expression

$$\mathbf{A}(\mathbf{r}) = \nabla(1/r) \times \boldsymbol{\mu}, \tag{7.2.8}$$

where $\boldsymbol{\mu}$ is the nuclear magnetic dipole moment,

$$\boldsymbol{\mu} = \frac{1}{2c} \int \mathbf{r}' \times \mathbf{J}(\mathbf{r}') d\tau'. \tag{7.2.9}$$

Since $\boldsymbol{\mu}$ transforms as a spherical tensor of the first rank, the *Wigner–Eckart* theorem justifies the replacement

$$\boldsymbol{\mu} \rightarrow g_n \mu_n \mathbf{I}, \tag{7.2.10}$$

where $\mu_n \equiv e\hbar/2Mc$ is the nuclear magneton (M is the mass of the proton), and the dimensionless nuclear g–value g_n is an empirical property of the nuclear ground state. The magnetic field intensity is then

$$\mathbf{H}(\mathbf{r}) = -\nabla \times \mathbf{A}(\mathbf{r}) = -g_n \mu_n \left[\frac{\mathbf{I}}{r^3} - \frac{3(\mathbf{I} \cdot \mathbf{r})\mathbf{r}}{r^5} \right]. \tag{7.2.11}$$

With these forms for the magnetic vector potential and the magnetic field, the fourth and fifth terms on the right–hand side of (7.2.2) can be combined by employing the transformation

$$\nabla(1/r) \times \boldsymbol{\mu} \cdot \mathbf{p} = \boldsymbol{\mu} \cdot \mathbf{r} \times \mathbf{p}/r^3 = \hbar \boldsymbol{\mu} \cdot \mathbf{l}/r^3. \tag{7.2.12}$$

The sixth term of (7.2.2) can also be transformed with the aid of (7.2.8) and (7.2.10) to:

$$\frac{1}{r^2} \frac{d}{dr} \frac{1}{[1 + (e\phi/2mc^2)]} g_e g_n \mu_B \mu_n \mathbf{I} \cdot \left[\mathbf{s} - \frac{\mathbf{r}(\mathbf{s} \cdot \mathbf{r})}{r^2} \right]$$

$$\cong \frac{8\pi}{3} \delta(\mathbf{r}) g_e g_n \mu_B \mu_n \mathbf{I} \cdot \mathbf{s}, \tag{7.2.13}$$

where a factor of 2 has again been replaced by g_e. The second form of this *Fermi* contact hyperfine interaction reflects its extreme locality, and is equivalent to neglecting the variation of the electronic wave function over the very small region enclosing the nucleus in which this term is appreciable.

We can combine the contributions of the three terms to obtain the hyperfine interaction of a many–electron system,

$$\mathcal{H}_{HF} = g_e g_n \mu_B \mu_n \sum_i \left[\frac{\mathbf{I} \cdot (\mathbf{l}_i - \mathbf{s}_i)}{r_i^3} + \frac{3(\mathbf{r}_i \cdot \mathbf{I})(\mathbf{r}_i \cdot \mathbf{s}_i)}{r_i^5} + \frac{8\pi}{3}\delta(\mathbf{r}_i)\mathbf{I} \cdot \mathbf{s}_i \right].$$

(7.2.14)

7.2.5 Quadrupole Interaction

The electric quadrupole interaction arises from the finite extension and aspherical shape of the nuclear charge distribution. The electric potential $\phi(\mathbf{r})$ at a point $\mathbf{r}$ outside the nucleus, obtained by solution of the Poisson equation, is given by

$$\phi(\mathbf{r}) = \int \frac{\rho(\mathbf{r}')}{|\mathbf{r} - \mathbf{r}'|} d\tau',$$

(7.2.15)

where $\rho(\mathbf{r}')$ is the charge density at point $\mathbf{r}'$ inside the nucleus. The factor $1/|\mathbf{r} - \mathbf{r}'|$ may be expanded in *normalized* spherical harmonics $Y_l^m(\theta, \phi)$ to obtain the expression

$$\phi(\mathbf{r}) = \sum_{l=0}^{\infty} \sum_{m=-l}^{l} \sqrt{\frac{4\pi}{2l+1}} r^{-l-1} Y_l^m(\theta, \phi) p_{lm},$$

(7.2.16)

where p_{lm} is a multipole moment given by [7.5]

$$p_{lm} = e \sum_{k=1}^{Z} \int |\Psi_{IM_I}(r_1 \dots r_N)|^2 \sqrt{\frac{4\pi}{2l+1}} r_k^l Y_l^{m^*}(\theta_k, \phi_k) d\tau_1 \dots d\tau_n.$$

(7.2.17)

The sum is over the Z protons in the nucleus, and the integral extends over the N nucleons. Since the nuclear ground state wave function Ψ_{IM_I} has definite parity and angular momentum, only multipole moments with even values of l which satisfy the inequality $l \leq 2I$ can be non–vanishing.

The monopole moment p_{00} is simply the total nuclear charge which gives rise to a spherically symmetrical Coulomb potential. The next surviving moments, p_{2m}, are responsible for the quadrupole interaction. Moments with $l > 2$ have a negligible effect.

The spherical harmonic $Y_l^{m^*}(\theta, \phi)$ in (7.2.17) is a special case of a spherical tensor operator. Application of the *Wigner–Eckart* theorem greatly simplifies the expression for p_{2m} by replacing the spherical harmonics $Y_2^{m^*}$ by second degree polynomials in the nuclear angular momentum operator $\mathbf{I}$, which transform as spherical tensors of the second rank. This substitution leads to the following form for the quadrupole interaction *Hamiltonian*:

$$\mathcal{H}_Q = -e \sum_i \phi(\mathbf{r}_i) = \frac{e^2 Q}{2I(2I-1)} \sum_i \left[\frac{I(I+1)}{r_i^3} - \frac{3(\mathbf{r}_i \cdot \mathbf{I})^2}{r_i^5} \right], \quad (7.2.18)$$

where the sum is over open shell electrons and Q is the nuclear quadrupole moment defined by

$$Q \equiv (2/e) p_{20} = \sum_{k=1}^{Z} \int |\psi_{II}(\mathbf{r}_1 \dots \mathbf{r}_N)|^2 (3z_k^2 - r_k^2) d\tau_1 \dots d\tau_N. \quad (7.2.19)$$

7.2.6 Total Hamiltonian

Two additional interactions are required in order to complete the *Hamiltonian*. The spin–spin interaction between the spin magnetic moments of two electrons is derived in analogy with the hyperfine interaction, and has the form

$$\mathcal{H}_{SS} = \frac{1}{2} \sum_{j \neq i} g_e^2 \mu_B^2 \left[\frac{\mathbf{s}_i \cdot \mathbf{s}_j}{r_{ij}^3} - \frac{3(\mathbf{r}_{ij} \cdot \mathbf{s}_i)(\mathbf{r}_{ij} \cdot \mathbf{s}_j)}{r_{ij}^5} - \frac{8\pi}{3} \delta(\mathbf{r}_{ij}) \mathbf{s}_i \cdot \mathbf{s}_j \right]. (7.2.20)$$

Finally, the nuclear *Zeeman* interaction involves the direct effect of the external magnetic field on the nuclear magnetic moment,

$$\mathcal{H}_{NZ} = -g_n \mu_n \mathbf{H} \cdot \mathbf{I}. \quad (7.2.21)$$

The total *Hamiltonian* for a system of electrons interacting with a single central nucleus plus a closed shell core and an external crystal field, can now be presented in the form

$$\mathcal{H} = \sum_i \left[\frac{\hbar^2}{2m} \nabla_i^2 - \frac{Ze^2}{r_i} + U(\mathbf{r}_i) + V_c(\mathbf{r}_i) \right] + \frac{1}{2} \sum_{i \neq j} \frac{e^2}{r_{ij}}$$

$$+ \mathcal{H}_{SO} + \mathcal{H}_{SS} + \mathcal{H}_Z + \mathcal{H}_{HF} + \mathcal{H}_Q + \mathcal{H}_{NZ}, \quad (7.2.22)$$

where the interactions listed in order of diminishing strength in the second line are defined by (7.2.3), (7.2.20), (7.2.6), (7.2.14), (7.2.18) and (7.2.21), respectively.

7.3 Central Ion Hyperfine Structure

In this section we consider the effects of hyperfine and quadrupole interactions with a central nucleus on the electronic structure of an impurity ion in a solid.

We begin with the electronic structure of the free ion, and then proceed to examine the modifications introduced by a crystal field. The *Hamiltonian* presented in (7.2.22), together with the equations which define its various terms, provides the basis for this discussion.

7.3.1 Free Ion Electronic Structure

The electronic structure of the free ion can be elucidated by sequentially examining the effects of successive terms in its *Hamiltonian* [7.6]. The first three terms in the *Hamiltonian* of (7.2.22), the kinetic energy and the central potentials of the nucleus and core electrons, comprise the central field approximation, and the fourth term vanishes in the absence of a crystal field. At this level of approximation, the *Hamiltonian* is invariant under independent rotations of the space and spin coordinates of individual electrons. The energy levels in the central field approximation correspond to configurations designated by the set of principal quantum numbers n_{l_i} and orbital angular momentum quantum numbers l_i. The degenerate eigenstates are distinguished by the quantum numbers m_{l_i} and m_{s_i}, respectively, denoting the projections of the orbital and spin angular momenta of the occupied spin orbitals. An additional restriction is imposed by the *Pauli* principle, which requires that the wave function be antisymmetric under exchange of the space and spin coordinates of any two electrons. The wave function is then the antisymmetrized product of these occupied spin orbitals (*Slater* determinant),

$$\psi \cong \frac{1}{\sqrt{N!}} \sum_p (-1)^p \mathcal{P} \prod_{i=1}^{N} \phi_{k_i}(j) \equiv \frac{1}{\sqrt{N!}} \det |\phi_{k_i}(j)|, \qquad (7.3.1)$$

where $\mathcal{P}$ is an operator which permutes the electron space and spin coordinates $\mathbf{r}_j, \mathbf{s}_j$, denoted concisely by j ; p is the number of pair–wise permutations (transpositions) into which $\mathcal{P}$ may be decomposed; N is the number of occupied spin orbitals; and k_i denotes the quantum numbers n_i, l_i, m_{l_i} and m_{s_i}. The spin orbitals $\phi_{k_i}(\mathbf{r}, \mathbf{s})$ are eigenfunctions of the one–electron *Hamiltonian*

$$h = -\frac{\hbar^2}{2m}\nabla^2 - \frac{Ze^2}{r} + U(r). \qquad (7.3.2)$$

The antisymmetrized wave function of (7.3.1) vanishes if any two spin orbitals share the same set of values of all four quantum numbers n, l, m_l and m_s; consequently, the spin orbitals are populated in order of increasing energy in the ground configuration. Finally, all of the functions belonging to the same configuration have a common parity determined by that of the component spin orbitals.

An open shell configuration is highly degenerate because of the large number of possible combinations of m_{l_i} and m_{s_i} which are consistent with the *Pauli* principle. When the fifth term, the mutual electrostatic interaction of the electrons ($\frac{1}{2}\sum_{i\neq j} e^2/r_{ij}$) is included, there is a sharp reduction in symmetry; the *Hamiltonian* is now invariant only under simultaneous rotations of the space and spin coordinates of all the electrons. As a consequence, much of the degeneracy is removed, and each configuration splits into a set of terms labeled by L and S, the total orbital and spin angular momentum quantum numbers, corresponding respectively to the operators $\mathbf{L} = \sum_i \mathbf{l}_i$ and $\mathbf{S} = \sum_i \mathbf{s}_i$. It should be noted that the coupling of spin angular momenta at this stage (*Russell–Saunders* coupling) is a consequence of the *Pauli* principle; the spin–spin interaction of (7.2.20) is weaker by four orders of magnitude. Terms are conventionally designated by the notation ^{2S+1}L, with numerical values $L = 0, 1, 2, 3, \ldots$ replaced by letters $S, P, D, F, \ldots$, respectively. Possible values of L and S are limited by the rules for coupling angular momenta, and additional severe restrictions on allowed combinations are imposed by the *Pauli* principle.

The corresponding wave functions are linear combinations of antisymmetrized product functions, which we represent by the kets $|\alpha L S M_L M_S\rangle$, where M_L and M_S are the respective projections of the total orbital and spin angular momenta, and α is a parameter which distinguishes terms with the same L, S values. The degeneracy of a term is $(2L+1)(2S+1)(2I+1)$, where I is the nuclear angular momentum quantum number.

The spin–orbit interaction $\mathcal{H}_{SO}$ in (7.2.22) reduces the symmetry further, and splits each term into fine–structure levels labeled by the total electronic angular momentum quantum number J, corresponding to the operator $\mathbf{J} = \mathbf{L} + \mathbf{S}$, in the notation $^{2S+1}L_J$. The corresponding eigenkets $|\alpha L S J M_J\rangle$ are linear combinations of the $|\alpha L S M_L M_S\rangle$, and are $(2J+1)(2I+1)$–fold degenerate. One can invoke the *Wigner–Eckart* theorem to express $\mathcal{H}_{SO}$ in a form which is valid within a single term,

$$\mathcal{H}_{SO} = \lambda(\alpha, L, S)\mathbf{L} \cdot \mathbf{S}. \tag{7.3.3}$$

Its eigenvalues, obtained by elementary means, are given by

$$E_J = \frac{1}{2}\lambda[J(J+1) - L(L+1) - S(S+1)]. \tag{7.3.4}$$

It follows from (7.3.4) that the interval between successive fine–structure levels J and $J-1$ is just λJ; this result is known as the *Landé* interval rule. The spin–spin interaction $\mathcal{H}_{SS}$ of (7.2.22) is two orders of magnitude weaker than the spin–orbit interaction, and produces no additional splitting of fine–structure levels. Its only effect on the electronic structure of the free ion is to upset the *Landé* interval rule. The interval rule is also upset by the proximity of other terms which have a fine–structure level with the same J value; since the spin–orbit interaction increases rapidly with atomic number, the latter effect is especially pronounced in lanthanide, actinide and heavy–metal impurities. As an example, consider neutral Pb, whose ground configuration is

6p^2 plus closed shells. Its multiplet structure consists of three terms: ^1D, ^3P and ^1S. The ^3P term splits further into fine–structure levels $^3P_0$, $^3P_1$ and $^3P_2$. Since the $^3P_2$ level is strongly repelled by the higher $^1D_2$ level, and the $^3P_0$ level is somewhat less strongly repelled by the highest $^1S_0$ level, the *Landé* interval rule is not even approximately satisfied, and the most that one can say is that the fine–structure levels at least occur in the expected order.

The *Wigner–Eckart* theorem can be invoked yet again to express the electronic *Zeeman* interaction $\mathcal{H}_Z$ as

$$\mathcal{H}_Z = g_J \mu_B \mathbf{H} \cdot \mathbf{J}, \tag{7.3.5}$$

$$g_J = \frac{1 + g_e}{2} + \frac{1 - g_e}{2} \frac{L(L+1) - S(S+1)}{J(J+1)}. \tag{7.3.6}$$

The next two terms of (7.2.22), $\mathcal{H}_{HF}$ and $\mathcal{H}_Q$, can also be transformed with the help of the *Wigner–Eckart* theorem; they can then be combined with the electronic and nuclear *Zeeman* interactions in a perturbation *Hamiltonian* $\mathcal{H}'$ given by

$$\mathcal{H}' = \mathcal{H}_Z + \mathcal{H}_{HF} + \mathcal{H}_Q + \mathcal{H}_{NZ}$$

$$= g_J \mu_B \mathbf{H} \cdot \mathbf{J} + a \mathbf{I} \cdot \mathbf{J} + 2b \mathbf{I} \cdot \mathbf{J} (2 \mathbf{I} \cdot \mathbf{J} + 1) + g_n \mu_n \mathbf{H} \cdot \mathbf{I}. \tag{7.3.7}$$

The coefficient b in the quadrupole interaction is given by

$$b = \frac{3e^2}{4I(2I - 1)J(2J - 1)}$$

$$\times \sum_{i=1}^{n} \int |\psi_{JJ}(r_1 \ldots r_n)|^2 \frac{(3z_i^2 - r_i^2)}{2r_i^5} d\tau_1 \ldots d\tau_n, \tag{7.3.8}$$

where $\psi_{JJ}(r_1 \ldots r_n)$ is the electronic wave function for $M_J = J$, and the sum is over electrons.

The coefficient a in the hyperfine interaction is different for each L, S term. For the special case of a single electron outside closed shells, it is given by

$$a = g_e g_n \mu_B \mu_n \frac{l(l+1)}{j(j+1)} <r^{-3}>, \quad l \neq 0, \tag{7.3.9}$$

$$a = g_e g_n \mu_B \mu_n \frac{2\pi}{j(j+1)} |\psi(0)|^2, \quad l = 0. \tag{7.3.10}$$

Note that the a and b coefficients defined here should not be confused with those introduced in Sect. 5.4, in connection with superhyperfine interactions, which have different meanings.

The effect of the perturbation $\mathcal{H}'$ is to split the fine–structure levels into *Zeeman* and hyperfine levels. The nature of this splitting depends on the relative strengths of the first two terms. In the absence of an external magnetic field, $\mathcal{H}'$ is diagonal in a coupled representation spanned by the eigenkets $|\alpha LSIJFM_F\rangle$, where $\mathbf{F} = \mathbf{I} + \mathbf{J}$. In a weak field, its eigenvalues are given approximately by

$$E_{FM_F} \cong \frac{1}{2}aK + bK(K+1) + g_F\mu_B HM_F, \qquad (7.3.11)$$

$$K = F(F+1) - I(I+1) - J(J+1), \qquad (7.3.12)$$

$$g_F = g_J\frac{F(F+1) - I(I+1) + J(J+1)}{2F(F+1)}$$

$$-\frac{m}{M}g_n\frac{F(F+1) + I(I+1) - J(J+1)}{2F(F+1)}. \qquad (7.3.13)$$

The opposite extreme, in which the electronic *Zeeman* interaction $\mathcal{H}_Z$ is much stronger than the hyperfine interaction $\mathcal{H}_{HF}$, is of greater interest in the present context. In that case, the eigenvalues of $\mathcal{H}'$ are given approximately by

$$\begin{aligned} E_{M_I M_J} = {}&g_J\mu_B HM_J + aM_I M_J \\ &+2b[(I(I+1)J(J+1) - I(I+1)M_J^2 \\ &-J(J+1)M_I^2 + 3M_I^2 M_J^2] - g_n\mu_n HM_I. \end{aligned} \qquad (7.3.14)$$

In intermediate coupling where $\mathcal{H}_Z$ and $\mathcal{H}_{HF}$ are comparable (magnetic fields of the order of 10^{-2} $T = 100$ Gauss), one must diagonalize the matrix of $\mathcal{H}'$ within each M_F block. The eigenvalues of $\mathcal{H}_Z + \mathcal{H}_{HF}$ can be expressed in closed form in the special case $J = \frac{1}{2}$ (*Breit–Rabi* formula), [7.7]

$$E_{M_F} = -\frac{\Delta W}{2(2I+1)} \pm \frac{\Delta W}{2}\sqrt{1 + \frac{4x M_F}{2I+1} + x^2}, \qquad (7.3.15)$$

$$\Delta W = \frac{1}{2}a(2I+1), \qquad (7.3.16)$$

$$x = \frac{g_J\mu_B H}{\Delta W}. \qquad (7.3.17)$$

7.3.2 Crystal Field Splitting

Interaction of the electronic system with an external crystal field is represented by the fourth term in (7.2.22), $\sum_i V_c(\mathbf{r}_i)$. As noted previously, the relative strength of this term depends principally on the position of the ion in the periodic table. Its symmetry, on the other hand, reflects the point symmetry of the impurity site, including any effects of local charge compensation. In a point–ion model of the crystal field, $V_c(\mathbf{r})$ is assumed to be a non singular solution of the *Laplace* equation; it can be expanded in spherical harmonics as

$$V_c(\mathbf{r}) = \sum_{k=0}^{\infty} \sum_{q=-k}^{k} B_k^q \, r^k \, Y_k^q(\theta, \phi), \qquad (7.3.18)$$

where the coefficients B_k^q are restricted by the requirement that the crystal potential has the symmetry of the defect site; i.e., that it transforms as a basis for the identity representation of the relevant crystallographic point group.

As a consequence of extremely compact open shell orbitals in the lanthanides and actinides, together with the shielding effects of valence orbitals, the crystal field term in the *Hamiltonian* is weaker than the relatively strong spin–orbit interaction appropriate to these heavy ions. By virtue of its symmetry, which is lower than that of the free ion, the crystal field partially removes the degeneracies of the fine–structure levels; J remains a good quantum number, but not M_j.

In the case of lighter ions, including transition metals of the iron, platinum and palladium groups, the crystal field term is much stronger than the spin–orbit interaction. For fields of intermediate strength, V_c is weaker than the electrostatic interaction $\frac{1}{2} \sum_{i \neq j} e^2/r_{ij}$, so L and S remain good quantum numbers, but not M_L, M_S, J or M_J, and the crystal field partially removes the degeneracies of the L, S terms. In the strong field case, V_c exceeds the electrostatic interaction. The l_i remain good quantum numbers, but not L and S, and V_c partially removes the degeneracies of the atomic configurations to yield strong crystal field configurations [7.8]. When the crystal field and electrostatic interactions are of comparable strength, one must diagonalize both operators simultaneously within the entire ground atomic configuration.

In the case of a single electron in an open d shell ($l = 2$), the distinction between strong and intermediate crystal fields disappears. The fivefold orbital degeneracy $(2l + 1)$ of the ground state is partially removed by an octahedral crystal field, which leaves an orbital triplet below an orbital doublet. The separation of these crystal field levels, conventionally designated 10 Dq, is a measure of the strength of the crystal field. In a point–ion model, 10 Dq is proportional to the ligand charge and inversely proportional to the fifth power of the impurity–ligand distance. Covalency effects, which must be included in a complete and rigorous description, are considered in Sect. 7.4, below. The crystal field splittings in the case of several electrons in an open d–shell are also functions of Dq.

7.3.3 Spin Hamiltonian

In the common circumstance that the crystal field leaves an orbital singlet lying lowest, the orbital angular momentum is "quenched" and makes no first order contribution to the energy. However, the combined effects of the spin–orbit interaction and the crystal field in second order substantially modify the

Zeeman, hyperfine and quadrupole interactions. These effects can be conveniently summarized in the form of a "spin *Hamiltonian*". The spin *Hamiltonian* operates only within the degenerate ground manifold of the unperturbed *Hamiltonian* [the first line of (7.2.22)] but is contrived so that its eigenvalues are the same, to second order, as those of the full perturbation *Hamiltonian* [the second line of (7.2.22). It has the general form [7.9]

$$\mathcal{H}_S = \mu_B \mathbf{H} \cdot \tilde{g} \cdot \mathbf{S} + \mathbf{S} \cdot \tilde{D} \cdot \mathbf{S} + \mathbf{S} \cdot \tilde{A} \cdot \mathbf{I} + \mathbf{I} \cdot \tilde{Q} \cdot \mathbf{I} - g_n \mu_n \mathbf{H} \cdot \mathbf{I} + \mathbf{H} \cdot \tilde{\alpha} \cdot \mathbf{I}. \tag{7.3.19}$$

The dominant contributions to each of the tensors in (7.3.19) are as follows:

$$\tilde{g} = g_e \tilde{1} - 2\lambda \tilde{\Lambda}, \tag{7.3.20}$$

$$\tilde{D} = -\lambda^2 \tilde{\Lambda} - \rho \tilde{1}, \tag{7.3.21}$$

$$\tilde{A} = -g_e g_n \mu_B \mu_n < r^{-3} > (\kappa \tilde{1} + 3\xi \tilde{l}), \tag{7.3.22}$$

$$\tilde{Q} = q' \tilde{l}, \tag{7.3.23}$$

$$\tilde{\alpha} = 4\mu_B^2 <r^{-3}> \tilde{\Lambda}, \tag{7.3.24}$$

$$\tilde{\Lambda} = \sum_{n \neq 0} \frac{\langle 0|\mathbf{L}|n\rangle \langle n|\mathbf{L}|0\rangle}{(W_n - W_0)}, \tag{7.3.25}$$

$$\tilde{l} = \frac{1}{2}\langle 0|(\mathbf{LL} + (\mathbf{LL})_c)|0\rangle - \frac{1}{3}L(L+1)\tilde{1}, \tag{7.3.26}$$

$$\xi = \frac{2l + 1 - 4S}{S(2l-1)(2l+3)(2L-1)}, \tag{7.3.27}$$

$$q' = \frac{\pm 3e^2 Q}{2I(2I-1)} \frac{2(2l-1-4S)}{(2l-1)(2l+3)(2L-1)} < r^{-3} >, \tag{7.3.28}$$

$$\rho = \frac{-4}{7(2L-l)} \left[\left(\frac{5}{S} - 4\right) p + \frac{1}{7}\left(62 - \frac{100}{S}\right) q \right]^2 \mu_B^2, \tag{7.3.29}$$

$$p = \int_0^\infty \frac{l}{r} R^2(r) dr \int_0^r r'^2 R^2(r') dr', \tag{7.3.30}$$

$$q = \int_0^\infty \frac{l}{r^3} R^2(r) dr \int_0^r r'^4 R^2(r') dr', \tag{7.3.31}$$

where the upper sign in (7.3.28) applies for a less than half filled shell, and the lower sign for more than half. The first term in the $\tilde{D}$ tensor, (7.3.21), arises from the second order effect of spin–orbit interaction and crystal field, which also accounts for the deviation of the $\tilde{g}$ tensor from its free electron value. The second term is the first order contribution of spin–spin interaction. The tensor $\tilde{A}$ incorporates both isotropic (contact) and anisotropic hyperfine

Table 7.1. Values of the proportionality constant α.*

$L \backslash \Gamma$	T_1	T_2
1	1	—
2	—	-1
3	-3/2	1/2
4	1/2	5/2

*Reference [7.9] (Appendix B, Table 4, p. 857).

interactions. The contact term is an indirect effect when the open shell orbitals vanish at the nucleus. It can then be described equivalently in terms
of an admixture of excited configurations or of the polarization of the occupied core orbitals by exchange interaction with electrons in the open shell;
these effects are represented by the empirical parameter κ. The quadrupole
interaction is included in the $\tilde{Q}$ tensor, and $\tilde{\alpha}$, the small correction to the
nuclear *Zeeman* term, arises from the second order effect of orbital *Zeeman*
and orbital hyperfine interactions. Several higher order contributions of cross
terms involving these interactions have been omitted from the equations.

A different situation arises when an orbitally degenerate state lies lowest.
This case is exemplified by an orbital triplet ground state, T_1 or T_2 in cubic
symmetry. The matrix of the orbital angular operator $\mathbf{L}$ within the triply
degenerate manifold can be represented by $\alpha \mathbf{l}'$, where $\mathbf{l}'$ is a fictitious orbital
angular momentum with $l' = 1$ and α is a numerical factor which depends on
L and on the irreducible representation Γ, as shown in Table 7.1.

The spin–orbit interaction is then represented by $\alpha \lambda \mathbf{l}' \cdot \mathbf{S}$. Fine–structure
levels are labeled by a fictitious total angular momentum quantum number
$\mathbf{J}'$ where $\mathbf{J}' = \mathbf{l}' + \mathbf{S}$, and, in analaogy with (7.3.4), their energies are given by

$$E(J') = \frac{1}{2}\alpha\lambda[J'(J'+1) - S(S+1) - 2]. \tag{7.3.32}$$

These fine–structure levels are split further by higher order effects of the
crystal field, as well as by *Zeeman*, hyperfine and quadrupole interactions.
Nevertheless, the quantum number J' corresponding to the lowest value of
$E(J')$ labels a set of states whose energies lie within a few cm^{-1} of one another, and which are well separated in energy, typically by hundreds of cm^{-1},
from higher excited states. The energies of these low lying states can also
be represented by the eigenvalues of a spin *Hamiltonian* which operates only
within their manifold, but with a major difference: The spin $\mathbf{S}$ is now an effective spin $\mathbf{S}'$, where $\mathbf{S}'$ is identified with $\mathbf{J}'$. The *Zeeman* term then has the

form $g'\mu_B \mathbf{H} \cdot \mathbf{S'}$, where, in analogy with (7.3.6), g' is given by

$$g' = \left(\frac{g'_l + g_e}{2}\right) + \left(\frac{g'_l - g_e}{2}\right)\left(\frac{l'(l'+1) - S(S+1)}{J'(J'+1)}\right), \qquad (7.3.33)$$

and g'_l is the orbital g factor, identified with the proportionality constant α in this case.

An alternative approach to the explicit inclusion of lower symmetry crystal field effects and of hyperfine and quadrupole interactions, correct to the second order of perturbation theory, is to adopt the most general form of spin *Hamiltonian* compatible with symmetry considerations, incorporating appropriate adjustable parameters [7.10, 11]. This form is a symmetry–adapted polynomial of degree $2S'+1$ in $\mathbf{S'}$ and $2I+1$ in $\mathbf{I}$, but is customarily limited to linear and quadratic terms in $\mathbf{H}$. It includes contributions from orders of perturbation theory higher than the second, but their adjusted coefficients tend to be correspondingly small. The generalized spin *Hamiltonian* has the advantage of permitting a more precise fit to spin–resonance spectra, but it also has the disadvantage that the physical interpretation of its various terms is obscured.

7.4 Covalency and Superhyperfine Interaction

In the preceding section, we treated the interaction of a substitutional paramagnetic impurity ion with its crystalline environment, in terms of an electrostatic potential $V_c(\mathbf{r})$, given by (7.3.18). A more rigorous description of the interaction of a paramagnetic ion with its neighbors (ligands) takes account of their electronic structure as well. This description is accomplished by consideration of a molecular complex, and leads to subtle modifications of the values of spin *Hamiltonian* parameters. However, a more significant manifestation of ion–ligand interactions is the phenomenon of superhyperfine structure. When magnetic interactions with ligands are taken into account, the spin *Hamiltonian* of (7.3.19) is augmented by superhyperfine, ligand quadrupole and ligand nuclear *Zeeman* terms, given by

$$\mathcal{H}_{shf} + \mathcal{H}_{LQ} + \mathcal{H}_{LNZ} = \sum_j (\mathbf{S} \cdot \tilde{A}_j \cdot \mathbf{I}_j + \mathbf{I}_j \cdot \tilde{Q}_j \cdot \mathbf{I}_j + g_{nj}\mu_n \mathbf{H} \cdot \mathbf{I}_j),$$

$$(7.4.1)$$

where the sums are over the nuclei of neighboring ions. These relatively weak interactions, which are sometimes apparent as superhyperfine structure of EPR spectra, provide the entire content of ENDOR spectroscopy. They are also intimately related to the central theme of this book, since they embody a wealth of information about the structures of point imperfections in solids. The present section is concerned with the formal description of superhyperfine and ligand quadrupole interactions in terms of a molecular orbital picture of the complex, often called ligand field theory.

7.4.1 Molecular Orbitals and Configuration Mixing

In order to describe covalency effects, bonding and antibonding molecular orbitals, ψ^b and ψ^a, respectively, are each constructed from a metal (central ion) orbital ϕ and a symmetry–adapted combination of ligand orbitals χ, as follows [7.12,13]:

$$\psi^a = (1 - 2\lambda S + \lambda^2)^{-1/2}(\phi - \lambda\chi), \tag{7.4.2}$$

$$\psi^b = (1 + 2\gamma S + \gamma^2)^{-1/2}(\chi + \gamma\phi), \tag{7.4.3}$$

$$S = \langle\phi|\chi\rangle, \tag{7.4.4}$$

$$\lambda = S + \gamma. \tag{7.4.5}$$

In principle, the covalency parameter γ is determined variationally by minimizing the total energy. All of the bonding orbitals are doubly occupied in the ground state; the unpaired spins of electrons in singly occupied antibonding orbitals are usually considered to be responsible for magnetic resonance, although a contrary view [7.14] has also been proposed. For example, the ground state wave function Ψ for a three–electron system can be represented in terms of the orthogonal molecular orbitals by a single *Slater* determinant,

$$\Psi = \frac{1}{\sqrt{3!}} \det |\psi_{\uparrow}^a, \psi_{\uparrow}^b, \psi_{\downarrow}^b|. \tag{7.4.6}$$

The $\gamma = 0$ case, corresponding to the absence of covalency, is designated the ionic model. In that case, the bonding orbital ψ^b is a pure ligand orbital; however, the antibonding orbital is not a pure metal orbital, but instead retains an admixture of ligand orbital as a consequence of the non orthogonality of the basis functions ϕ and χ.

Covalency can be described alternatively in terms of configuration mixing. The ionic model for a three–electron system can be represented by the single *Slater* determinant

$$\Psi_G = \frac{1}{3!(1 - S^2)} \det |\phi_{\uparrow}, \chi_{\uparrow}, \chi_{\downarrow}|. \tag{7.4.7}$$

An improved ground state wave function Ψ can be constructed by admixture of an excited charge–transfer configuration Ψ_E, .

$$\Psi = \frac{1}{(1 + 2\gamma S + \gamma^2)} \cdot (\Psi_G + \gamma\Psi_E), \tag{7.4.8}$$

$$\Psi_E = \frac{1}{3!(1 - S^2)} \det |\phi_{\uparrow}, \chi_{\uparrow}, \phi_{\downarrow}|, \tag{7.4.9}$$

$$\gamma = \frac{\langle\Psi_G|\mathcal{H}|\Psi_E\rangle - \langle\Psi_G|\Psi_E\rangle\langle\Psi_G|\mathcal{H}|\Psi_E\rangle}{[\langle\Psi_G|\mathcal{H}|\Psi_G\rangle - \langle\Psi_E|\mathcal{H}|\Psi_E\rangle]}. \tag{7.4.10}$$

Note that the charge is transferred *from* the ligand *to* the metal in the charge–transfer configuration Ψ_E. (7.4.6) and (7.4.8) are alternative expressions for precisely the same wave function, Ψ. Thus, in this case, the description in terms of configuration mixing is completely equivalent to the molecular orbital description; however, it has the advantage of greater generality, since other excited configurations are possible whose admixture could not be reduced to a molecular orbital description. Configuration mixing can also facilitate the description of hyperfine interactions which are transferred to more remote ligands by intervening ions.

7.4.2 Superhyperfine Interaction

The unpaired spin density associated with the wave function of (7.4.6) is

$$\rho_\uparrow - \rho_\downarrow = \frac{|\phi|^2 - \lambda(\phi^*\chi + \chi^*\phi) + \lambda^2|\chi|^2}{(1 - 2\lambda S + \lambda^2)}. \tag{7.4.11}$$

The distribution of unpaired spin density between metal and ligand orbitals can be summarized by coefficients f_m and f_l, defined by

$$f_m = \frac{1}{(1 - 2\lambda S + \lambda^2)}, \tag{7.4.12}$$

$$f_l = \frac{\lambda^2}{(1 - 2\lambda S + \lambda^2)}. \tag{7.4.13}$$

Note that overlap increases the unpaired spin density on both metal and ligand ions, while covalency further increases it on the ligands, but diminishes it on the metal. There is also unpaired spin density associated with the overlap term in (7.4.11), but this component makes a negligible contribution to the hyperfine interaction.

The coefficient f_m appears as an additional multiplying factor in the central ion hyperfine term in the spin *Hamiltonian*; e.g., (7.3.22). The coefficient f_l is a measure of the strength of that component of the superhyperfine interaction which is associated with unpaired spin in ligand orbitals. The nature of this interaction depends on the type of ligand orbital involved: A ligand s–orbital contributes a contact hyperfine interaction proportional to its spin density at the nucleus, while a ligand p–orbital has both an anisotropic hyperfine interaction and an indirect contact interaction via core polarization. The molecular orbital can be generalized to allow linear combination of several symmetry–adapted ligand orbitals χ_i, with one metal orbital ϕ.

The molecular orbital description can be further extended to include more than one singly occupied antibonding orbital. As a consequence of the *Wigner–Eckart* theorem, the contribution to the spin *Hamiltonian* of the superhyperfine interaction with ligand j is given by

$$\mathcal{H}_{shf_j} = [a_j \mathbf{S}' \cdot \mathbf{I}_j + b_j(2S'_z I_{jz} - S'_x I_{jx} - S'_y I_{jy})], \tag{7.4.14}$$

where the z direction is along the line joining the metal ion to ligand j. Note that the coefficients a_j and b_j should not be confused with the a and b introduced in Sect. 7.3.1, in connection with the hyperfine structure of the free ion, which have different meanings. These coefficients are determined by calculating and equating matrix elements of (7.4.14) with those of the superhyperfine operator

$$
\mathcal{H}_{shf_j} = g_e g_n \mu_B \mu_n \sum_i \left[\frac{\mathbf{I}_j \cdot (\mathbf{l}_i - \mathbf{s}_i)}{r_i^3} + \frac{3(\mathbf{r}_i \cdot \mathbf{I}_j)(\mathbf{r}_i \cdot \mathbf{s}_i)}{r_i^5} \right.
$$
$$
\left. + \frac{8\pi}{3} \delta(\mathbf{r}_i) \mathbf{I}_j \cdot \mathbf{s}_i \right] \tag{7.4.15}
$$

in a common reference state. This calculation is facilitated by the fact that $\mathcal{H}_{SHF_j}$ is a one–electron operator. In the special case that a state of maximum effective spin projection can be represented by a single *Slater* determinant, e.g., for an orbitally non degenerate *Hund's* rule ground state, the coefficients are given by

$$
a_j = \frac{g_e g_{nj} \mu_B \mu_n}{2S'} \left[\frac{8\pi}{3} f_{sj} |s_j(0)|^2 - \kappa_j (f_{\sigma j} + 2f_{\pi j}) <r^{-3}>_{pj} \right], \tag{7.4.16}
$$

$$
b_j = \frac{g_e g_{nj} \mu_B \mu_n}{2S'} \left[\frac{2}{5} (f_{\sigma j} - f_{\pi j}) <r^{-3}>_{pj}) \right] + \frac{g_e g_{nj} \mu_B \mu_n}{R^3}, \tag{7.4.17}
$$

$$
f_{\alpha j} = \sum_\zeta f_{\alpha j \zeta}, \tag{7.4.18}
$$

where g_{nj} is the nuclear g–value appropriate to the nucleus of ligand j, ζ denotes a particular row of a particular irreducible representation, and the sum is over all values of ζ, for which ψ_ζ^a is singly occupied. The orbitals $p_{\sigma j}$ and $p_{\pi j}$ are directed parallel and perpendicular, respectively, to the line joining the metal with ligand j, and the total wave function is assumed to have axial symmetry about this line. The spin–density coefficient $f_{\alpha j \zeta}$ of ligand atomic orbital $\chi_{\alpha j}$ in antibonding orbital ψ_ζ^a is the appropriate generalization of f_l in (7.4.13).

The quantities $|s_j(0)|^2$ and $<r^{-3}>_{pj}^2$ play a ubiquitous role in the theory of superhyperfine and quadrupole interactions. Some typical values of these quantities calculated from free ion orbitals [7.15] for several common constituents of ionic crystals are listed in Appendix H. These free ion values are expected to be only semi–quantitatively reliable for application to ions in crystals. Two caveats are in order: (1) Analytic atomic orbitals which have been optimized for energy calculations in either atoms or molecules are not necessarily suitable for the computation of these quantities whose values depend critically on details of the wave function very close to the nucleus. Contracted *Gaussian* basis functions are particularly unsuitable. (2) For heavy ions, the calculated values are also subject to substantial relativistic corrections. Although theoretical formulas for these correction factors are available [7.16] a

s(e_g)

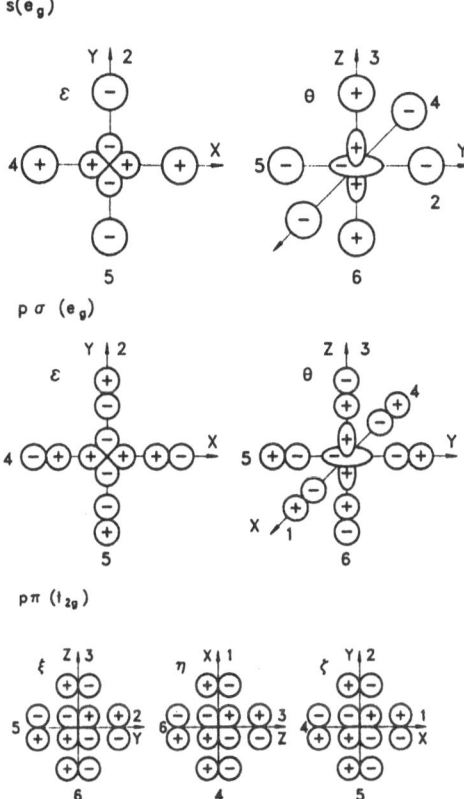

$p\sigma$ (e_g)

$p\pi$ (t_{2g})

Fig. 7.1. Symmetry–adapted metal and ligand orbitals for an octahedrally coordinated transition metal ion

particularly simple and convenient empirical formula for the correction factor γ for the spin density at the nucleus, $|s(0)|^2$, has been proposed in the form [7.17]

$$\gamma = 1 + (3.76 \times 10^{-6})Z^3, \tag{7.4.19}$$

where Z is the atomic number of the nucleus.

As an example of the application of these formulas, consider an MX_6 octahedral complex, where M is a transition metal ion and the ligand X is an anion such as F^- or Cl^-. Symmetry–adapted combinations of ligand s, $p\sigma$ and $p\pi$ orbitals are illustrated in Fig. 7.1, and antibonding molecular orbitals constructed by combining them with metal d–orbitals are listed in Table 7.2. The conditions assumed in the derivation of (7.4.16–18) are satisfied for configurations d^3 and d^8, as well as for d^5 in the high–spin state, $S = 5/2$, which occurs for a sufficiently small value of Dq. Coefficients f_{sj} and $f_{\sigma j}$ vanish

Table 7.2. Symmetry–adapted antibonding orbitals of an MX_6 octahedral complex

$$e_g: \quad \psi_\theta = N_\sigma\{d_{z^2} - (12)^{-1/2}\lambda_s(2s_3 + 2s_6 - s_1 - s_4 - s_2 - s_5)$$
$$-(12)^{-1/2}\lambda_\sigma(-2z_3 + 2z_6 + x_1 - x_4 + y_2 - y_5)\}$$

$$\psi_\epsilon = N_\sigma\{d_{x^2-y^2} - 2^{-1/2}\lambda_s(s_1 + s_4 - s_2 - s_5)$$
$$-2^{-1/2}\lambda_\sigma(-x_1 + x_4 + y_2 - y_5)\}$$

$$N_\sigma = [1 + 4\lambda_\sigma<d_{x^2-y^2}|x_1> - 4\lambda_s<d_{x^2-y^2}|s_1> + \lambda_\sigma^2 + \lambda_s^2]^{-1/2}$$

$$t_{2g}: \quad \psi_\xi = N_\pi\{d_{yz} - \frac{1}{2}\lambda_\pi(z_2 - z_5 + y_3 - y_6)\}$$

$$\psi_\eta = N_\pi\{d_{zx} - \frac{1}{2}\lambda_\pi(x_3 - x_6 + z_1 - z_4)\}$$

$$\pi_\zeta = N_\pi\{d_{xy} - \frac{1}{2}\lambda_\pi(y_1 - y_4 + x_2 - x_5)\}$$

$$N_\pi = [1 - 4\lambda_\pi<d_{xy}|y_1> + \lambda_\pi^2]^{-1/2}$$

in a d^3 configuration (e.g., V^{2+}, Cr^{3+}), since only π antibonding orbitals are singly occupied.

Similarly, $f_{\pi j}$ vanishes in a d^8 configuration (e.g., Ni^{2+}, Cu^{3+}) in which only σ antibonding orbitals are singly occupied. All of the coefficients are non–vanishing in a d^5 configuration (e.g., Mn^{2+}, Fe^{3+}) and have the values $f_s = \frac{1}{3}N_\sigma^2\lambda_s^2$, $f_\sigma = \frac{1}{3}N_\sigma^2\lambda_\sigma^2$ and $f_\pi = \frac{1}{4}N_\pi^2\lambda_\pi^2$, in terms of the wave function parameters defined in Table 7.2.

The last term in (7.4.17) arises from the direct dipolar interaction of the total electron–spin magnetic moment on the metal ion with the ligand nuclear magnetic moment. The isotropic superhyperfine interaction ordinarily grossly predominates over the anisotropic; however, for ligands sufficiently far removed from the metal ion that overlap and covalency effects that are negligible, this dipolar term is the only surviving contribution to the superhyperfine interaction. Its diagnostic importance lies in the information it conveys concerning the metal–ligand distance R. As a practical matter, the inequality $a < b$ is the criterion for applicability of the dipolar approximation. The dipolar approximation may be improved by taking account of the spatial distribution of the unpaired spin (multipole correction) [7.18,19].

7.4.3 Ligand Core Polarization

The second term in (7.4.16) arises from polarization of the ligand core by its valence p–orbital. Polarization of ligand cores can arise not only from admixtures of excited configurations involving higher ligand orbitals, but also those involving higher metal orbitals. The latter mechanism was investigated theoretically by *Adrian et al.* [7.20]. In terms of the present nomenclature, the dominant indirect contribution to contact hyperfine interaction has the form

$$\Delta a_j = \left(\frac{8\pi}{3} g_e g_{nj} \mu_B \mu_n\right) \left(\frac{-2e^2}{\epsilon}\right) \langle \chi_\zeta | \delta(\mathbf{r}_j) | \chi_\zeta \rangle$$

$$\times |\langle \chi_\zeta | \phi_\zeta \rangle|^2 \langle \phi_\zeta \phi_\eta | r_{12}^{-1} | \phi_\eta \phi_\zeta \rangle, \tag{7.4.20}$$

where ϵ is an average excitation energy. The exchange integral involving open shell orbitals, ϕ_ζ and ϕ_η, survives only for parallel spins, and thus mediates a different admixture of ligand orbitals χ_ζ for up and down spins, with consequent core polarization. This mechanism was invoked in a rather different context, in which the "metal" is an O^- ion with the hole in a $p\pi$ orbital, and the "ligand" is a substitutional Al^{3+} impurity ion [7.21].

In recent years, computational technology has progressed to the point that molecular clusters can be modeled by *ab–initio* self–consistent field, molecular orbital calculations [7.22]. The computational method is based on the variational principle, in which the energy is given approximately by

$$E_t = \frac{\langle \Psi_t | \mathcal{H} | \Psi_t \rangle}{\langle \Psi_t | \Psi_t \rangle}, \tag{7.4.21}$$

where Ψ_t is a trial function. This expression provides an upper bound on the ground state energy whose error is quadratic in the wave function error; consequently, the approximation can be optimized by varying Ψ_t to minimize E_t. In the case of a closed shell configuration, one adopts a trial function Ψ_t in the form of a single *Slater* determinant in molecular spin orbitals. In the standard spin–restricted *Hartree–Fock* (RHF) method, the molecular spin orbitals are further constrained by associating each molecular orbital with two spin functions. Variation of the molecular orbitals $\psi_k(\mathbf{r})$, subject to the constraint that they remain orthonormal, leads to the *Hartree–Fock* equations,

$$\mathcal{F}\psi_k(\mathbf{r}) = \left[-\frac{\hbar^2}{2m}\nabla^2 + V(\mathbf{r}) + 2e^2 \sum_i \int \frac{|\psi_i(\mathbf{r}')|}{|\mathbf{r} - \mathbf{r}'|} d\tau'\right]\psi_k(\mathbf{r})$$

$$-e^2 \sum_l \int \frac{\psi_l^*(\mathbf{r}')\psi_t(\mathbf{r}')}{|\mathbf{r} - \mathbf{r}'|} d\tau' \psi_l(\mathbf{r})$$

$$= \epsilon_k \psi_k(\mathbf{r}), \tag{7.4.22}$$

where the *Fock* operator $\mathcal{F}$ plays a role analogous to that of the one–electron *Hamiltonian* h of (7.3.2), and the diagonal *Lagrange* multiplier ϵ_k serves in some sense as its energy eigenvalue. $V(\mathbf{r})$ in (7.4.22) is the nuclear potential energy, and the integrals embody the *Coulomb* and exchange interactions with other electrons.

In the linear combination of atomic orbitals (LCAO) method, in which each molecular orbital $\psi_k(\mathbf{r})$ is expressed as a linear combination of non orthogonal basis functions, "atomic" orbitals $\chi_j(\mathbf{r})$, the expansion coefficients C_{jk} satisfy the *Roothaan* equation [7.23]

$$\sum_j F_{ij} C_{jk} = \sum_j S_{ij} C_{jk} \epsilon_k, \qquad (7.4.23)$$

where $S_{ij}(= \langle \chi_i | \chi_j \rangle)$ is an overlap integral. Since the elements of the *Fock* matrix $F_{ij}(= \langle \chi_i | \mathcal{F} | \chi_j \rangle)$ depend implicitly on the occupied molecular orbitals, (7.4.23) must be solved by iterating to self consistency.

The *Hartree–Fock* equations can be modified to accomodate open–shell configurations of interest in the present context. *Ab–initio* electronic structure calculations directed toward the explanation of hyperfine structure, including core polarization, have been performed in recent years. These include spin–unrestricted *Hartree–Fock* (UHF) calculations which incorporate different molecular orbitals for different spins in a single *Slater* determinant. Equations similar to (7.4.23) are obtained for each spin projection, and must be solved simultaneously. The UHF method yields states which are not eigenfunctions of the total spin, and consequently tends to exaggerate core polarization. A refinement of the method, projected unrestricted *Hartree–Fock* (PUHF), corrects for spin contamination, with improved results [7.24].

The most recent efforts have been concentrated on multi–configuration, self–consistent field (MCSCF) calculations. In this approach, the trial function is expressed as a linear combination of *Slater* determinants from an appropriately chosen mixture of configurations, and both orbitals and mixing coefficients are optimized for minimum energy. This method appears to provide satisfactory predictions of hyperfine constants in calculations on small radicals [7.25, 7.26].

7.4.4 Ligand Quadrupole Interaction

The quadrupole interaction *Hamiltonian* $\mathcal{H}_Q$ of (7.2.18), which is expressed as a function of the electron coordinates, comprises the potential energy of the electrons in the field of the nuclear quadrupole moment. Alternatively, one can view the same interaction as the energy of orientation of a nuclear quadrupole moment in the electric field gradient of the electrons. In the principal axis coordinate system, the expectation value of the quadrupole *Hamiltonian* in the electronic state Ψ_e is given by [7.27]

$$\langle \Psi_e | \mathcal{H}_Q | \Psi_e \rangle = \frac{e^2 q Q}{4I(2I-1)} \left[3I_z^2 - I(I+1) + \frac{1}{2}\eta(I_+^2 + I_-^2) \right], \quad (7.4.24)$$

$$eq \equiv \frac{\partial^2 V}{\partial z^2}, \quad (7.4.25)$$

$$\eta \equiv \frac{\frac{\partial^2 V}{\partial x^2} - \frac{\partial^2 V}{\partial y^2}}{\frac{\partial^2 V}{\partial z^2}}. \quad (7.4.26)$$

The advantage of this alternative viewpoint is that one can include additional sources of the electric field gradient.

The contribution of the ion's own electronic structure to the electric–field gradient is determined by evaluating the left–hand side of (7.4.24), with $\mathcal{H}_Q$ given by (7.2.18), and with the help of *Wigner–Eckart* theorem. The resulting expression for eq is then a function of the quantum numbers which characterize the ground electronic state. The electric field gradient arises from the total charge distribution and not just from the unpaired spin density; however, completely filled shells make no direct contribution, since their charge density is spherically symmetric. An indirect contribution arising from the distortion of closed shells is discussed below.

Ligand quadrupole interaction reflects not only the extra electronic charge on the ligand associated with antibonding orbitals, but also depletion of charge on the ligand associated with transfer of charge from the ligand to the metal in the bonding orbitals. For example, the total electron density associated with the ground state wave function of (7.4.2,3) is given by

$$\rho_\uparrow + \rho_\downarrow = |\psi^a|^2 + 2|\psi^b|^2 = \frac{|\phi|^2 - \lambda(\phi^*\chi + \chi^*\phi) + \lambda^2|\chi|^2}{(1 - 2\lambda S + \lambda^2)}$$
$$+ 2\left[\frac{|\chi|^2 + \gamma(\chi^*\phi + \phi^*\chi) + \gamma^2|\phi|^2}{(1 + 2\gamma S + \gamma^2)} \right]. \quad (7.4.27)$$

Thus, the fractional change in ligand charge introduced by overlap and covalency effects is given by $f_l - 2h_l$, where the coefficient h_l is defined by

$$h_l = -\frac{1}{(1 + 2\gamma S + \gamma^2)}. \quad (7.4.28)$$

Note that h_l and f_l are approximately equal in the limit $S \ll \gamma$ [7.28].

The value of q for a single electron in an atomic orbital with angular momentum quantum numbers l and m is

$$q_{lm} = \frac{2[l(l+1) - 3m^2]}{(2l-1)(2l+3)} \langle r^{-3} \rangle_l. \quad (7.4.29)$$

It follows that there is no contribution from s–orbitals, and that the values of q for $p\sigma$ and $p\pi$ orbitals are given by

$$q_{p\sigma} = -2q_{p\pi} = -\frac{4}{5}<r^{-3}>_p. \qquad (7.4.30)$$

In the special case that a state of maximum effective spin projection can be represented by a single *Slater* determinant, the contribution of the ligand quadrupole interaction to the spin *Hamiltonian* is

$$\mathcal{H}_{LQj} = Q_{j\parallel}\left[I_{jz}^2 - \frac{1}{3}I_j(I_j+1)\right], \qquad (7.4.31)$$

$$Q_{j\parallel} = \frac{3e^2Q}{5I_j(2I_j-1)}[f_{\pi j} - f_{\sigma j} - 2(h_{\pi j} - h_{\sigma j}]<r^{-3}>_p, \qquad (7.4.32)$$

$$h_{\alpha j} = \sum_\zeta h_{\alpha j\zeta}, \qquad (7.4.33)$$

where $h_{\alpha j\zeta}$ is the appropriate generalization of h_l in (7.4.28), and the sum over ζ is over all doubly occupied bonding orbitals. Note that the quadrupole contribution to the spin *Hamiltonian* (7.4.31), vanishes unless $I_j \geq 1$.

A secondary contribution to the electric field gradient arises from the charge distribution of neighboring ions. This contribution should be inherently weaker, since the ionic charges are substantially further removed from the ligand nucleus than is its own electronic charge. However, it is amplified appreciably through distortion of the closed shell core orbitals by the ionic charge distribution. The contribution of the open shell electrons is similarly modified by distortion of the core. Thus, the total electric field gradient is

$$eq = eq_c(1 - \gamma_\infty) + eq_v(1 - R_q), \qquad (7.4.34)$$

where q_c is the crystal field contribution, q_v is the contribution of valence electrons, and γ_∞ and R_q are *Sternheimer* antishielding factors. Calculations show that γ_∞ is negative and may be as large as -100, whereas R_q may have either sign and is much smaller, $|R_q| < 0.2$. The latter parameter may be absorbed in an effective radial integral $< r_q^{-3} >$, defined by

$$<r_q^{-3}>\equiv<r^{-3}> (1 - R_q). \qquad (7.4.35)$$

7.4.5 Pseudopotentials

A frozen–core approximation greatly facilitates molecular orbital calculations on clusters containing relatively heavy ions and, as will be seen, defect calculations by other methods as well. In this approximation, occupied atomic core orbitals are assumed to be known and fixed, independent of valence orbitals

and their crystalline environment. Exploitation of the frozen–core approximation in the orthogonalized–plane–wave method of electronic band–structure
calculation inspired a re–formulation of the problem by *Phillips* and *Kleinman*
[7.29] known as pseudopotential theory. The *Schroedinger* equation satisfied
by the valence orbitals,

$$\mathcal{H}\psi_v = (T + V)\psi_v = E_v\psi_v, \tag{7.4.36}$$

can be transformed by introducing a smooth pseudo–wave function ϕ such
that

$$\psi_v = N(1 - \mathcal{P})\phi, \tag{7.4.37}$$

$$\mathcal{P} = \sum_c |\psi_c\rangle\langle\psi_c|, \tag{7.4.38}$$

where N is a normalization constant and ψ_c is an occupied core orbital,
assumed to be known. It is always possible to construct ϕ in the form

$$\phi = \frac{1}{N}\left(\psi_v + \sum_c a_c\psi_c\right). \tag{7.4.39}$$

Substitution of (7.4.37) into (7.4.36) leads to an eigenvalue equation for the
pseudo–wave function,

$$[\mathcal{H} + (E_v - \mathcal{H})\mathcal{P}]\phi \equiv (T + V_p)\phi = E_v\phi. \tag{7.4.40}$$

The exact pseudopotential V_p, given by

$$V_p = V + (E_v - \mathcal{H})\mathcal{P}, \tag{7.4.41}$$

is a relatively intractable non–local operator, and (7.4.40) is satisfied by any
pseudo–wave function of the form (7.4.39). However, model pseudopotentials
incorporating varying degrees of energy and angular momentum dependence
have been successfully employed in approximate calculations.

A particular form of local model pseudopotential has found extensive application in molecular orbital calculations[7.30,31]. One proceeds by construc-
ting a smooth pseudo–wave function ϕ^R with no internal radial nodes from
the exact one–electron orbitals for a reference state R of an atom. The corresponding model pseudopotential, called an effective core potential (ECP),
is derived by inverting the *Schroedinger* equation,

$$V_p^{ECP}(\mathbf{r}) = \frac{(E_v^R - T)\phi^R(\mathbf{r})}{\phi^R(\mathbf{r})}. \tag{7.4.42}$$

A distinct reference state is chosen and an effective core potential constructed
for each value of the angular momentum quantum number l.

The molecular energy levels calculated with effective core potentials are excellent approximations to those determined by all–electron calculations. However, the corresponding molecular orbitals are actually pseudo–wave functions, which lack much of the structure of the true molecular orbitals within the atomic cores. An important caveat is that the hyperfine and superhyperfine interactions must be calculated with the true molecular orbitals ψ_v, recovered by orthogonalizing the pseudo–molecular orbitals ϕ to the occupied core orbitals as indicated in (7.4.37,38). Hyperfine interactions calculated from ϕ alone may be in error by orders of magnitude. This caveat will be emphasized in subsequent sections as well.

7.4.6 Lattice Dynamical Effects

In the *Born–Oppenheimer* approximation, the electronic wave function adjusts to the instantaneous positions of both metal and ligand ions. Consequently, both bonding and antibonding orbitals are modulated by lattice vibrations, primarily through the parameters S and γ. Since phonon lifetimes are short compared with the lifetimes of electronic states and with the time constants involved in magnetic resonance experiments, this modulation is manifest in average values of the parameters S and γ, which differ significantly from static values appropriate to the average ion positions by virtue of the extreme nonlinearity of their dependence on metal–ligand distance. Its effects are apparent primarily in the temperature dependence of superhyperfine interactions and in relaxation of symmetry restrictions [7.32]. An example of the latter effect is the bonding of t_{2g} metal orbitals with s ligand orbitals, enabled by transient asymmetrical displacements, with consequent enhancement of contact hyperfine interactions.

7.5 Orthogonalized Envelope Functions

As noted previously, the substitutional transition–metal impurity ions considered above are characterized by compact wave functions, and their superhyperfine interactions are limited to a few close neighbors. At the opposite extreme are shallow impurity states in semiconductors, with exceedingly diffuse wave functions. Color centers in insulators and deep–level impurities in semiconductors occupy intermediate positions; their wave functions are comparable in extent with the lattice parameter. Descriptions appropriate to these more diffuse systems are considered in this section.

7.5.1 Wannier's Theorem and Effective–Mass Theory

The diffuse wave functions associated with shallow impurity states strongly reflect the electronic structure of the host crystal; it is, therefore, convenient to start with host crystal band functions. These are constrained by translational symmetry to be *Bloch* functions,

$$\psi^{(\mu)}(\mathbf{k}, \mathbf{r}) = \exp(i\mathbf{k} \cdot \mathbf{r})u^{(\mu)}(\mathbf{k}, \mathbf{r}), \tag{7.5.1}$$

$$u^{(\mu)}(\mathbf{k}, \mathbf{r} + \mathbf{R}) = u^{(\mu)}(\mathbf{k}, \mathbf{r}), \tag{7.5.2}$$

where $\mathbf{R}$ is a direct lattice vector and $\mathbf{k}$ is a wave vector in the first *Brillouin* zone of the reciprocal lattice, consistent with periodic boundary conditions. The index μ denotes a particular energy band with energy levels $E^{(\mu)}(\mathbf{k})$.

A *Wannier* function $a^{(\mu)}(\mathbf{r} - \mathbf{R})$, defined by

$$a^{(\mu)}(\mathbf{r} - \mathbf{R}) \equiv \frac{1}{\sqrt{N}} \sum_{\mathbf{k}} \exp(-i\mathbf{k} \cdot \mathbf{R})\psi^{(\mu)}(\mathbf{k}, \mathbf{r}), \tag{7.5.3}$$

can be associated with each lattice point. *Wannier* functions are localized near lattice points, and those belonging to different bands or different lattice points are orthogonal. Although these functions are associated with specific bands, they are not energy eigenfunctions. Consider a slowly varying perturbing potential $U(\mathbf{r})$. Expansion of the perturbed wave function $\psi(\mathbf{r})$ in terms of *Wannier* functions,

$$\psi(\mathbf{r}) = \frac{1}{\sqrt{N}} \sum_{\mu} \sum_{\mathbf{R}} a^{(\mu)}(\mathbf{r} - \mathbf{R})\phi^{(\mu)}(\mathbf{R}), \tag{7.5.4}$$

leads to the differential equation [7.33]

$$[E^{(\mu)}(-i\nabla) + U(\mathbf{R}) - E]\phi^{(\mu)}(\mathbf{R}) = 0, \tag{7.5.5}$$

where $\mathbf{R}$ is treated as a continuous variable. For the case of an excess electron in a semiconductor or insulator, $E^{(\mu)}(-i\nabla)$ may be expanded about the conduction band minimum, $\mathbf{k} = \mathbf{k}_0$, to obtain the approximate form

$$E^{(\mu)}(-i\nabla) \cong \frac{\hbar^2}{2}\tilde{m}^{*-1}\nabla\nabla, \tag{7.5.6}$$

where $\tilde{m}^{*-1}$ is a reciprocal effective–mass tensor defined by

$$\tilde{m}^{*-1} \equiv \frac{1}{\hbar}\frac{\partial^2 E^{(\mu)}(\mathbf{k})}{\partial\mathbf{k}\partial\mathbf{k}}\bigg|_{\mathbf{k}=\mathbf{k}_0}. \tag{7.5.7}$$

(7.5.4–7) express the essential content of *Wannier's* theorem.

7.5.2 Continuum Models

Consider a semiconducting crystal with cubic symmetry for which the conduction band minimum lies at the center of the first *Brillouin* zone (Γ-point). (7.5.4–7) can be specialized to a shallow donor impurity at the origin, $\mathbf{R} = 0$,

$$\left[-\frac{\hbar^2}{2m^*}\nabla^2 - \frac{e^2}{\kappa R} - C\delta(\mathbf{R}) - E \right] \phi(\mathbf{R}) = 0. \tag{7.5.8}$$

The periodic potential is absorbed in the effective–mass m^*, leaving an effective interaction with the ionized donor impurity. The long–range interaction is represented by the Coulomb potential of a point charge in a dielectric continuum with dielectric constant κ, and the central–cell correction by a model pseudopotential, $-C\delta(\mathbf{R})$. Solutions of (7.5.8) in the absence of the small central–cell correction, are hydrogen–atom wave functions with scales increased by the factor $\kappa m/m^*$ and binding energies diminished by the factor $m^*/m\kappa^2$.

It is instructive to consider an approximate solution in the presence of the central–cell correction. For this purpose, we employ the variational principle with a hydrogenic ground state trial function ϕ_t given by

$$\phi_t = \frac{\alpha^3}{\pi}\exp(-\alpha R), \tag{7.5.9}$$

where α is an adjustable parameter. The corresponding expression for the energy is

$$E_t = \frac{\hbar^2}{2m^*}\alpha^2 - \frac{e^2}{\kappa}\alpha - \frac{C\alpha^3}{\pi}. \tag{7.5.10}$$

This expression has a local minimum for a finite value of α provided that C satisfies the inequality

$$C < \frac{\pi\hbar^4\kappa}{12m^{*2}e^2}. \tag{7.5.11}$$

It follows that (7.5.11) is the condition for a shallow donor state; larger values of the central–cell correction lead to the collapse of the wave function to a much more compact scale appropriate to a deep–level impurity state, and thus, to violation of the effective–mass approximation on which (7.5.8) is based. From this point of view, a given defect state is either shallow or not; there is no reason to expect a continuum of intermediate cases.

It should be noted that the shallow donor states in silicon and germanium are rather more complicated than this discussion suggests, since the conduction bands in these materials have several equivalent minima away from zone center. Consequently, the effective–masses are anisotropic and the shallow donor ground state displays a degeneracy which is partially removed by the central–cell correction (valley–orbit splitting) [7.34].

Continuum models have also been employed extensively for color centers in ionic crystals. The approximation is demonstrably inconsistent for F center $1s$ ground states, since the majority of the unpaired spin is associated with the central–cell. On the other hand, an abundance of evidence supports the view that the relaxed excited $2s$ and $2p$ states of F centers are truly shallow states. A modification called the semi–continuum model, which has enjoyed considerable success, employs a square–well potential in the central–cell and a Coulomb tail in a dielectric continuum outside.

The relevant point about continuum models for the purpose of the present discussion is that the solution $\phi(\mathbf{R})$ of (7.5.8) is *not* the wave function, but is rather an envelope function whose values at lattice points are coefficients of *Wannier* functions in the expansion of the wave function $\psi(\mathbf{r})$, as indicated in (7.5.4). It is imperative that hyperfine and superhyperfine interactions be calculated with the true wave function $\psi(\mathbf{r})$ rather than with the envelope function $\phi(\mathbf{R})$. The essential difference between them arises from the requirement that conduction band *Wannier* functions be orthogonal to all of the valence band *Wannier* functions. Consequently, orthogonalization of the envelope function to occupied core orbitals provides a useful approximation of the true wave functions.

7.5.3 Point–Ion Model and Ion–Size Corrections

Another approach to the description of color centers in ionic crystals involves an extension of the crystal field concept. In the point–ion model of *Gourary* and *Adrian* [7.35], each excess electron moves in an electrostatic potential $V_{PI}(\mathbf{r})$ derived from an array of point–ions which represent the crystal, including defects. Since the electronic wave function of a color center is more extensive than those of a transition–metal ion, (7.3.18) for $V_c(\mathbf{r})$ must be extended to include solutions of the *Laplace* equation, beyond the first sphere of ions, which are singular at the origin,

$$V_{PI}(\mathbf{r}) = -\sum_{L=0}^{\infty} \sum_{M=-L}^{L} Y_L^M(\theta, \phi) \left[e_{LM}\, r^L + \sum_{s<} f_{LMs} \left(\frac{r_s^L}{r^{L+1}} - \frac{r^L}{r_s^{L+1}} \right) \right],$$

$$(7.5.12)$$

where the sum over $s_<$ means only over spherical shells of ions whose radius r_s is less than r. The point–ion potential has been interpreted as a model pseudopotential [7.36].

A correction for finite ion–size effects was introduced by *Bartram, Stone-ham* and *Gash* (BSG) [7.37] as a model pseudopotential of the form

$$V_p^{BSG} = V_{PI} + \sum_j C_j \delta(\mathbf{r} - \mathbf{R}_j), \qquad (7.5.13)$$

$$C_j = A_j + B_j(E - U_j), \qquad (7.5.14)$$

where U_j is the potential at the site of ion j due to the remaining ions, and the coefficients A_j and B_j are properties of the ion at site j. The ion–size correction is explicitly energy dependent. [In the original BSG formulation, $\bar{V}(= \langle \phi | V_p | \phi \rangle)$ appeared in (7.5.14) in place of E, which has been used in subsequent work; these forms are approximately equivalent for a smooth pseudo–wave function, but the energy–dependent form is more tractable.]

Since both V_{PI} and V_p^{BSG} are pseudopotentials, the corresponding solutions, $\phi(\mathbf{r})$, of the *Schroedinger* equation are pseudo–wave functions which must be orthogonalized to occupied core orbitals in order to recover the true wave function, $\psi_v(\mathbf{r})$, for the purpose of calculating hyperfine interactions.

7.5.4 Green's Function Method for Deep–Level Impurities

The continuum model for shallow impurity states has the advantageous feature that properties of the host lattice are incorporated in the effective mass tensor and the *Wannier* functions. The envelope function then reflects only the change in potential introduced by the impurity. Rapid progress has been made in recent years in developing a comparable description of deep–level impurity states. This description, which is based on a *Green's* function method introduced by *Koster* and *Slater* [7.38], has been elaborated by several investigators. The bound state wave functions satisfy the integral equation

$$\psi(\mathbf{r}) = \int G_E(\mathbf{r}, \mathbf{r}') U(\mathbf{r}') \psi(\mathbf{r}') d\tau', \qquad (7.5.15)$$

$$G_E(\mathbf{r}, \mathbf{r}') = \sum_\mu \sum_\mathbf{k} \frac{\psi^{(\mu)}(\mathbf{k}, \mathbf{r}) \psi^{(\mu)*}(\mathbf{k}, \mathbf{r}')}{[E - E^{(\mu)}(\mathbf{k})]}, \qquad (7.5.16)$$

where $U(\mathbf{r})$ is the change in potential due to the impurity, and $\psi^{(\mu)}(\mathbf{k}, \mathbf{r})$ is defined by (7.5.1,2). Note that the sum must be extended over several bands μ for a deep–level state. In the formulation of *Baraff* and *Schlüter* [7.39], $\psi(\mathbf{r})$ is expanded in a finite set of basis functions $\chi_i(\mathbf{r})$ such that

$$U(\mathbf{r})\psi(\mathbf{r}) \cong U(\mathbf{r}) \sum_i C_i \chi_i(\mathbf{r}). \qquad (7.5.17)$$

The expansion coefficients C_i and energy E are derived from a related variational principle,

$$\sum_i [N_{ij} - D_{ij}(E)] C_j = 0, \qquad (7.5.18)$$

$$N_{ij} = \int \chi_i^*(\mathbf{r})U(\mathbf{r})\chi_j(\mathbf{r})d\tau, \tag{7.5.19}$$

$$D_{ij}(E) = \int\int \chi_i^*(\mathbf{r})U(\mathbf{r})G_E(\mathbf{r},\mathbf{r}')U(\mathbf{r}')\chi_j(\mathbf{r}')d\tau d\tau'. \tag{7.5.20}$$

However, the expansion $\sum_i C_i\chi_i(\mathbf{r})$ is a good approximation to $\psi(\mathbf{r})$ only in the region where $U(\mathbf{r})$ is appreciable. For a neutral deep–level impurity, this region is comparable with a single unit cell. Superhyperfine interactions, on the other hand, are governed by values of the wave function outside this region, where $\psi(\mathbf{r})$ must be recovered from (7.5.15). In practice, the Green's function $G_E(\mathbf{r},\mathbf{r}')$ is calculated from pseudo–*Bloch* functions $\phi^{(\mu)}(\mathbf{k},\mathbf{r})$ derived from a pseudopotential representation of the atomic cores, plus a local approximation of exchange and correlation effects as in the X_α (*Hartree–Fock–Slater*) [7.40] and density functional [7.41] approximations. Consequently, (7.5.15) actually yields a pseudo–wave function $\phi(\mathbf{r})$ which must be orthogonalized to occupied core orbitals to recover the true wave function $\psi(\mathbf{r})$.

7.5.5 Orthogonalization to Core Orbitals

All of the methods cited in this section involve an envelope function which must be orthogonalized to occupied ion–core orbitals before they are used to calculate superhyperfine interactions. An instructive approximation to the contact hyperfine interaction is obtained by neglecting the variation of the pseudo–wave function $\phi(\mathbf{r})$ over the ion–core orbitals, and by replacing N^2 with unity, in (7.4.37,38):

$$a_j \cong \frac{8\pi}{3}g_e g_{nj}\mu_B\ \mu_n A_j|\phi(\mathbf{r}_j)|^2, \tag{7.5.21}$$

$$A_j = 1 - 2\sum_c \chi_{jc}(\mathbf{r}_j)\int \chi_{jc}(\mathbf{r})d\tau$$

$$+ \sum_c\sum_{c'}\chi_{jc}(\mathbf{r}_j)\chi_{jc'}(\mathbf{r}_j)\int \chi_{jc}(\mathbf{r})d\tau\int \chi_{jc'}(\mathbf{r}')d\tau', \tag{7.5.22}$$

where $\mathbf{r}_j$ is the position of the nucleus of ion j, and $\chi_{jc}(\mathbf{r})$ is an atomic core orbital centered on ion j. The amplification factor [7.42] A_j, which may exceed unity by several orders of magnitude, reflects the difference between the spin density of the true wave function at the nucleus and that of the envelope function. Only occupied s–orbitals contribute to the amplification of the contact hyperfine interaction. An analogous amplification factor for the anisotropic hyperfine interaction associated with the gradient of the envelope function can be derived from occupied p–orbitals. Experimental data on the hyperfine constants, together with theoretical estimates of amplification factors, provide a detailed map of the envelope function which challenges theoretical interpretation.

Although the amplification factor approximation is illuminating, it is too crude to be quantitatively reliable. Equations (7.4.37,38) entail evaluation of $\langle \psi_c | \phi \rangle$ where the core orbitals ψ_c are eigenfunctions of the same *Hamiltonian* as ψ_v, and are, therefore, also inherently delocalized. In the *Schmidt* orthogonalization procedure, the ψ_c are re–interpreted as atomic core orbitals χ_j:

$$\psi_v = N(\phi - \sum_j \langle \chi_j | \phi \rangle \chi_j), \qquad (7.5.23)$$

$$N = (1 - \sum_j \langle \chi_j | \phi \rangle^2)^{-1/2}. \qquad (7.5.24)$$

If the envelope function has the form of a diffuse atomic orbital centered on the defect site, the two–center overlap integrals can be evaluated by standard methods [7.43].

Schmidt orthogonalization is a substantial improvement over the amplification factor approximation, but an additional improvement can be effected by correcting for the mutual overlap of atomic core orbitals on different centers. This correction can be accomplished by employing a procedure for symmetrical orthogonalization of core orbitals, proposed by *Löwdin* [7.44]. In this procedure, the orbitals χ_j in (7.5.23,24) are replaced by normalized, orthogonalized core orbitals φ_j, defined by the series expansion

$$\varphi_j \equiv \chi_j - \frac{1}{2} \sum_{k \neq j} \chi_k \langle \chi_k | \chi_j \rangle + \frac{3}{8} \sum_{k \neq l} \sum_{l \neq j} \chi_k \langle \chi_k | \chi_l \rangle \langle \chi_l | \chi_j \rangle - \dots \quad (7.5.25)$$

These symmetrically orthogonalized functions are essentially identical with *Wannier* functions in the extreme tight–binding approximation. The dominant overlap integrals in (7.5.25) are typically of order 0.1.

Successful electronic structure calculations have also been performed with variational trial functions constructed by explicitly orthogonalizing envelope functions to core orbitals in the frozen–core approximation [7.45]. Hyperfine interactions are then calculated directly from these wave functions [7.46].

7.6 Simple Approximations and Illustrations for Interpretation of shf and Quadrupole Interactions

In the preceding sections, guidelines for the theoretical interpretation of shf and quadrupole interactions were presented. The actual theoretical calculation of the interaction parameters is, in most cases, a formidable and difficult task. There are not many cases in the literature where a satisfactory quantitative interpretation of the experimental shf and quadrupole interaction could be achieved. Often, however, a crude estimate of the interaction parameters can be very helpful for the determination of a defect structure model and to obtain a rough idea about its electronic structure. As pointed out in Chaps. 5 and 6, the ENDOR analysis yields precise information on the symmetry and size of the shf and quadrupole tensors, but no information on the distance between a ligand shell and the defect center. This information, however, is necessary to establish an unambiguous defect model. What is needed is information on the relative distances of ligand shells from the center obtained from the relative magnitude of their interaction constants. For example, if it is known from simple theoretical arguments that the envelope function falls off in a monotonic way as a function of distance from the center, then it is straightforward to establish the sequence of ligand shells from their decreasing interactions. For this purpose simple approximations of calculating the shf interactions are often sufficient.

In this section, some simple approximations are discussed and illustrated with actual examples which may be helpful. They are applicable, in particular, to defects in ionic crystals. However, the approximations discussed below become less and less reliable the more covalent the host crystal. This is particularly true for covalent semiconductors, where simple approximations cannot be used any more.

7.6.1 Point Dipole–Dipole Interaction

For many defects one observes that for neighbor nuclei beyond the second or third shell, the isotropic shf constant a becomes very small so that the measured shf interaction is predominantly determined by the anisotropic part. Furthermore, the orientation of the shf tensor points to the "center" of the defect. If this is observed, the shf interaction can be approximated by the point dipole–dipole interaction, in which the unpaired electron is replaced by a point dipole at the site of the atom which carries the unpaired electron. This observation was made even for next nearest neighbors of highly localized defects. The anisotropic shf interaction then becomes

$$\frac{b}{h} = \frac{\mu_0}{8\pi} g_n \mu_n g_e \mu_B \cdot \frac{2}{R_\alpha^3}, \tag{7.6.1}$$

$$\frac{b}{h} = 95.625 \frac{g_n}{R_\alpha^3} \quad [\text{MHz}], \tag{7.6.2}$$

where R_α is the distance between the neighbor nucleus and the defect center (in atomic units). In fact, (7.6.1) is obtained by integrating (3.5.7) with ψ being a compact envelope function. The lack of importance of the overlap transfer terms is seen by the disappearance of the isotropic shf constant a at the more distant neighbors. The isotropic shf constant a is much more sensitive to overlap contributions than b because of the high value of the amplitude of the s-functions at the nucleus, compared to the average $< r^{-3} >$ over an outer p–orbital; see also (7.4.17) in Sect. 7.4.2 and Appendix H. This

Table 7.3. Superhyperfine interactions of five Li^+ shells surrounding Fe^{3+} impurities in $LiTaO_3$ and the classical point dipole–dipole shf interaction constants b_{dd} calculated assuming that Fe^{3+} is on a substitutional Li^+ site (in MHz). The experimental uncertainty is ± 0.015 MHz. (After [7.47])

Shell	a	b	b_{dd}
1	0.118	0.58	0.58
2	-0.015	0.21	0.22
3		0.19	0.19
4		0.12	0.12
5		0.09	0.09

simple observation of the point dipole–dipole interaction being the dominant interaction for more distant shells can sometimes be used to establish a defect model if the symmetries of the experimentally determined shf tensors allow for several possibilities, or to determine details of a center model. A recent example is that of an ENDOR investigation of Fe^{3+} impurity centers in congruent $LiTaO_3$, where the most important question was on which site Fe^{3+} was located [7.47]. Figure 7.2 shows the possible sites on the **c**-axis of the crystal which are compatible with the symmetry of the EPR and ENDOR spectra. The Li^+ shells are numbered according to increasing distance from the assumed Fe^{3+} site. Fe^{3+} could either be on the Li site (Fig. 7.2a) or on the Ta^{5+} site (Fig. 7.2b). Only the Li^+ ions of the lattice are shown; the oxygens are omitted for the sake of clarity. Experimentally, the shf interactions with five shells of ^7Li neighbors could be resolved. Figures 7.3a and b show the ENDOR angular dependence for rotation of the magnetic field in the x–y plane (dots). Also shown as solid lines are the expected angular dependences for the two sites, for the assumption that the ^7Li shf interaction is entirely due to the point dipole–dipole interaction. Clearly, the Ta^{5+} site cannot explain the data, while the assumption of the Li site gives a very good explanation for most shells. Inclusion of a small isotropic constant for shell 1 explains the experimental data very well, as seen from Table 7.3, which contains a comparison of the experimentally determined shf interactions by fitting the dot chains and those calculated with (7.6.2) for Fe^{3+} being on the Li^+ site (g_n (^7Li) = 2.1707).

Another recent example where the point–dipole approximation gave valuable information on the structure of an aggregate defect is that of $F_H(OH^-)$ centers in KCl, in which an OH^- molecule is aggregated to an F center in a next nearest (200) position. Its position could be concluded from the (100)

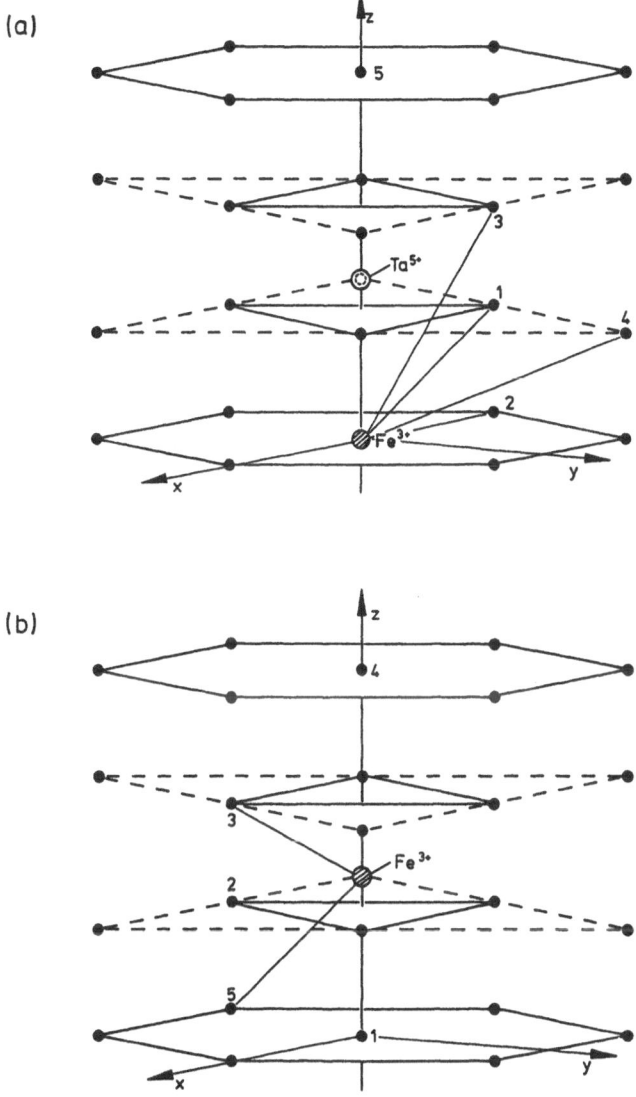

Fig. 7.2. a Fe^{3+} on a Li^+ site on the c–axis of a $LiTaO_3$ lattice with ^7Li neighbor shells ordered according to increasing distance from Fe^{3+}. Oxygen atoms are not shown. **b** Fe^{3+} on a Ta^{5+} site on the c–axis of a $LiTaO_3$ lattice with ^7Li neighbor shells ordered according to increasing distance from Fe^{3+}. Oxygen atoms are not shown

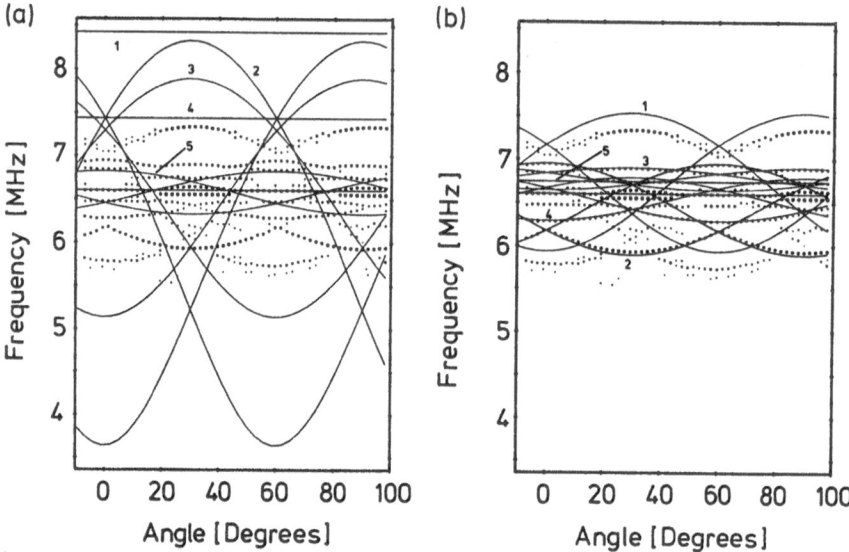

Fig. 7.3. a Experimental angular dependence (X–band) of ^{7}Li ENDOR lines for rotation of the magnetic field in the x–y plane (Fig. 7.2). The solid lines are calculated assuming a point dipole–dipole interaction for Fe^{3+} being on a Ta^{5+} site on the **c**–axis of $LiTaO_3$. (After [7.39]). **b** Experimental angular dependence (X–band) of ^{7}Li ENDOR lines for rotation of the magnetic field in the x–y plane (Fig. 7.2). The solid lines are calculated assuming a point dipole–dipole interaction for Fe^{3+} being on a Li^+ site on the **c**–axis of $LiTaO_3$. (After [7.47])

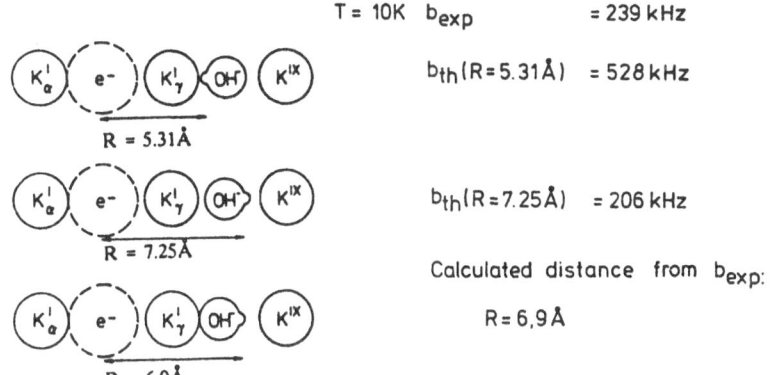

$T = 10\,K \quad b_{exp} \qquad\qquad = 239\,kHz$

$b_{th}(R = 5.31\,\text{Å}) \quad = 528\,kHz$

$b_{th}(R = 7.25\,\text{Å}) \quad = 206\,kHz$

Calculated distance from b_{exp}:

$R = 6.9\,\text{Å}$

Fig. 7.4. Details of the structure of $F_H(OH^-)$ centers in KCl for various positions of the OH$^-$ dipoles. (After [7.48])

symmetry of the ENDOR spectra of the proton (Chap. 5). What could not be concluded from the ENDOR analysis, however, was the orientation of the OH$^-$ dipole. It could, in principle, be such that the proton is next to a nearest K$^+$ neighbor or opposite (Fig. 7.4). Of course, from an electrostatic point of view, the latter would be expected. For the proton it was found experimentally that $a \approx 0$, $b/h = 239$ kHz. Assuming that the O$^-$ occupies the Cl$^-$ site, R would be 5.31 Å for the near position of the proton, and 7.25 Å for the other one. The respective b values calculated with (7.6.2) are 528 kHz and 206 kHz. Clearly, the dipole is oriented with the proton opposite, as expected. Since OH$^-$ replaces the larger Cl$^-$ ion, it seems to be attracted towards the nearest K$^+$ neighbor. From the experimental b value of the proton, one estimates with (7.6.2) a distance of $R_\alpha = 6.9$ Å, which is approximately what one expects from ionic radii, allowing the nearest K$_\gamma^I$ neighbor to relax about 10% outwards (Fig. 7.4). Its shf interaction is smaller compared to the other first shell K$^+$ neighbors [7.48]. The actual position of an impurity atom is seen most "directly" through the point dipole–dipole interaction from distant neighbors. They can be used more reliably to actually determine the impurity site, compared to nearer neighbors where the uncertainties in the interpretation of the shf interactions are larger. As an example, reference is made to the case of off–center Ni$^+$ defects in CaF$_2$, the position of which could be located best from the 7th shell F$^-$ neighbors [7.49]. Estimates using (7.6.2) are certainly possible in ionic crystals. To what extent this is also possible for deep–level defects in semiconductors seems an open question at present.

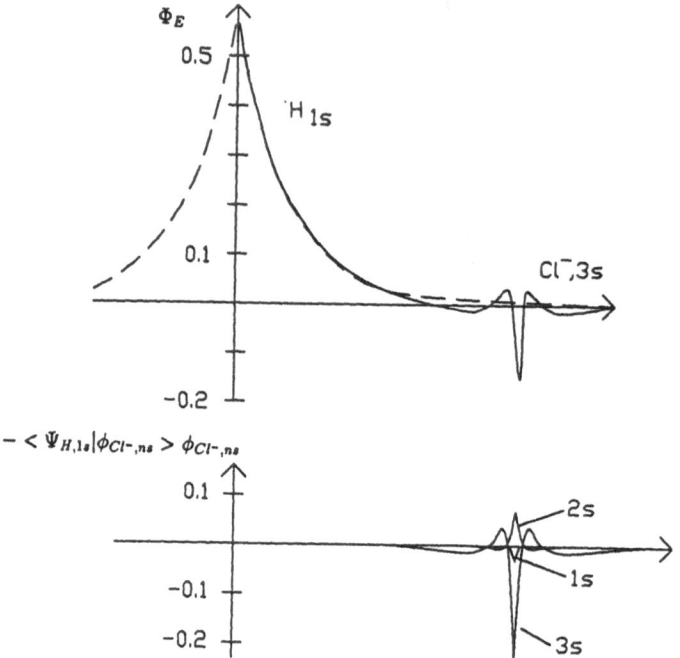

Fig. 7.5. a Hydrogen $1s$ function orthogonalized to a Cl^- $3s$ orbital at a distance of $R = 5.1$ a.u. **b** All Cl_{ns}^- admixtures obtained by orthogonalizing a hydrogen $1s$ function to a Cl^- ligand at $R = 5.1$ a.u.

7.6.2 Calculation of Isotropic shf Constants with Orthogonalized Envelope Function

As outlined in Sect. 7.5.5, the envelope function ϕ for calculating the shf constants must be orthogonalized to the ligand core orbitals χ_j

$$\psi = N\left(\phi_e - \sum_j \langle\chi_j|\phi\rangle\chi_j\right), \tag{7.6.3}$$

where the overlap integrals are the admixture coefficients of occupied orbitals of neighboring lattice atoms. The shf interactions of atomic hydrogen centers in alkali halides were successfully interpreted by taking the atomic hydrogen $1s$ function as envelope function [7.50,51].

Figure 7.5a shows the wave function according to (7.6.3) for the example of a hydrogen atom at a distance of 5.1 a.u. from a Cl^- ion, which represents a ligand in the case of interstitial atomic hydrogen centers in KCl [7.50]. The amplitude of $\phi = \psi_{H_{1s}}$ at the site of the Cl nucleus can be neglected in comparison to the amplitudes of $S \cdot \chi$ (Cl_{3s}^-) with $S = \langle\chi_{H_{1s}}|\chi_{Cl^-}\rangle$. In fact, the order of magnitude of the measured isotropic shf constant is really given by the square of the admixture of neighbor orbital amplitudes into the envelope function. However, for its calculation, one must not neglect the inner Cl^- orbitals which are admixed with alternate signs. Although the overlap integrals become smaller for the inner shells, the amplitudes of the inner shell orbitals become bigger (Appendix H). Therefore, the decrease in overlap integrals is partly compensated, and the admixture contributions from all shells are of the same order of magnitude. This is demonstrated in Fig. 7.5b where all the Cl^- s–admixtures $S_i \chi_i$ are shown to scale. Taking only the $3s$ admixture, one would calculate with (7.6.3) $a/h = 22.5$ MHz, while one obtains $a/h = 15.1$ MHz, taking into account all Cl^- s–orbitals. (The normalization constant N in (7.6.3) was neglected). It is a common observation, that the neglect of the inner orbitals overestimates the isotropic constant a by about 30%. Note that $a/h = g_n \cdot 801.1 |\psi(0)|^2$ a.u. [MHz].

The inclusion of a covalency [(7.4.3) in Sect. 7.4.1] can approximately be done by replacing the overlap integrals to the outermost ligand orbitals by $(\langle \phi | \chi_j \rangle + \lambda_j)$ in (7.6.3) where λ_j is a covalency parameter, which is mostly determined empirically from the shf data. The shf interactions with many neighbor shells of interstitial atomic hydrogen centers in alkali halides can successfully be explained by the assumption of just one small covalency parameter to the outer p–orbitals of the nearest halogen ions [7.51]. It should be noted that the method of the orthogonalized envelope function is a one-particle approximation which always yields positive shf constants. An experimentally observed negative isotropic shf constant, or b–values smaller than predicted by the point–dipole approximation (7.6.1), indicate that this approximation breaks down. In this case, polarization effects have to be considered in a multi–particle approach (Sect. 7.4.3).

7.6.3 Transferred shf Interactions

The mutual nonorthogonality of lattice ions or atoms leads to transfer effects of the shf interactions, in which unpaired spin density is transferred from nearest to next nearest neighbors. When applying the *Löwdin* transformation (7.5.25) to orthogonalize the lattice orbitals, and when orthogonalizing an envelope function to the orthogonalized set of lattice orbitals, the resulting transfer effect depends on the geometrical configuration of the ligands with respect to the paramagnetic atom carrying the unpaired electron. In order to illustrate this, let us assume the geometrical configuration of Fig. 7.6, and estimate the admixture coefficients of s–orbitals of the nearest neighbors A and B, and the next nearest neighbor C. For simplicity, let us consider only outer s–orbitals of the neighbors, and an envelope function ψ_E of s–character. When neglecting the mutual overlap of the cores of A, B and C, the isotropic constants would be given by the following expressions, neglecting covalency effects and the amplitude of ψ_E at the neighbor sites:

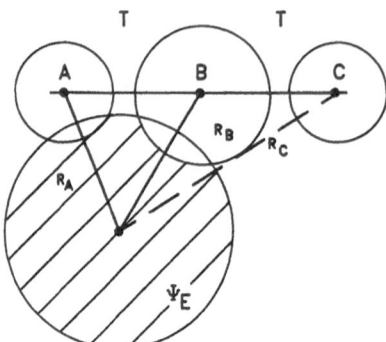

Fig. 7.6. Schematic representation of a paramagnetic impurity with envelope function and first and second shell neighbors for the calculation of superhyperfine transfer effects

$$a(A) \approx c|\langle \psi_E | \chi_A \rangle \chi_A(0)|^2, \qquad (7.6.4)$$

$$a(B) \approx c'|\langle \psi_E | \chi_B \rangle \chi_B(0)|^2, \qquad (7.6.5)$$

$$a(C) \approx c''|\langle \psi_E | \chi_C \rangle \chi_C(0)|^2 \approx 0. \qquad (7.6.6)$$

where c, c' and c'' contain the nuclear g factors, etc.; see (7.5.21). $\chi(0)$ are the amplitudes of the ligand s–orbital at the nuclear sites. For a compact wave function ψ_E, $a(C)$ will be very small due to the comparatively small overlap integral $\langle \psi_E | \chi_C \rangle$, because of the larger distance of C compared to A and B from the center. Consideration of the mutual core overlaps according to the *Löwdin* procedure, changes the admixture coefficients. When retaining only the leading terms, one obtains for the admixture of χ_C

$$a(C) \approx c''|\langle \psi_E | \chi_B \rangle \langle \chi_B | \chi_C \rangle|^2 |\chi_C(0)|^2 \qquad (7.6.7)$$

Thus, unpaired spin density in the near neighbor B is transferred through the transfer integral $T = \langle \chi_B | \chi_C \rangle$ to the second nearest neighbor. This is the transferred shf interaction. However, for the near neighbors A and B the transfer effect between A and B has a different consequence: here, the transfer integral T reduces the admixture into ψ_E. One obtains:

$$a(A) \approx c[\langle \psi_E | \chi_A \rangle - \langle \psi_E | \chi_B \rangle \langle \chi_B | \chi_A \rangle]^2 |\chi_A(0)|^2, \qquad (7.6.8)$$

$$a(B) \approx c'[\langle \psi_E | \chi_B \rangle - \langle \psi_E | \chi_A \rangle \langle \chi_A | \chi_B \rangle]^2 |\chi_B(0)|^2. \qquad (7.6.9)$$

The two terms in (7.6.8,9) may be of the same order of magnitude, thus reducing drastically the admixture one would have calculated neglecting the mutual overlap between the cores of A and B.

If $\langle \psi_E | \chi_B \rangle > \langle \psi_E | \chi_A \rangle$, as indicated by the size of the circles in Fig. 7.6, then $a(A)$ will be small, while $a(B)$ may not be affected much. Thus, one can observe a sort of quantum interference effect in the transferred shf interaction. In actual calculations, one has, of course, to include $p-$ and $d-$functions and take into account their orientations when calculating the transfer integrals. Covalency effects will be transferred constructively to outer neighbors but destructively to near neighbors, depending on the geometrical situation.

A situation similar to the one outlined above was found for interstitial hydrogen centers in alkali halides which showed a very anisotropic spin density distribution. In fact, the isotropic shf interaction of the nearest K neighbors was almost cancelled to zero by this effect in KBr, while the nearest Br neighbors at the same distance showed a large shf interaction. The isotropic shf interaction of the second shell K neighbors exceeded that of the first shell due to this effect. They experienced a transfer effect according to (7.6.7) [7.51]. For the explanation of the very low spin density at the nearest F neighbors of $F(F^-)$ centers in BaFCl and BaFBr, this destructive interference was also essential [7.52, 53].

For more compact wave functions of the paramagnetic atom or ion, the overlap admixtures of nearest neighbor orbitals are less important compared to covalent admixtures as discussed in Sect. 7.4. In this case, in (7.6.7), $\langle \psi_E | \chi_B \rangle$ is essentially replaced by the covalency parameter λ of (7.4.2). Again, the transfer integral T is responsible for the transfer effect, and the geometrical considerations apply in an analogous way.

7.6.4 Calculation of Anisotropic shf Constant b with Orthogonalized Envelope Function

The anisotropic constant b for axial symmetry is given by

$$\frac{b}{h} = N^2 \cdot C \int \frac{3\cos^2(\theta) - 1}{r^3} \left(\psi_E - \sum_i \langle \psi_E | \chi_i \rangle \chi_i \right)^2 dV, \qquad (7.6.10)$$

with

$$C = \frac{\mu_0}{8\pi} g_n \mu_n g_e \mu_B / h.$$

In the wave function only $p-$ and $d-$orbital ligand admixtures must be considered, since the integral vanishes for $s-$functions. The integral in (7.6.10) contains several parts, which are conveniently calculated separately. As will be seen below, some contributions are normally only small and can be neglected, in particular if the envelope function is not known too well, so that a rather crude estimate of b is all one would like to calculate.

In the following, only the admixture of $p-$orbitals is taken into account. If $d-$orbitals are to be included, the calculation must be extended in an analogous way. In order to illustrate the calculation, a compact envelope function and a neighbor with two $p-$orbitals $n\ p$ and $n'\ p$ (e.g., $n = 3, n' = 2$) is assumed.

(1) From the envelope function one obtains

$$\frac{b^{dd}}{h} = N^2 C \int \frac{3\cos^2(\theta) - 1}{r^3} |\psi_E|^2 dV. \tag{7.6.11}$$

For neighbors beyond the first shell this can usually be approximated by

$$\frac{b^{dd}}{h} = N^2 C \cdot \frac{2}{R_\alpha^3}, \tag{7.6.12}$$

where R_α is the distance between the center of the defect and the neighbor nucleus (Sects. 7.4.2, 6.1).

(2) From the admixture of the np orbital one obtains

$$\frac{b^{np\ np}}{h} = N^2 C |\langle \psi_E | \chi_{np} \rangle|^2 \int \frac{3\cos^2(\theta) - 1}{r^3} |\chi_{np}|^2 \, dV \tag{7.6.13}$$

$$= \frac{4}{5} N^2 C |\langle \psi_E | \chi_{np} \rangle|^2 \int_0^\infty \frac{1}{r^3} |P_{np}(r)|^2 \, dr. \tag{7.6.14}$$

Here, $P(r)$ equals $rR(r)$ of the neighbor orbital χ. The angular integration is, of course, the same for all p–functions and yields the factor $4/5$. The integrals $< r^{-3} >_p = \int_0^\infty \frac{1}{r^3} |P_p(r)|^2 dr$ are one–center integrals and can be calculated, e.g., from the *Clementi–Roetti* tables [7.15]. It is sufficient to consider only the outermost p–orbital, since the decrease in the square of the overlap integral to the inner p–orbitals by far outweighs the increase in the quantity $< r^{-3} >_{np}$.

(3) The most important cross term is

$$\frac{b^{np\ n'p}}{h} \approx 2 \cdot \frac{4}{5} \cdot N^2 \cdot C \langle \psi_E | \chi_{np} \rangle \langle \psi_E | \chi_{n'p} \rangle \int_0^\infty \frac{1}{r^3} P_{np} P_{n'p} dr. \tag{7.6.15}$$

The integral is also a one–center integral.

(4) It is more difficult to calculate the cross term with the envelope function:

$$\frac{b^{\psi_E\ np}}{h} = N^2 C \langle \psi_E | \chi_{np} \rangle \int \psi_E \frac{3\cos^2(\theta) - 1}{r^3} \chi_{np} dV. \tag{7.6.16}$$

For an estimate, one can expand ψ_E about the neighbor nucleus. One obtains an expression of the form

$$\frac{b^{\psi_E\ np}}{h} \propto \frac{\partial \psi_E}{\partial r}\Big|_{R_\alpha} \langle \psi_E | \chi_{np} \rangle \int_0^\infty \frac{1}{r} P_{np}(r) \, dr. \tag{7.6.17}$$

For a smoothly varying envelope function which falls off monotonically, this term can normally be neglected.

For illustration as a quantitative example, the terms 1 – 4 are given for atomic hydrogen at a distance of 5.1 a.u. from a Cl$^-$ neighbor, assuming only *Schmidt* orthogonalization:

$b^{dd} = 0.374$ MHz, $b^{3p\ 3p} = 3.962$ MHz, $b^{2p\ 2p} = 0.019$ MHz,
$2b^{2p\ 3p} = -0.513$ MHz, $2b^{H_{1s}\ 3p} = 0.021$ MHz.

Thus, it is usually sufficient to calculate the following terms for p–admixtures:

$$\frac{b}{h} \approx N^2 B \Big\{ \frac{5}{4} \frac{2}{R_\alpha^2} + \sum_{np} |\langle \psi_E | \chi_{np} \rangle|^2 \int_0^\infty \frac{1}{r^3} |P_{np}(r)|^2 dr$$

$$+ 2 \sum_{np,n'p} \langle \psi_E | \chi_{n'p} \rangle \langle \psi_E | \chi_{np} \rangle \int_0^\infty \frac{1}{r^3} P_{np}(r) P_{n'p}(r) dr \Big\},$$

$$(7.6.18)$$

$$\frac{b}{h} \approx 38.25 g_n \cdot N^2 \cdot \{\ \ \} \text{ a.u. [MHz].} \qquad (7.6.19)$$

where the $\{\ \ \}$ is that of (7.6.18). It is generally sufficient to consider only the two outermost p–orbitals.

If a covalency is to be considered, then $\langle \psi_E | \chi_{np} \rangle$ is to be replaced by $(\langle \psi_E | \chi_{np} \rangle + \lambda_{np})$, where λ_{np} in the covalency parameter describing the covalent admixture of ligand np orbital into the envelope function. The consideration of a covalency also somewhat influences the normalization constant. See Sect. 7.4.2 and e.g., [7.51] for a detailed discussion of atomic hydrogen centers in alkali halides.

7.6.5 Dynamical Contributions to shf Interactions

Localized vibrations of the impurity atom, or of the neighbor atoms with large enough amplitudes, cause a dynamical part of the shf interactions, which can be quite substantial. This may show up in a temperature dependence of the shf interactions or in an isotope effect, if one can substitute the impurity atom by an isotope with different mass but identical electronic configuration.

If there is a vibration, usually with frequencies of the order of 10^{12} – 10^{13} Hz, then with EPR/ENDOR one sees only the averaged effect of these vibrations. A dynamical part arises because all overlap integrals and covalency parameters depend non linearly (mostly exponentially) on the distance between ψ_E and the neighbor atom cores and, therefore, the interactions depend on the actual relative position between the paramagnetic impurity and the neighbors (Sect. 7.4.6):

$$\bar{a} = \int a(\mathbf{r}) |\varphi(\mathbf{r} - \mathbf{r}_0)|^2 \, dV, \qquad (7.6.20)$$

$$\bar{b} = \int b(\mathbf{r}) |\varphi(\mathbf{r} - \mathbf{r}_0)|^2 \, dV. \qquad (7.6.21)$$

In these integrals, $\mathbf{r}_0$ describes the equilibrium distance between neighbor and impurity, φ stands for the nuclear motions. To illustrate the essential effect

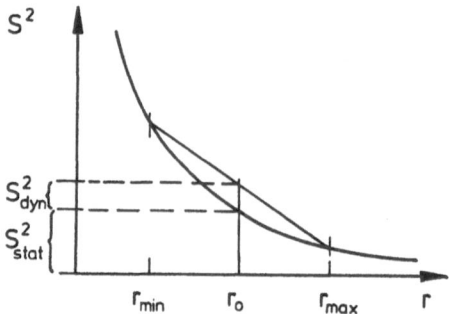

Fig. 7.7. Schematic representation of static and dynamical contributions to the superhyperfine interactions

for the example of a vibrating paramagnetic atom, the isotropic shf constant of a nearest neighbor may be approximated as

$$\bar{a} \approx A|\psi_{ns}(0)|^2 \int S^2(r)\varphi^2(\mathbf{r} - \mathbf{r_0})dV, \tag{7.6.22}$$

where $\mathbf{S}(r)$ is the admixture coefficient of the outer s-orbital of the neighbor, and may either be an overlap integral $\langle\psi_E|\chi_{ns}\rangle$ or also contain a covalency parameter. Also, the latter will depend on the distance between the impurity and the neighbor. The overlap integral can be approximated about the equilibrium distance $\mathbf{r_0}$ by

$$S(r) = S(r_0)\exp[-\beta(r - r_0)]. \tag{7.6.23}$$

For simplicity it is assumed that only the impurity atom vibrates. Figure 7.7 shows schematically the dependence of $S^2(\mathbf{r})$. The static value $S^2(r_0)$ leads to a shf constant which is too small since, in the experiment, one averages over the vibrations $\varphi^2(r - r_0)$, which is indicated by a straight line between $S^2(r_{min})$ and $S^2(r_{max})$ in Fig. 7.7. Of course, in order to calculate the effect properly, one has to calculate the average of $S^2(r)$ with the proper wave function $\varphi(\mathbf{r} - \mathbf{r_o})$ describing the motion of the impurity.

An isotope effect was observed for the shf interactions of nearest neighbors of interstitial hydrogen atoms in KCl. For example, a of the nearest ^{35}Cl neighbors was observed to be 23.74 MHz for H centers, and 23.13 MHz for D centers [7.54]. A calculation of the dynamical effect according to (7.6.20) assuming a zero point vibration in a harmonic potential

$$\varphi(r - r_0) = \left(\frac{\alpha}{\pi}\right)^{3/4}\exp[-\alpha(r - r_0)^2/2], \tag{7.6.24}$$

with the potential parameter

$$\alpha = \frac{1}{h}\sqrt{k \cdot m}, \tag{7.6.25}$$

where k is a force constant and m the mass of the hydrogen atom, is done with the model of the orthogonalized hydrogen $1s$–function as envelope function, taking only the Cl^- $3s$–function into account. For this simple model the integration (7.6.20) can be done analytically when $S(r)$ is assumed to follow (7.6.23). One obtains

$$\bar{a} = a_{static}(1 - \frac{\beta}{\alpha r_0}) \exp(\frac{\beta^2}{\alpha}). \qquad (7.6.26)$$

For D centers β is the same, but $\alpha(D) = \alpha(H)\,2^{1/2}$ and, thus, a is smaller. The effect is more pronounced for large values of β, i.e., for compact wave functions. For the anisotropic constants analogous effects were observed. However, the integration (7.6.21) must be done numerically [7.54].

For a vibrating impurity with increasing temperature, one expects the following influence on the shf interactions:

(i) The lattice will expand with increasing temperature and, therefore, the overlaps will decrease, which will result in a decrease in shf interactions.

(ii) If the potential parameter of the vibration is sufficiently small, then higher vibrational modes can be populated thermally with the result of increasing vibrational amplitudes and, therefore, increasing shf interactions.

(iii) The vibrational amplitudes of the nearest neighbor vibrations will also increase and contribute to an increase in shf interactions.

Which of the effects dominates, depends on the defect. For atomic hydrogen on anion and cation sites in alkali halides, a thermally activated increase of nearest neighbor shf interactions was observed (Fig. 7.5) [7.55–57]. Recently, similar effects for $F_H(F^-)$ centers and for $F_H(OH^-)$ centers in KCl could be observed [7.48, 58] where the motion of heavier atoms (K) and molecules (OH^-) showed up in an increase of ENDOR frequencies with temperature. The high precision of ENDOR frequency measurements enables the study of these dynamical effects. Small temperature changes of a few kHz for the interaction constants can safely be detected (Sect. 5.7).

7.6.6 Quadrupole Interactions

The quadrupole interaction measures the electrical field gradient experienced at the nucleus. The quadrupole tensor elements are given by

$$Q_{ik} = \frac{eQ}{2I(2I-1)} \frac{\partial^2 V}{\partial x_i \partial x_k}\Big|_{r=r_l}, \qquad (7.6.27)$$

where Q is the nuclear quadrupole moment, V the electrostatic potential (Sect. 7.4.4).

The calculation of the electrical field gradient at the nuclear site r_l is a formidable task. The origin of the electrical field gradient is the charge distribution of the unpaired electron (which is not necessarily the same as the spin density distribution), the lattice distortion, the distortion of neighbor core orbitals, covalency effects and the orthogonality admixtures of p– or d– orbitals into an envelope function. A rigorous theoretical interpretation of the quadrupole interactions is hard to achieve, and was hardly attempted.

However, the calculation of a crude estimate can be helpful to assess the charge state of a defect, by considering two contributions, which often are the major sources of the electrical field gradient. For axial symmetry one has

$$q = \frac{eQ}{4I(2I-1)} V_{zz}(r_l). \tag{7.6.28}$$

The contribution of a point charge is simple to calculate:

$$V_{zz}(p) = \pm \frac{2|e|}{R_\alpha^3}(1 - \gamma_{\infty,\alpha}), \tag{7.6.29}$$

where R_α is the distance of the nucleus from the charge, $|e|$ the magnitude of the charge and $(1 - \gamma_{\infty,\alpha})$ the *Sternheimer* anti–shielding factor for a charge outside the core of the atom or ion α under consideration [7.59,60]. The other important contribution to V_{zz} is from the unpaired spin density moving in p– (and d–) orbitals. This contribution is calculated in the same way as is the anisotropic shf constant

$$V_{zz}(b) \propto \int \frac{3\cos^2(\theta) - 1}{r^3} |\psi|^2 \, dV \tag{7.6.30}$$

where ψ is the wave function of the defect containing the neighbor orbital admixtures. Comparison with (7.6.10) shows, that for a particular neighbor nucleus

$$V_{zz}(b_\alpha) \propto b_\alpha(p_\sigma), \tag{7.6.31}$$

where $b_\alpha(p_\sigma) = b_\alpha \exp(-b^{dd})$. That part of ψ which gives rise to the point dipole–dipole contribution b^{dd} of b does not contribute to $V_{zz}(b)$. One can, therefore, establish a relation between $b(p)$ and the quadrupole constant $q(p)$ which arises from the p_σ–admixtures [7.28].

$$q(p_\sigma) = \frac{Se^2Q(1-\gamma)}{2I(2I-1)\mu_B g_n \mu_n} b(p_\sigma) \tag{7.6.32}$$

Here, $(1 - \gamma)$ is the atomic anti–shielding factor. The spin S enters since it is taken into account in the shf term, but not in the quadrupole term in the spin *Hamiltonian*. Therefore, as a crude estimate one has

$$q \approx q(p) + q(p_\sigma)$$

$$\approx \frac{e^2Q}{4I(2I-1)} \left[\pm \frac{2}{R_\alpha^3}(1 - \gamma_\infty) + \frac{2S(1-\gamma)}{\mu_B g_n \mu_n} b(p_\sigma) \right] \tag{7.6.33}$$

   As an application of the simple arguments given, an ENDOR investigation of Ni^{3+} in GaP shall be mentioned. Experimentally it was found that there is a Ga ligand with a quadrupole constant $q/h = 0.018$ MHz with almost [111] axial symmetry. Ni^{3+} could either be substituting for Ga^{3+} or be in a tetrahedral interstitial site, according to the ENDOR analysis of the shf tensors. Which place it occupied could not be decided from the analysis of the spectra. However, were Ni^{3+} on the interstitial site, one would calculate from the first term in (7.6.33) $q(p) = 0.45$ MHz ($R_\alpha = 4.71$ Å, $(1-\gamma_\infty) = 24$), which is a factor of 25 too large. On the other hand, for substitutional Ni^{3+} there would be no extra charge, and the small q be predominantly due to the small amount of unpaired p_σ-orbitals, due to the second term in (7.6.33) [7.62].

   In an ENDOR investigation of Mn^{2+} substituting for Ga^{3+} in GaP, it was found there that for the same type of Ga neighbor the quadrupole interaction was much bigger. The diagonal elements of Q_{ik} were $Q_{xx} = -1.0$ MHz, $Q_{yy} = 0.40$ MHz, $Q_{zz} = 0.60$ MHz. Contributions were estimated to be due to the charge mismatch (Mn^{2+} instead of Ga^{3+}), and the lattice deformation of the nearest P neighbors due to the difference in ionic radii. All contributions were of the same order of magnitude which made an accurate estimate difficult [7.61]. A discrepancy factor of 2 between experimental and model calculation seems hardly meaningful.

   An estimate according to (7.6.33) can often be very helpful to determine the charge state of a defect, since the point charge contribution is often the largest one. Such an estimate could, for example, decide that the As interstitial in the As_{Ga}–As_i pair defect (paramagnetic EL2 defect) in GaAs, must carry either a positive or negative charge, since the order of magnitude of the observed quadrupole interaction of the nearest As neighbors of the As interstitial could not have been understood otherwise [7.62].

8. Technology of ENDOR Spectrometers

One straightforward approach to building an ENDOR spectrometer is to take a conventional EPR spectrometer, add the ability to produce a sufficiently high radio–frequency magnetic field in the microwave cavity, modulate this field, and then try to detect the change of the EPR signal, the ENDOR signal, with a lock–in amplifier. However, this will usually not work satisfactorily.

The experimental constraints for ENDOR experiments, in particular for the investigation of defects in semiconductors, are, in most cases quite different from those for EPR. If the performance of an ENDOR spectrometer is pushed to the limits, then almost any part of the apparatus is affected by these constraints. A good ENDOR spectrometer, on the other hand, usually exhibits only mediocre performance for EPR.

It is not even possible to simultaneously optimize an ENDOR spectrometer for the investigation of all defect systems in solids. As an example, the wide range of electronic and nuclear spin–lattice relaxation times calls for very different values of temperature, microwave power, radio–frequency field amplitude, and modulation frequency. This cannot be handled by a single experimental set up under the condition of optimum signal–to–noise–ratio (SNR). Therefore, in this chapter, general guidelines for the design of ENDOR spectrometers are discussed, rather than detailed instructions given for a special purpose.

The situation is somewhat different in the case of optical detection of magnetic resonance. Once an EPR signal has been detected optically with sufficient SNR, from a technological point of view, the addition of the ENDOR facilities is comparatively straightforward. The discussion of spectrometers for optical detection of ENDOR concentrates, therefore, primarily on the technology of optical detection of magnetic resonance in general, rather than on the radio–frequency technology for optically detected ENDOR (Chap. 9).

8.1 Experimental Constraints for Conventional ENDOR

8.1.1 Modulation Frequency

In an ENDOR experiment, signals must be detected in the audio and even in the sub–audio frequency range. This requires a technology for the entire ENDOR spectrometer which differs substantially from that of typical EPR spectrometers. Due to the relatively long nuclear spin–lattice relaxation times involved in the ENDOR process, modulations can only be applied at frequencies as low as about 10 Hz to several kHz. 10 kHz, for example, has been found to be too high in all cases measured so far. In this low frequency range many sources of noise and interferences are significant, which are almost completely unknown in typical EPR experiments, where signals are detected at about 100 kHz due to the modulation of the static magnetic field B_0.

8.1.2 Sensitivity

The intensity of an ENDOR signal is typically two orders of magnitude lower than the corresponding EPR signal. Therefore, the highest possible sensitivity is imperative. In all practical cases the time necessary for an ENDOR measurement is strongly related to the achievable SNR. An improvement of a factor of two in sensitivity will, for example, increase the time for the experiment by a factor of four. In an ENDOR experiment very many data must be collected in order to determine the structure of a defect. Therefore, measurement time becomes the limiting factor in the case of weak signals. A continuous measurement time of about three weeks for an ENDOR investigation is not unusual. It is, therefore, highly desirable to make the spectrometer completely accessible for computer control. For a lengthy unattended spectrometer operation under optimum conditions, more peripheral parameters of the spectrometer like the cavity coupling, the microwave phase setting and the helium flow and pressure in the cryostat, must also be automatically controlled.

8.1.3 Temperature

The magnitude of an ENDOR signal depends strongly on the relaxation times for the electronic system and the neighbor nuclei, and on the various cross relaxation times which must all have suitable values to obtain sufficiently high signal intensity. These relaxation processes depend on temperature in a very unpredictable way. Temperature is, therefore, an important parameter which has to be optimized experimentally for the best ENDOR signals. The range needed is from about 1.5 K, typical for defects with electronic ground states with high angular momentum, up to about 100 K for s–like systems which are easier to saturate. About 10 K turned out to be a temperature at which

ENDOR signals of most defects are at least visible. Sometimes the temperature range for optimum signals is remarkably small, e.g., about 10 K. The cryostat must therefore allow for lengthy experiments at any temperature, at least in the range from 3.5 K to about 100 K, which can be achieved with suitable flow cryostats.

8.1.4 Microwave and Radio–Frequency Field Intensities

The microwave magnetic field, B_1, at the location of the sample in the microwave cavity must be high enough to allow for a partial saturation of the EPR transition used for ENDOR. However, the application of too high a microwave field for the EPR saturation of defects with very short electron spin–lattice relaxation times is not helpful, because the radio–frequency (rf) magnetic field at the sample, B_2, must be able to partially counterbalance the transition rate induced by B_1. As a rule of thumb, it is sufficient for the field B_2 to induce an NMR transition rate, which is about 1% of the EPR transition rate induced by B_1. Hence it is assumed that only about 1% of the electronic relaxation takes place via the nuclear levels which are saturated by the B_2 field. For the same reason, the ENDOR effect does not exceed 1% of the EPR signal in most cases. This situation is typical for defects in solids for which the electron spin has substantial interactions with many neighbor nuclei due to the extended wave functions of these defects. From an experimental point of view, it is much easier to produce sufficiently high B_1 fields rather than high B_2 fields. Therefore, the limiting factor is predominantly the available B_2 field. On the other hand, for most defects in solids the ENDOR effect tends to saturate upon increasing the B_2 field beyond a specific level. Further enhancement of the B_2 level usually broadens the ENDOR lines with no substantial improvement of the signal intensity. A possible explanation for this behaviour is the assumption of a weak relaxation path in series with the "short circuit" provided by the induced NMR transition rate. In this case, the optimum value of the B_1 field and the most efficient value of the B_2 field are defined by the series relaxation path. If this path is inefficient, the ENDOR signals can be very weak or undetectable, even though the EPR transitions are easily saturated.

Again, from experience, rf fields in the order of 0.3 – 1 mT rms, and maximum microwave power levels of 1 – 10 mW are sufficient. There are also defect systems from which maximum ENDOR signals are obtained for a microwave power of 10^{-7} W applied to a cylindrical TE_{011} cavity with a loaded quality factor Q of about 10,000. Depending on the type of cavity used for the ENDOR experiments, the microwave power necessary to produce a specific B_1 field at the sample can vary within an order of magnitude.

Due to the hyperfine enhancement effect [8.1], ENDOR line intensities are not affected if the rf field, B_2, decreases proportionally to $1/f_{ENDOR}$, where

f_{ENDOR} is the frequency of the rf field. From an experimental point of view, it is very difficult, if not impossible, to retain the $1/f_{ENDOR}$ relation for B_2 down to the lowest frequencies. One has to select a "corner frequency" up to which B_2 is nearly constant, and above which it starts to fall off proportionally to $1/f_{ENDOR}$. As an example, 10 MHz is a good value for this corner frequency. The necessary values for B_2 stated above refer to frequencies below about 10 MHz.

8.1.5 Microwave and ENDOR Frequency

The magnetic field, and consequently the microwave frequency, should be as high as possible. The higher the magnetic field, the better the frequency separation of ENDOR lines for nuclei with different nuclear g values. High magnetic fields can, therefore, greatly disentangle complicated ENDOR angular dependences with many overlapping lines due to different isotopes of neighbor nuclei. The separation of EPR lines due to electron g value anisotropies is proportional to the magnetic field. At high fields, it is therefore easier to measure the ENDOR spectra for different defect orientations separately. This, again, particularly helps in the case of low–symmetry defects in high–symmetry hosts to reduce the degree of complexity of the ENDOR spectra. At the same time, higher order effects in the spin *Hamiltonian* are reduced. This facilitates the interpretation of ENDOR spectra. In particular, multiple splittings of ENDOR lines due to indirect nuclear–nuclear couplings of neighbor nuclei with high interaction energy are less pronounced for high electron *Zeeman* energies.

However, from a spectrometer technology point of view, there are serious drawbacks for ENDOR experiments at high microwave frequencies. With increasing microwave frequency, the Q of the microwave cavities decreases. At the same time, both the frequency noise and the amplitude noise of microwave generators tend to increase considerably. Microwave detectors and amplifiers, if available at all, also have worse performances with higher frequencies. The same is true for all other microwave components, the prices of which can get very high for higher microwave frequencies. It is usually unavoidable that the sensitivity of ENDOR spectrometers decreases with increasing microwave frequency. Also, the more complex and expensive superheterodyne technology must be applied (Sect. 8.2).

Apart from the sensitivity of the spectrometer, the geometry of the samples strongly influences the SNR achievable for different microwave frequencies. If the samples cannot be made big enough to achieve an optimum filling factor in large, low frequency resonators, the improvement of the filling factor in a smaller, high frequency cavity can by far overcompensate for the loss in spectrometer sensitivity. If, on the other hand, the sample size can easily be adjusted at a constant center concentration to cavities for different frequencies, the selection of a higher frequency yields no advantage but only the disadvantage of the reduced spectrometer performance. The X–band frequency range (about 9.5 GHz) is still the best choice for high sensitivity requirements. However, the rapid progress in microwave technology will presumably enable the design of almost as sensitive K–band spectrometers (about 24 GHz) in the near future.

The rf frequency range necessary for ENDOR experiments depends on the magnitude of the hf, shf and quadrupole interactions to be investigated. The lower end of this range should be somewhat lower than the smallest value of the nuclear *Zeeman* frequencies. For example, 0.5 MHz is a sufficiently low value for X–band spectrometers. As an upper limit, 30 – 40 MHz suffices for ligands of point defects in silicon investigated so far. However, for defects in III–V semiconductors, up to 300 MHz may be necessary to measure the interactions with nuclei of the first neighbor shell.

The minimum frequency step size in an ENDOR spectrum may be about 1 kHz. However, in special ENDOR–induced EPR investigations a frequency resolution of up to 100 Hz may be desirable.

8.1.6 Static Magnetic Field

The requirements for the static magnetic field in an ENDOR experiment are almost the same as for the EPR experiment. The homogeneity is less important in the case of ENDOR, since the frequency of an ENDOR line depends only weakly on the magnetic field through the nuclear *Zeeman* term. However, the short term stability of the field is of great importance. Due to EPR rapid passage effects, even small, low frequency fluctuations of the field value can cause big interference signals with *Fourier* components of the frequency of the detected ENDOR signal. It is characteristic for these effects that the noise signal is present only if the magnetic field is adjusted to an EPR line. The noise contribution of magnetic field instabilities has been found to be important only in cases for which ENDOR signals have to be detected at low temperatures in the regime of relatively long electron spin–lattice relaxation times.

8.1.7 Modulation of Parameters

For the different ENDOR measurement techniques several parameters have to be modulated in a suitable way. In conventional ENDOR experiments the frequency or the amplitude of the rf field is usually modulated to achieve an effect–modulation for high sensitivity detection of ENDOR signals. Both of these methods have their advantages and disadvantages. Frequency modulation has the advantage that the detection of otherwise large background signals is minimized. However, for high resolution measurements of multiple overlapping lines, a small frequency modulation amplitude is required, which decreases the signal intensity, and, thus the SNR. The best choice is, therefore, the amplitude modulation, as long as the background signal problems can be satisfactorily dealt with. In any case, a square wave modulation ("on–off modulation") is superior to a modulation with a sine function, as long as harmonics of the modulation frequency cause no problems. However, such problems have not yet been observed. The advantage of the square wave is that its fundamental *Fourier* component, as detected by a lock–in amplifier, is higher by about 30% than its amplitude . This results in a 30% signal improvement. A square wave frequency modulation is equal to a "frequency jump modulation" which can, however, only be realized by special synthesizers ("direct" synthesizers).

The frequency of the rf field modulation should be variable between about 5 Hz and 5 kHz. The noise spectrum of the spectrometer sometimes exhibits pronounced peaks at several frequencies due to microphonic effects. The modulation frequency must, therefore, be continuously variable to allow adjustment to a suitable noise minimum.

In order to adjust the frequency of the microwave source automatically to the resonant frequency of the cavity, automatic frequency control (AFC), the microwave frequency is usually modulated with a frequency of, e.g., 80 kHz, and a suitable frequency control signal is produced by a lock–in amplifier. A much better performance of the AFC is, however, achieved by the special "dc–AFC" technique as described below.

8.2 ENDOR Spectrometer Design

Figure 8.1 shows a block diagram of a conventional ENDOR spectrometer. The spectrometer consists of the microwave source (1), the microwave attenuator (2), the circulator (3), the cavity (4) with elements (4.1) to produce the rf magnetic field at the sample (4.2) and with additional elements (4.3) to modulate the magnetic field generated by the magnet (5), the cryostat system (6), the rf generator (7), the microwave detector system (8) which delivers the AFC signal and the low frequency ac signal containing the ENDOR information for further amplification by the pre–amplifier (9). The detector system needs a microwave reference provided by the reference arm (10) with facilities for microwave phase shifting (11). The rf amplitude or the frequency of the rf generator can be modulated by the reference signal provided by a lock–in amplifier (12), which demodulates the ac signal delivered by the signal preamplifier (9). The output of the lock–in amplifier is a dc signal obscured by noise. Finally, this dc signal, is the sum of the ENDOR signal and a background signal.

For a better understanding of the function of the different parts of the spectrometer, it is useful to discuss some relations which describe the detected

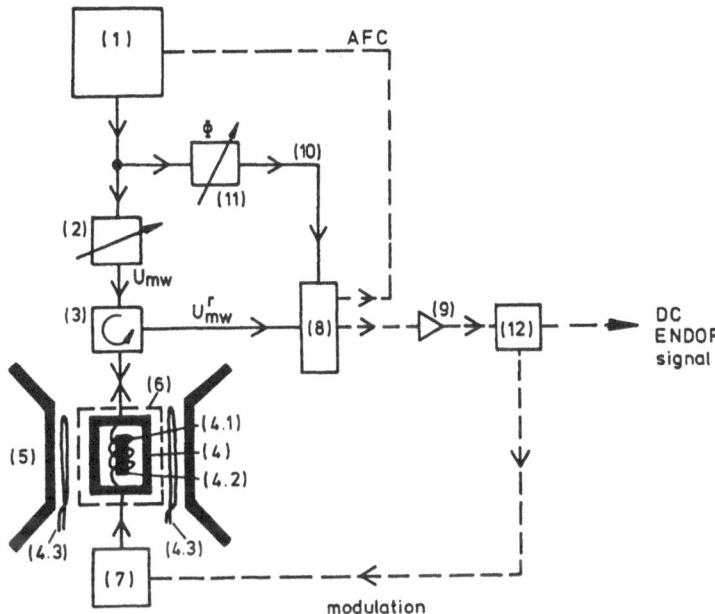

Fig. 8.1. Simplified block diagram of an ENDOR spectrometer consisting of: (1) microwave source; (2) microwave attenuator; (3) circulator; (4) cavity; (4.1) element in the cavity to produce rf field B_2; (4.2) sample; (4.3) coil for modulation of static magnetic field; (5) magnet; (6) cryostat system; (7) rf generator; (8) detector system; (9) signal preamplifier; (10) microwave reference arm; (11) microwave phase shifter; (12) lock–in amplifier

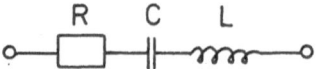

Fig. 8.2. Equivalent circuit for a microwave cavity

signal in terms of different parameters of the spectrometer: The microwave source (1) generates a microwave with frequency Ω. A fraction of this microwave passes the attenuator (2), and flows through the circulator (3) to the cavity (4) with an amplitude U_{mw}. Another fraction flows through the reference arm (10) to the detector system (8). A portion of the microwave is reflected at the cavity and flows back through the circulator to the detector system (8). The reflected microwave, U_{mw}^r, differs in amplitude and in phase from the incoming microwave U_{mw}. It is convenient to describe this by a complex amplitude of the reflected microwave, where the real part of U_{mw}^r, $\mathrm{Re}(U_{mw}^r)$, is in phase with U_{mw}, and where the imaginary part of U_{mw}^r, $\mathrm{Im}(U_{mw}^r)$, is out of phase by 90 degrees, with respect to U_{mw}. The signal U_{mw}^r can easily be calculated by:

$$U_{mw}^r = U_{mw}\frac{R_c - Z}{R_c + Z}, \tag{8.2.1}$$

where R_c is the (complex) impedance of the cavity and Z is the impedance of the microwave transmission line, usually a rectangular wave guide or a coaxial cable. The electrical properties of the cavity can be described by an equivalent circuit consisting of a resistor, R, a capacitor, C, and an inductance, L, as sketched in Fig. 8.2. This resonance circuit has a resonance frequency, Ω_{res}:

$$\Omega_{res} = \frac{1}{\sqrt{LC}}. \tag{8.2.2}$$

The quality factor, Q, of the resonance circuit is defined by:

$$Q = \Omega_{res}\frac{L}{R}. \tag{8.2.3}$$

With these expressions, the impedance, R_c, of the resonance circuit is obtained as:

$$R_c = R + i[Q \cdot R(\Omega/\Omega_{res}) - L(\Omega_{res}^2/\Omega)], \tag{8.2.4}$$

where i is the imaginary unit. If the cavity contains a sample in which a magnetic resonance transition is induced, then the inductance, L, has to be replaced by an "effective" inductance, L':

$$L' = L[1 + f(\chi' - i \cdot \chi'')], \tag{8.2.5}$$

where χ' is the real part and χ'' is the imaginary part of the magnetic susceptibility of the sample. It is assumed here that only a fraction f of the effective volume of the cavity is filled by the sample. The factor f is called the *filling factor*. Insertion of (8.2.5) in (8.2.4) and the result, in turn, into (8.2.1) yields a very complicated expression for the reflected microwave signal, U_{mw}^r, which is detected by the detector system. This expression, however, can be greatly simplified by some approximations:

(a) χ' and χ'' are very small, therefore, only terms linear in χ' and χ'' are considered.

(b) It is assumed that $R = Z$, which means that the cavity coupling has been adjusted perfectly to minimum reflection of microwave power under a condition where no magnetic resonance takes place.

(c) Only small deviations of the microwave frequency, Ω, from the resonance frequency, Ω_{res}, are considered. Higher order terms of $(\Omega_{res} - \Omega)$ are neglected.

The relative deviation, δ, of the resonance frequency Ω_{res} of the cavity from the microwave frequency Ω, is defined as:

$$\delta = \frac{(\Omega_{res} - \Omega)}{\Omega}. \tag{8.2.6}$$

The reflected signal U_{mw}^r then amounts to:

$$U_{mw}^r = U_{mw} Q^2 \delta^2 - U_{mw} f Q \frac{\chi''}{2} - U_{mw} f Q^2 \chi' \delta$$

$$+ i U_{mw} Q \delta - i U_{mw} f Q \frac{\chi'}{2} \tag{8.2.7}$$

The second term is the EPR absorptive signal and the fifth term the dispersive signal. The dependence of these terms on $U_{mw} f Q$ is somewhat misleading in the case of ENDOR, since one deals with saturated EPR signals. This will be considered later in more detail together with the discussion of microwave cavities. The fourth term indicates a direct signal contribution of the frequency deviation to the dispersive signal. This contribution can be efficiently used for the realization of the AFC which will be treated together with the discussion of the microwave detector system. The first term is very important for the SNR of the spectrometer. It shows in which way a frequency deviation δ produces an error signal which is detected as an offset to the absorptive signal. Such a frequency deviation may be due to, for example, the frequency noise of the microwave generator. Details will be discussed in the section dealing with the properties of microwave sources. The third term in (8.2.7) is very small, and its influence on the noise properties of the spectrometer can be neglected in practical cases.

If one assumes that the AFC works perfectly ($\delta = 0$), then one can derive an expression for the consequences of a misadjustment of the cavity coupling under similar approximations as for (8.2.7). Let a cavity coupling parameter, λ, be defined by:

$$\lambda = \frac{(Z - R)}{Z}.$$

(8.2.8)

A perfect coupling corresponds to $\lambda = 0$. Neglecting higher order terms in λ, one obtains for the reflected signal U_{mw}^r:

$$U_{mw}^r = U_{mw} \left(\frac{\lambda}{2} - \frac{\lambda^2}{4}\right) - U_{mw} f Q \chi'' \left(\frac{1}{2} - \frac{\lambda^2}{8}\right)$$

$$- i U_{mw} f Q \chi' \left(\frac{1}{2} - \frac{\lambda^2}{8}\right)$$

(8.2.9)

In practical cases, λ is always small ($|\lambda| \ll 1$). The effect of a slight misadjustment of the coupling on the detected resonance signals is, therefore, negligible. The main effect of a wrong coupling is the offset to the absorptive signal as expressed by the first term in (8.2.9). This offset signal can be directly used as the control signal for a circuit for automatic readjustment of the coupling. An absolute stability of λ is imperative, since any fluctuation of λ adds large amounts of noise to the absorptive signal.

Finally, it is of interest whether a mutual influence of frequency and coupling inaccuracies exists. This could be the case if δ and λ have finite values at the same time. It can, however, be verified that to a very good approximation this does not happen. The effects of $\delta \neq 0$ and of $\lambda \neq 0$, with respect to the reflected signal U_{mw}^r, are additive. By combining (8.2.7) and (8.2.9) and neglecting small terms, one obtains as a good approximation of the reflected signal U_{mw}^r:

$$U_{mw}^r = U_{mw} Q^2 \delta^2 + U_{mw} \frac{\lambda}{2} - U_{mw} f Q \frac{\chi''}{2}$$

$$+ i U_{mw} Q \delta - i U_{mw} f Q \frac{\chi'}{2}$$

(8.2.10)

So far, it has been assumed that the microwave circulator (3), Fig. 8.1, is an ideal microwave component. There is, however, always a certain amount of leakage of the circulator which adds a fraction of the incoming microwave amplitude U_{mw} directly to the signal U_{mw}^r. Reflections of microwave power at some points of the transmission line from the circulator to the cavity due to discontinuities of the line impedance, have a similar effect. As can be easily seen from (8.2.10), the real part and the imaginary part of this leakage can be completely counterbalanced by readjusting the cavity coupling λ (8.2.8) and the frequency deviation δ (8.2.6), accordingly. This will be achieved automatically by the cavity coupling device and by the AFC. Good circulators exhibit better than 20 db isolation, which causes no noticeable degradation of the SNR. There is, however, an important point to be observed: Since the leakage microwave amplitude is at least partially counterbalanced by the cavity coupling, any instability of the amount of leakage can add tremendous amounts of noise to the detected signal U_{mw}^r in the same way as instabilities of the cavity coupling do. Therefore, the transmission line between the cavity and the circulator must be of the highest possible quality, and any possible vibration of this line must be avoided. A waveguide should be preferred for this interconnection, and it should be as short as possible with a minimum number of flanges. Any other microwave transmission line in the entire spectrometer is far less critical in this respect, and very compact high performance spectrometers can be built by making use of semi–rigid coaxial microwave cables with SMA connectors and the corresponding coaxial microwave devices.

8.3 Components of ENDOR Spectrometer

The main difficulty in the design of a conventional ENDOR spectrometer arises from the fact that ac signals must be detected with the highest possible sensitivity in the frequency range from about 5 Hz – 5 kHz. There are many sources of noise and interferences in this frequency range. Many of the design features of the spectrometer can, therefore, be considered initially as noise countermeasures. The most important sources of noise are the following (compare Fig. 8.1):

(a) the signal pre–amplifier

(b) the microwave detector

(c) amplitude and frequency noise of the microwave generator

(d) microphonic interference generated by the cavity

(e) interferences from rf leakage

(f) microphonic interference from microwave leakages

First of all, the different parts of the spectrometer will be discussed from the aspect of how to achieve the highest possible performance with respect to an optimum SNR.

8.3.1 Signal Pre–Amplifier

It is the purpose of this component to provide sufficient amplification of the signal delivered by the microwave detector to overcome the input noise level of the lock–in amplifier. It goes without saying that the input noise of

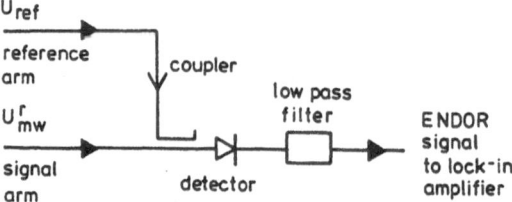

Fig. 8.3. Block diagram for a microwave detector system using one microwave detector diode

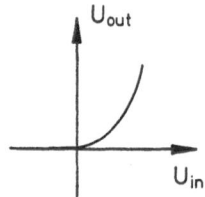

Fig. 8.4. Transfer function of the detector system shown in Fig. 8.3

the pre–amplifier should be considerably lower than the noise contribution of the microwave detector. This is not a major problem with modern operational amplifiers in the range of higher frequencies (above 10 kHz) for EPR applications. For low frequencies, the $1/f$ noise behavior of the amplifier becomes significant. For a given impedance of the microwave detector, a careful selection of the operational amplifier, considering voltage noise and current noise, is necessary. The amplification factor of the amplifier should not be much higher than dictated by noise considerations. Otherwise, the lock–in amplifier can be overloaded by noise spikes too easily. Amplification factors of 10 – 100 are usual. The lower end of the amplifier band width should be as high as possible with respect to the lowest signal frequency of interest. This is, again, to prevent overloading by interference signals which usually exhibit high amplitudes, predominantly in the low frequency range. Some applications, however, require a dc amplifier. If the amplifier is also used for EPR applications and for pre–amplifying the AFC signal, then the upper limit of the band width should extend to at least about 200 kHz.

8.3.2 Microwave Detector

The most popular and very simple microwave detector circuit is sketched in Fig. 8.3. It consists of the microwave detector diode, a directional coupler to add the reference arm microwave power to the signal arm, and a low pass filter behind the diode with a cut off frequency of, for example, 1 MHz. The transfer function of the single diode is shown schematically in Fig. 8.4. Neglecting higher order terms, the amplitude of the output signal, U_{out}, depends on the amplitude of the input signal, U_{in}, for $U_{\text{in}} \geq 0$ as:

$$U_{\text{out}} = a\, U_{\text{in}}^2, \tag{8.3.1}$$

where a is a characteristic constant of the diode. The amplitude of the microwave input signal U_{in} is equal to:

$$U_{\text{in}} = U_{\text{mw}}^r(t) \sin(\Omega t) + U_{\text{ref}} \sin(\Omega t + \phi), \tag{8.3.2}$$

where $U_{\text{mw}}^r(t)$ is the time dependent amplitude of the microwave signal containing the ENDOR information, Ω is the microwave frequency, U_{ref} is the microwave amplitude of the reference arm, and ϕ is the phase of the reference arm relative to the signal arm. The effect of the low pass filter on the output signal can be simulated in the calculation by averaging the output voltage, $\hat{u}_{\text{out}}$, over one microwave period. It follows for the output voltage, behind the low pass filter, $\hat{u}_{\text{out}}$:

$$\hat{u}_{\text{out}} = U_{\text{mw}}^r(t)\, U_{\text{ref}}\, a\, \cos(\phi) + U_{\text{mw}}^r(t)^2 \frac{a}{2} + U_{\text{ref}}^2 \frac{a}{2}. \tag{8.3.3}$$

Equation (8.3.3) shows the disadvantages of this detector: The output voltage is linearly dependent on the signal amplitude $U_{\text{mw}}^r(t)$ only in the limit of small signals. The output voltage depends on the amplitude of the reference arm. Any amplitude noise of the microwave source is, therefore, directly transferred to the signal channel. The microwave power of the reference arm adds a dc offset to the output voltage. In the case of large signals $U_{\text{mw}}^r(t)$ the output voltage does not even depend on the phase ϕ. A considerable improvement can be achieved by the use of two diodes in a bridged configuration. However, these relatively complex arrangements are far out performed by modern double balanced mixers. These are small, integrated devices with four diodes and three terminals. There is one input for the signal channel and the reference arm, respectively, and one output for the output signal U_{out}. The output voltage behind the low pass filter, $\hat{u}_{\text{out}}$, is given by the relation:

$$\hat{u}_{\text{out}} = U_{\text{mw}}^r(t)\, a \cos(\phi). \tag{8.3.4}$$

Again, a is a device constant reflecting the conversion loss which amounts to about $4 - 6$ db. There is only a negligible dc offset caused by the reference arm, and the output voltage is almost independent of the microwave amplitude of the reference arm, U_{ref}. The linearity for $U_{\text{mw}}^r(t)$ is excellent up to microwave power levels of about 10 mW. However, a microwave power of at least 10 mW for the reference arm is required.

    The technology of the double balanced mixers offers a very convenient way to obtain the dc AFC signal, which can be directly used to control the frequency of the microwave source. Usually, the AFC signal is obtained by

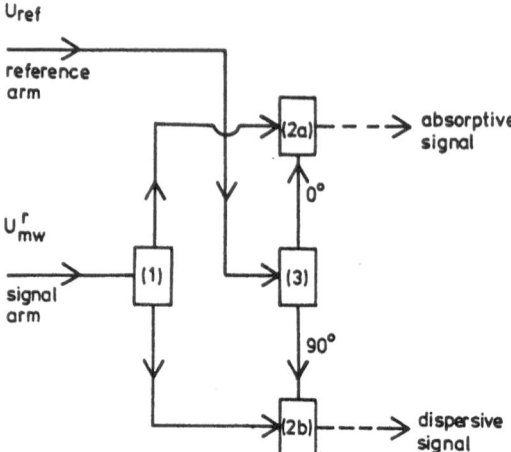

Fig. 8.5. Comfortable detector system for simultaneous detection of absorptive and dispersive signal: (1) microwave power splitter; (2a, 2b) double balanced mixers with dc output; (3) 90° hybrid

modulating the frequency of the microwave generator by a frequency of, e.g., 80 kHz, detecting the microwave absorptive signal due to this modulation [first term in (8.2.10)], and obtaining the dc AFC signal by a lock–in amplifier. It is, however, much easier to detect the dispersive and absorptive microwave signals simultaneously by the introduction of a second double balanced mixer. The circuitry is shown in Fig. 8.5. The signal arm carrying the signal U^r_{mw} is split into two lines by the power divider (1). The microwaves of these two lines are in phase with each other. Each of these lines is connected to the signal input of a separate double balanced mixer (2a and b). The microwave reference arm is also split into two lines by the application of a 90° hybrid (3), and these two lines are 90° out of phase with each other. These lines are connected to the corresponding reference inputs of the mixers. If the phase shifter, (11), Fig. 8.1, is properly set, the output of one mixer delivers the absorptive signal, whereas the second mixer provides the dispersive signal. This signal itself is already the dc signal necessary to control the frequency of the microwave source. Formally, the AFC operation is now provided by the fourth term in (8.2.10). Frequency modulation of the microwave source is no longer necessary. The band width of the signal delivered by the dispersive double balanced mixer can be made considerably larger than the band width of the AFC lock–in amplifier. This, in turn, allows a much larger loop amplification of the AFC circuitry with the consequence of a superb stability of the AFC.

So far, the detector noise observed on top of the output voltage $\hat{u}_{out}$ has not been considered. In the frequency range above about 10 ... 30 kHz, the noise level of the detector is approximately equal to its conversion loss, about

6 db. However, this level increases approximately proportionally to $1/f$ for lower frequencies f. It is, therefore, essential to provide some low noise pre–amplification of the signal before it can be obscured by the detector noise. This pre–amplification can be achieved quite easily with a low noise microwave amplifier [8.2]. The amplifier is simply inserted into the signal path between the circulator (3) and the detector system (8) (Fig. 8.1). For proper operation of the amplifier in the presence of microwave power reflected back from the detector system, a small microwave isolator should be installed directly behind the amplifier. The microwave amplifier should have an amplification factor of at least 20 db. For signal detection at very low frequencies (5 Hz), as much as 50 db may be necessary to overcome the noise of all the elements of the spectrometer behind the amplifier. Suitable GaAs–FET amplifiers are commercially available for frequencies up to about 40 GHz (Q–band). The noise levels of these amplifiers are very low for the X–band microwave range (e.g., 1.6 db for $8.5 - 9.6$ GHz) and rise to about 6 db for Q–band frequencies. As for all microwave devices, the progress in technology is very rapid and, therefore, much better amplifiers will certainly be available in the near future, especially for the high frequency microwave bands. It is apparent from the amplifier noise levels mentioned, that the insertion of an amplifier will also significantly improve the sensitivity of EPR spectrometers [8.2].

There is an easy way to check whether the overall noise figure of the spectrometer is exclusively determined by the input noise of the microwave amplifier, which is absolutely the lowest noise level of the spectrometer one can achieve:

This is done by suitably terminating the microwave input of the amplifier and reading the rms noise output of the lock–in amplifier, (12) in Fig. 8.1, for given values of amplification, band width, and reference frequency of the lock–in amplifier. If the power supply voltage of the microwave pre–amplifier is then switched off, the rms noise level should fall to a considerably lower value due to the noise of the spectrometer without the microwave amplifier. If this second noise level is not considerably lower than the first one, then either the amplification of the amplifier is too low or the noise level of the detector system is too high. The challenge in spectrometer design is to maintain the noise level observed with the amplifier terminated and switched on (first measurement) under the normal operating conditions of the spectrometer during the experiments.

Given the great potential for noise reduction by the application of microwave pre–amplification, one might assume that this technique is always straightforward. There are severe drawbacks. Due to high microwave–amplification factors, even the smallest amount of microwave power reflected from the cavity due to some instability, for example, of the cavity coupling, tends to overload the amplifier output or the detector system. It is exactly this problem which makes the application of microwave pre–amplification difficult. This is particularly true if one needs both high spectrometer sensitivity

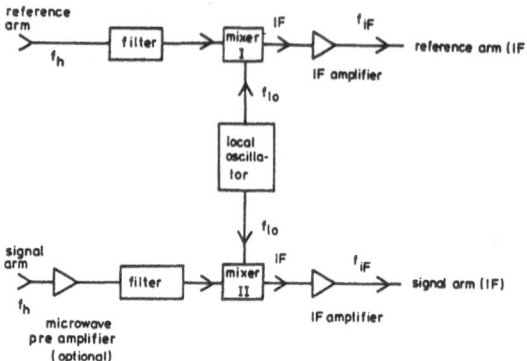

Fig. 8.6. Block diagram of essential parts of a superheterodyne spectrometer. For explanations see text

and relatively high microwave power levels. Because of stability problems, the amplification should never be higher than dictated by the required spectrometer sensitivity. In high performance spectrometers the amount of microwave pre–amplification is, therefore, adjustable. This can, for instance, be achieved by cascading a selected number of amplifiers.

At very high microwave frequencies, direct amplification of the microwave ENDOR signal is no longer feasible because of the lack of suitable amplifier technology. In this situation it is necessary to make use of the superheterodyne technique. In order to circumvent amplification and detection of the high frequency microwave, the microwave frequency of the signal arm and of the reference arm is downconverted, (10) in Fig. 8.1, by the application of two microwave mixers and an additional microwave oscillator, the so–called *local oscillator*. This is shown in Fig. 8.6. The downconverted lower frequency of the microwave is termed the *intermediate frequency* (IF). One selects an IF frequency low enough to be easily handled by available microwave technology. For a given high microwave frequency, f_h, the value of the IF frequency, f_{IF}, is determined by the fixed frequency local oscillator frequency, f_{lo}, by:

$$f_{IF} = |f_h - f_{lo}|. \tag{8.3.5}$$

It follows that for fixed values of the local oscillator frequency, f_{lo}, and of the IF frequency, f_{IF}, there are two possible values for the microwave frequency f_h, $f_h^{(1)}$ and $f_h^{(2)}$, which are always detected simultaneously:

$$f_h^{(1)} = |f_{lo} - f_{IF}| \tag{8.3.6}$$

and

$$f_h^{(2)} = f_{lo} + f_{IF}. \tag{8.3.7}$$

Indeed, only one of these two microwave frequencies can be equal to the frequency of the main microwave generator and of the cavity resonance and can, therefore, carry the ENDOR signal information. At the second microwave frequency, only the noise contributions of the main generator or of other components in front of the mixers are converted down into the two IF channels, the signal arm and the reference arm. To avoid the detection of these additional noise contributions, the use of suitable high frequency microwave filters is recommended. This is shown in the block diagram of the superheterodyne microwave spectrometer in Fig. 8.6. For proper operation of the filters, the two frequencies $f_h^{(1)}$ and $f_h^{(2)}$ should not be too close to each other, this means the IF frequency must not be too low.

For example, for a cavity and a main microwave generator designed for operation in the frequency range of $f_h = 50 \ldots 54$ GHz, the local oscillator selected operates at the fixed frequency $f_{lo} = 42$ GHz, and consequently, the IF frequency is in the range of $f_{IF} = 8 \ldots 12$ GHz. The passband for the filters should be selected to fall in the range of $45 \ldots 60$ GHz. These filters stop noise contributions of, for example, the main generator which could be detected in the range of $30 \ldots 34$ GHz.

The question now arises under what circumstances must the relatively complex superheterodyne technology be used. It is presently required for spectrometers operating at microwave frequencies above about 40 GHz. Also, if high microwave amplification factors are necessary for ENDOR signal detection at very low frequencies (e.g., 5 Hz), the corresponding microwave amplifiers for frequencies above about 20 GHz are still relatively expensive. The same is true for high performance double balanced mixers and other microwave components operating at frequencies above 20 GHz. Therefore, it is probably less expensive to provide the main part of the amplification and detection of absorptive and dispersive signals at a low IF frequency, and to use only a small amplifier in front of the mixer for the signal channel. This amplifier then needs an amplification factor just high enough to overcome the relatively low noise level of the mixer plus that of the IF amplifier (about 7 – 8 db). If the spectrometer is intended for operation at different microwave bands, then the superheterodyne technique offers an economical solution, since the same IF microwave components and related parts of the spectrometer can be used for all bands.

8.3.3 Microwave Sources

The most popular microwave source used in almost any EPR spectrometer is the reflex klystron. Despite the development of high performance microwave oscillators using solid state devices, for ENDOR applications the klystron can still offer the lowest values of both frequency and amplitude noise. There are, however, some drawbacks of klystrons which are independent of the noise figures of the device itself.

Klystrons need relatively complex power supplies. In order to achieve a very good frequency stability in the time scale from 1 s to about 0.1 ms, the voltages of the power supply must be extremely "clean" with respect to ripple, noise and line interferences down to the low frequency range. This is not easily achieved. In addition, it is very difficult to transfer the low dc voltage for frequency control in the presence of the high reflector and resonator voltages, under conditions of very high band width and minimum introduction of noise. Finally, klystrons are unattractive from an economic point of view due to their limited lifetime.

The only disadvantage of solid state devices, on the other hand, is their higher noise level. The microwave power level of these devices is sufficient for the ENDOR applications discussed here (low power ENDOR for solid state defects as opposed to high power ENDOR for applications in chemistry).

The interesting question now arises as to how much the noise level of the microwave source really influences the total noise figure of the spectrometer under normal working conditions. This is strongly dependent on the conditions for the particular ENDOR experiment. The most relevant type of noise of the microwave source is the frequency noise. Frequency instabilities are converted via the first term in (8.2.10) to amplitude noise, which will be noticeable when this contribution exceeds the total noise of the microwave detector system. The importance of the first term in (8.2.10) depends linearly on the microwave amplitude U_{mw}. Therefore, the microwave generator noise is generally not a problem if the ENDOR experiment can be carried out at sufficiently low microwave power levels. This is the case for many defect systems which need only low microwave power to saturate the EPR transitions. At higher power levels, the influence of the generator frequency noise can be greatly reduced by a properly working AFC system. In this case, the AFC must cancel any frequency instability in a time scale considerably shorter than that defined by the inverse of the modulation frequency. For example, if, in an ENDOR experiment the rf field is modulated at a rate of 1 kHz and, consequently, the ENDOR signal lock–in amplifier operates at this frequency, the AFC system must be able to respond to a frequency instability at least within 0.1 ms. Otherwise, *Fourier* components of the frequency noise can be present significantly in the frequency range around 1 kHz and will be detected. This condition is, however, still not sufficient to achieve an optimum suppression of frequency noise by the AFC. Due to the square law dependence of the first term in (8.2.10) on the frequency deviation δ, high frequency components of frequency instabilities are downconverted into the frequency range of the modulation. In a simple example, let 100 kHz be the upper limit of the AFC band width. This means the loop amplification and, thus, the operation of the AFC circuitry is no longer relevant at frequencies above 100 kHz. There are, however, frequency noise contributions of the microwave generator, even in the MHz range. For example, consider two spectral lines of this noise spectrum, one at $\omega_1 = 1$ MHz with an amplitude $\Delta\Omega_1$, and the other $\omega_2 = 1.001$ MHz with an amplitude $\Delta\Omega_2$. The frequency deviation parameter δ then yields:

$$\delta = \frac{\Delta\Omega_1 \sin(\omega_1 t) + \Delta\Omega_2 \sin(\omega_2 t)}{\Omega}, \tag{8.3.8}$$

and one obtains for the low frequency components of δ^2:

$$\delta^2 = \frac{\Delta\Omega_1 \, \Delta\Omega_2 \, \cos[(\omega_2 - \omega_1)t] + \frac{1}{2}\Delta\Omega_1^2 + \frac{1}{2}\Delta\Omega_2^2}{\Omega^2}. \tag{8.3.9}$$

Thus, a noise signal is added to the absorptive signal via the first term in (8.2.10) at a frequency of $\omega_2 - \omega_1 = 1$ kHz and with an amplitude, U_{noise}, and this is not cancelled by the AFC:

$$U_{\text{noise}} = U_{\text{mw}} \, Q^2 \, \frac{\Delta\Omega_1 \cdot \Delta\Omega_2}{\Omega^2}. \tag{8.3.10}$$

In order to estimate the total amount of noise converted down to frequencies around 1 kHz, one has to integrate over all frequencies ω_2 above 100 kHz in this example. Qualitatively it follows from this simplified example that the noise level decreases as the band width of the AFC circuit increases, provided the AFC circuit itself exhibits a sufficiently low noise level. At first glance it is surprising that the high frequency noise contributions beyond the cut off frequency of the AFC, rather than the low frequency instabilities of the microwave generator frequency, predominantly determine the noise level of the spectrometer. The amplitudes $\Delta\Omega_1$ and $\Delta\Omega_2$ of the frequency excursions at a high frequency rate, the so–called *noise bottom* of the generator's frequency noise, are usually higher for solid state devices than for klystrons. As already mentioned above, in many of the actual ENDOR experiments the microwave amplitude U_{mw} can be made low enough so as to avoid relevant noise contributions according to (8.3.10), and this is also the case for solid state microwave oscillators, particularly for spectrometers with high performance AFC circuits. The most suitable solid state oscillators for ENDOR spectrometers currently available are the so called YIG–tuned *Gunn* or transistor oscillators. Devices with bipolar transistors offer better performance than devices with GaAs–FET transistors. The frequency of these oscillators is determined by magnetic resonance of a Yttrium Iron Garnet sphere rather than by a resonant cavity or a dielectric material resonator. This offers the convenience of wide band, electrical tunability, just by controlling the current through an internal coil which determines the magnetic field for the resonance. These devices are available for frequencies up to about 40 GHz. In general, it is extremely difficult to theoretically predict which device will offer the best noise level for a given application. The specifications provided by the manufacturers are usually not detailed enough to make estimations according to (8.3.10).

In the case of an AFC circuit with a rapid response time, it is no longer possible to detect the ENDOR dispersive signal. It will be cancelled by the AFC in the same way as a frequency instability of the microwave generator. From a point of view of noise problems, the detection of dispersive signals is, therefore, very difficult. If, for some reason, their detection is desired, then the frequency of the microwave source must be additionally stabilized by some other means. One possibility is to stabilize the frequency of the microwave generator by a harmonic of a crystal oscillator with the help of a high band width special AFC. Alternatively, an additional high Q cavity can also be used for this purpose. However, in both cases, frequency instabilities of the ENDOR cavity due to microphonic effects are no longer cancelled and may cause considerable noise contributions. Whenever possible, it is best to look for high sensitivity ENDOR measurement techniques based completely on the detection of absorptive signals.

The noise problems discussed so far arise from the frequency noise of the microwave generator which is the most important source of noise. There are, however, also conditions under which the amplitude noise of the generator may contribute significantly to the total noise level. One might expect that the ENDOR signal depends only weakly on B_1, and thus on U_{mw}, since the EPR signal is saturated. Amplitude fluctuations of U_{mw} should therefore play no major role. However, the effect of the imaginary part χ'' of the static susceptibility of the sample is not negligible in cases where the EPR transition does not easily saturate and the ENDOR effect is small. The large dc EPR absorptive signal during the ENDOR experiment, in this case, will then be counterbalanced by the cavity coupling device which adjusts the coupling parameter λ, (8.2.8) accordingly (also 8.2.10). Thus, there are two large components of the absorptive signal which cancel each other. However, the signal due to χ'' depends in a nonlinear way on U_{mw} because of the partial saturation of the EPR, whereas the signal due to the setting of the coupling depends linearly on U_{mw}. Consequently, the resulting signal does depend on amplitude fluctuations of U_{mw}. This effect can only be minimized by selection of a microwave generator with low amplitude noise. Solid state devices with *Gunn* diodes or with bipolar transistors exhibit better performance than GaAs–FET transistor oscillators; a klystron may exhibit an even better performance provided the power supply is excellent.

Amplitude fluctuations of the microwave of the reference arm, (10), Fig. 8.1, are no longer important if double balanced mixers are used as microwave detectors.

8.3.4 ENDOR Microwave Cavities

The microwave cavity is a very critical part of the spectrometer. It influences the performance of the spectrometer to a very large extent. There are many different design approaches to ENDOR cavities. However, an ideal cavity, or even a "very good" one, does not yet exist. The constraints for ENDOR cavities are more complex than for EPR cavities since, in addition to the microwave field, B_1, a sufficiently high rf magnetic field, B_2, has to be produced at the sample. The sample must usually be cooled to liquid helium temperatures with almost no generation of microphonic noise; the cavity coupling must be adjusted with extremely high precision, and rf interference effects in the cavity must be avoided. The capability of varying the resonance frequency of the cavity by up to 10% is also desirable in order to determine nuclear g values of isotopes by magnetic field shift measurements. Finally, the product of the filling factor f and the quality factor Q, fQ, must be as high as possible for high sensitivity (Sect. 8.2.). It is quite clear that this optimization will strongly depend on the type of sample to be investigated. Critical parameters of the sample are its geometry and dimensions, its dielectric constant, and its microwave and rf loss behavior. Samples of solids may be very different with respect to these parameters and, therefore, only general cavity design guidelines will be discussed below.

The first important point is that ENDOR experiments must always be carried out at a given field, B_1, which is optimized experimentally. The performance of different cavities must, therefore, be compared under the condition of a constant field, B_1, rather than constant microwave amplitude, U_{mw}, applied to the cavity. Consider, for example, a sample with a given volume, V. Then the product of the cavity Q and the microwave power applied to the cavity, $Q\,P_{\mathrm{mw}}$, is proportional to:

$$Q\,P_{\mathrm{mw}} \propto \frac{B_1^2 V}{f}, \tag{8.3.11}$$

where it is assumed that B_1 is essentially constant over the sample volume V. The microwave power P_{mw} is proportional to the square of the amplitude U_{mw}. According to (8.2.10) the detected absorptive signal, $\mathrm{Re}\,(U_{\mathrm{mw}}^r)$, is then proportional to:

$$\mathrm{Re}\,(U_{\mathrm{mw}}^r) \propto B_1 \sqrt{\frac{V}{fQ}}\,Q^2\,\delta^2$$

$$- B_1 \sqrt{V f Q}\,\frac{\chi''}{2} \tag{8.3.12}$$

If the microwave field B_1 is small, then the contribution of the first term in (8.3.12) to the total noise of the spectrometer is negligible compared to the noise of the detector system which is independent of any other spectrometer parameter. In this case, according to the second term in (8.3.12), the SNR is proportional to:

$$\text{SNR} \propto \sqrt{V f Q}\, \chi'',\qquad(8.3.13)$$

which exhibits only a relatively weak dependence on the filling factor and on the quality factor of the cavity. If, on the other hand, B_1 is large, the noise may be determined primarily by the first term in (8.3.12), and the SNR is according to (8.3.12), proportional to:

$$\text{SNR} \propto \left(\frac{f}{Q}\right)\chi''.\qquad(8.3.14)$$

This set of conditions favors cavities with lower Q. These simple approximations are useful in examining trends; however, they must be interpreted cautiously. The SNR, according to (8.3.14), does not diverge for very low values of Q because this will result in other problems; for example, the AFC will no longer work properly. As mentioned above, one tries to avoid the condition assumed in (8.3.14) by the selection of a high performance microwave source and an efficient AFC circuit.

Another problem encountered in designing ENDOR cavities is the need for low sample temperatures. One standard solution is to keep the cavity at room temperature while the sample is cooled by cold helium gas which flows through an evacuated quartz dewar inside the cavity. Those flow systems are available commercially. The main disadvantage of such a solution is that unavoidable turbulance in the gas stream causes microphonic effects which can significantly exceed other noise sources. This effect is more serious at lower temperatures, where the density and velocity of the gas increase. Lowering the gas pressure reduces the turbulence, but at the same time decreases the cooling power. Bath cryostats are unacceptable because one needs the ability to vary the temperature over a wide range. A more successful approach is to cool the sample by direct mechanical contact with a cold finger or sample holder which is cooled itself by some means external to the cavity. One then needs only one quartz tube inside the cavity to provide the necessary vacuum for heat insulation [8.3]. The problem with this system is that the temperature of the sample will not be well–defined, particularly for those samples with low heat conductivity. It is generally difficult to achieve sample temperatures below about 10 K with this technique.

The most advanced solution to the problem is a construction where the entire cavity is at the temperature of the sample, and the sample is cooled by contact with a gas atmosphere inside the cavity [8.4]. The cavity body, in turn, is cooled from the outside by a flow cryostat system, but with no gas flowing through the inner part of the cavity. This method of cooling the sample requires a much more sophisticated cryogenic technique compared to the solutions discussed above. In order to keep the cavity at very low temperatures

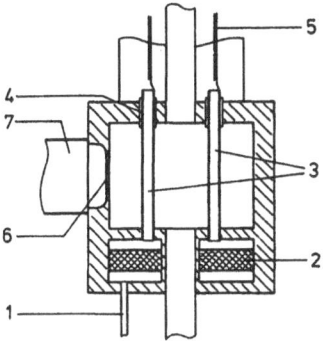

Fig. 8.7. Schematic representation of an ENDOR TE_{011} cavity with integrated Helium flow cryostat: (1) liquid He supply; (2) heat exchanger; (3) silver rods for B_2 field generation, cooled by He gas flow; (4) PTFE insulators; (5) rf transfer lines; (6) coupling iris of cavity; (7) wave guide

without excessive consumption of liquid helium, a very efficient pre–cooling of all interconnecting parts which could transfer heat to the cavity, and of the room temperature radiation shield around the cavity, is essential. This pre–cooling is achieved by the cold He gas leaving the flow cryostat. No liquid nitrogen is used. The waveguide to the cavity is made of stainless steel, which is an excellent thermal insulator at liquid He temperatures, and a special rf transfer line is necessary, which will be described later. By such methods the liquid He consumption under normal operation at a cavity temperature of 4 K can be reduced to about 1 l/h maximum; see Fig. 8.7 for the cooling system of the cavity.

No matter what type of microwave resonator is used, any resonator needs a suitable coupling device, and this is a very critical part, with respect to the introduction of noise. In order to vary the amount of coupling, some part, such as a plunger, has to be displaced mechanically. Particularly if a high microwave pre–amplification is used in the signal channel, this displacement has to be achieved with sub–micrometer accuracy and stability. A high precision motor–driven micrometer can be used for automatic coupling control. The coupling has to be readjusted almost constantly during the measurements. For example, even a slight variation of the gas pressure in the cavity makes a readjustment of the coupling necessary. The motor must not transfer vibrations to the coupling device; it should, therefore, run at a very low speed. If piezo devices are used for additional precise adjustment of the coupling, the output voltage of the driver must be free of ripple. Strong noise contributions to the absorptive signal were observed due to otherwise unimportant small harmonics of the line frequency on top of the piezo control voltage. If the cavity is at low temperature, the motion of the micrometer must be transferred to the plunger at the cavity, e.g., via a stainless steel tube.

The main technical problem in the design of an ENDOR cavity is the generation of the rf magnetic field, B_2. This field is produced by an rf current flow through suitable conductors in the cavity. These conductors, however, must not disturb the microwave mode of the cavity. They must not be parallel to directions of the microwave electric field. It follows that ideal shapes of rf conductors are, in many cases, incompatible with the desired cavity construction and vice versa. Two basic designs are widely used:

(i) The B_2 field is generated by a cylindrical coil in the center of a cylindrical cavity [8.5]. The sample is inserted into the center of the rf coil. The cavity mode is of the TM type, usually TM_{110}. This construction has the advantage that only a rather small volume defined by the dimensions of the coil is filled by the rf field. Strong B_2 fields can, therefore, be efficiently generated by relatively small rf currents and voltages. This construction is a good choice where the generation of high B_2 fields is of prime interest. This is, however, usually not the case for ENDOR investigations of defects in solids. The Q of the cavity is lower than achievable for other designs by about a factor of two. The position of the coil in the cavity is somewhat critical; it needs some means for mechanical stabilization. Also, extra elements are necessary for the modulation of the static magnetic field.

(ii) The B_2 field is generated by an rf current flow through four straight rods parallel to the axis of a cylindrical TE_{011} cavity [8.6]. The sample is, again, in the center of the cylinder. The rods are arranged in a way to form a *Helmholtz* pair of coils, with one loop per coil for optimum rf field homogeneity. The same rods can be used simultaneously for the modulation of the static magnetic field. The rods can be made very solid, providing an excellent mechanical stability. The cavity Q is at least twice as high as in the case (i), up to 15,000. The B_1 field along the cylindrical axis has a broad maximum in the center of the cavity and is, thus, not very homogeneous. However, if one uses ENDOR only as a tool for defect analysis, and is not interested in investigating ENDOR mechanisms, this inhomogeneity of the field is rather advantageous. Since samples of solids often cannot be made very long, the B_1 maximum provides a better filling factor in the case of the TE_{011} cavity compared to the TM_{110} cavity.

A considerable disadvantage of the B_2 generation by rods is their low rf impedance. About 10 Amps of rf current per rod are needed to generate a B_2 field of about 0.5 mT rms in a typical X–band TE_{011} cavity. The power of the rf generator needed to produce a given B_2 field by the rod arrangement is about four times higher than needed to produce the same field by the coil design. However, the rf problems can be solved and one can exploit the advantages of the TE_{011} cavity. The basic design principle of a TE_{011} cavity with integrated flow cryostat is shown in Fig. 8.7. Liquid He is supplied through the capillary (1) from the cryostat. The main part of the cooling system is the heat exchanger (2) which consists of small bronze metal spheres of 50 micrometer diameter sintered together to form a plate of 5 mm thickness. All the helium vaporizes when passing through this plate. Not shown are facilities for temperature measurement and control. The cold He gas then flows through four silver tubes (3) with 3 mm outer diameter (only two are shown). These tubes act as rods for the B_2 field generation. Silver has a high heat conductivity at low temperatures and, thus, there is no noticeable temperature gradient along these rods. The He gas atmosphere inside the cavity is very efficiently cooled by these rods and there is no flowing gas inside the cavity. The rods are insulated from the cavity top by little PTFE rings (4). The He gas flowing out of the tubes provides pre–cooling, in particular, of the rf transfer lines (5). Also shown schematically are the coupling iris (6) and the wave guide (7). The rf line must be able to carry about 20 Amps of rf current, it should not be too lossy, and it should only have low heat conductivity. This can be achieved over the entire temperature range from 3.5 K – 300 K. PTFE stripes are used, with about 0.5 mm thickness and about 40 mm width, which are coated on both sides with a copper layer of about 15 micrometer thickness. These two layers form a low impedance rf transmission line which operates in the frequency range from 0.5 – 200 MHz.

Very many different designs of ENDOR cavities are possible [8.7]. There is, however, one additional general point which has to be observed for proper operation of the resonator for ENDOR purposes. This is the effect of nonlinear microwave and rf losses. Any insulating material and any metal surface has some loss for microwaves and for radio frequency. As experience shows, these losses depend on the microwave amplitude and on the rf amplitude in a nonlinear way. If some material in the cavity is exposed to microwave and rf at the same time, then there is a mutual influence of different losses: The amount of the microwave loss of the cavity, which is measured by the absorptive signal, depends on the rf amplitude. If the rf is modulated, this effect can give rise to very large and unstable background signals. The only way to minimize this effect is the use of low loss materials in the cavity and for the rf transmission lines. The cavity must be designed in a way that along any metal surface in a good approximation, there flows either a microwave current or an rf current, never both at the same time.

There are strong effects which can give rise to large interference signals. These are *radar effects* due to microwave leakages of the cavity or of other parts of the spectrometer. The microwave power applied to the cavity is of the order of mW, the minimum microwave power detected in an actual experiment is in the order of 10^{-20} W. The slightest microwave leakage of the spectrometer causes a microwave field in the laboratory, which is finally coupled to the detector system via the same leakages. The microwave field in the laboratory is, however, not stable. Any vibrating part will modulate this field and will

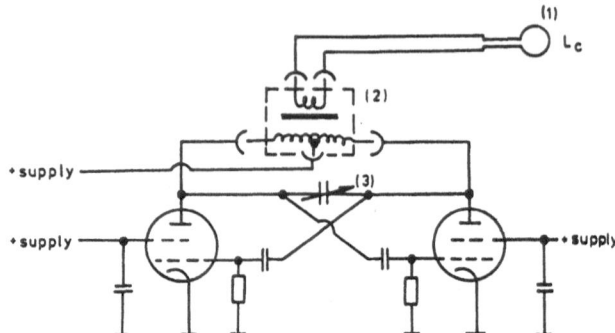

Fig. 8.8. Simple rf power oscillator using electron tubes. (1) cavity element with inductance L_c to generate a B_2 field; (2) interchangable transformer unit determining the frequency range; (3) variable capacitor (After [8.8])

show up as a corresponding dispersive or absorptive signal. These interference signals have almost no *Fourier* components in the frequency range around 100 kHz, and, therefore, cause no problems for EPR experiments; however, the situation is completely different for detection in the audio–frequency range.

8.3.5 Radio–Frequency Generators

From an rf technology point of view, it is very inconvenient that rf currents with very different frequencies have to flow through elements in the cavity with no possibility to adjust these elements according to the frequency. A very simple approach to solve this problem is sketched in Fig. 8.8. The inductance of the cavity elements for B_2 generation, L_c (1), is transformed to a considerably higher value by the output transformer (2) of a power oscillator. The variable capacitor (3) forms, together with this transformed inductance, a resonance circuit which determines the frequency of the oscillator. The frequency is variable by the capacitor by about a factor of 2.5. Different frequency ranges can be obtained by different transformers which must be exchanged accordingly [8.8]. An advantage of this approach is that the power of the oscillator can be relatively small (30–50 W), since the current through the cavity elements is multiplied by the quality factor of the resonance circuit. It is, however, very difficult to achieve frequencies above about 50 MHz with this concept. Sufficiently convenient computer control of the frequency is almost impossible. Triple resonance experiments are also impossible because one oscillator with one resonance circuit cannot produce two different frequencies at a time.

   A much better solution is offered by a broad band technique without any resonant elements. In this case, the rf is generated by a computer controlled synthesizer and amplified by a broad band power amplifier. The only problem

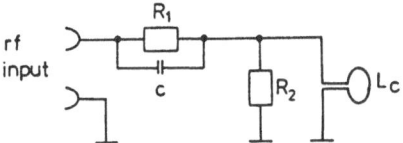

Fig. 8.9. Circuit to match the rf cavity element with inductance L_c to the output of a broadband power amplifier

is to couple the output of the amplifier to the cavity elements. For triple reso-nance experiments two synthesizers are used. Their output rf signals are added and amplified simultaneously by the broadband amplifier. Power amplifiers are always designed to deliver power to a resistive load, usually 50 Ohms. The cavity elements to produce the B_2 field, however, behave (to a good approxi-mation) like an inductance, L_c, with a purely imaginary frequency dependent impedance. If one simply connects this inductance to the amplifier output, many amplifiers will withstand this, however, the performance, with respect to frequency response and generation of harmonics, is too poor for most ap-plications. The problem can be solved by the circuit shown in Fig. 8.9. The resistors R_1 and R_2 both have a value, R, of e.g., $R = R_1 = R_2 = 50$ Ohms. It can be easily seen that for very low frequencies, the capacitor, C, and the resistor, R_2, have only a negligible influence. The total impedance seen by the amplifier is $R_1 = R$. For very high frequencies, on the other hand, the influence of R_1 and of L_c is negligible, and the impedance for the amplifier is $R_2 = R$. For any rf frequency, Ω_{rf}, the total impedance of the circuit, R_{tot}, seen by the amplifier is equal to:

$$R_{tot} = \frac{1}{\left(\frac{1}{R} + i\,\Omega_{rf}\,C\right)} + \frac{1}{\left(\frac{1}{i\,\Omega_{rf}\,L_c} + \frac{1}{R}\right)}. \qquad (8.3.15)$$

If one now selects the capacitor C as:

$$C = \frac{L_c}{R^2}, \qquad (8.3.16)$$

then it can be easily verified that the frequency dependent imaginary part of (8.3.15) vanishes for all frequencies, and it follows:

$$R_{tot} = R \text{ (for any frequency)}. \qquad (8.3.17)$$

Thus, it is possible to operate the broad band power amplifier under optimum conditions without tuning any element as a function of the actual frequency.

9. Experimental Aspects of Optically Detected EPR and ENDOR

In this chapter some of the essential features of spectrometers for optical detection of EPR and ENDOR are described. Unless necessary, we will not distinguish between ODEPR and ODENDOR, but simply refer to ODMR. Basically, there are two kinds of ODMR spectrometers. In one, the ODMR effect is measured as a microwave- or rf–induced change of the fluorescence or phosphorescence light intensity. It is often achieved by adding optical components to an ordinary EPR spectrometer and providing it with a special cavity (emission–type spectrometer). Since such spectrometers were described elsewhere previously [9.1–3], they will be dealt with only briefly here. The other type of ODMR spectrometer is based on a spectrometer to measure the magnetic circular dichroism of the absorption (MCDA) or the magnetic circular polarization of emitted light (MCPE), be it fluorescence or phosphorescence, to which the microwave and rf components are added (MCDA–type spectrometer). This type of spectrometer was described earlier [9.4,2] for the special purpose of investigating excited states of F centers in alkali halides. It has since been developed further for a more general use to study EPR and ENDOR of ground and excited states of many types of defects, and will, therefore, be described here in more detail. Several of the more critical components of the MCDA–type spectrometer are discussed in view of experience gathered in the Paderborn group over the last ten years. The MCDA–type spectrometer requires a higher degree of precision for the optical components compared to the emission–type spectrometer. It is usually not used for conventional detection of EPR, although this would be possible by including a microwave bridge. The emission–type spectrometer is usually also operated as a conventional EPR spectrometer.

9.1 Sensitivity Considerations

In Sect. 4.12 the basic sensitivity of ODMR was discussed for both the emission and MCDA methods. The discussion in this section focuses on the experimental conditions which should be met in order to optimize the sensitivity and signal–to–noise–ratio (SNR), respectively.

9.1.1 Magnetic Circular Dichroism of Absorption

As pointed out in Chap. 4, the noise is limited by the shot noise. Thermal noise is not a limiting factor except in the infrared. The SNR is given by the intensity of the light leaving the sample and arriving at the detector. This is dependent on the power of the light source (see below), on the attenuation of the image–forming optical components, including monochromators, filters, etc., and the total absorption and reflection of the sample. Of course, the SNR also depends on the quantum efficiency of the detector, which can be a serious problem, particularly in the infrared range. Normally, the light intensity is so high that the quantum noise is larger than the noise from detectors and the electronics. However, the light intensity may be limited by photochemical reactions of the defects under study, particularly in the MCDA method.

The MCDA effect is proportional to the occupation difference between the magnetic sublevels in thermal equilibrium, disregarding optical pumping effects (Sect. 4.8). Therefore, in principle, it is advantageous to work at very low temperature and high magnetic fields. A temperature of 1.5 ... 1.7 K is attainable by pumping on liquid He. Although a lower temperature than that could be realized by the use of a ^{3}He cryostat, it must be borne in mind that the thermal load by optical, microwave and rf absorptions by the sample will exceed the thermal load such a cryostat is normally capable of withstanding.

A large number of defects with an intense absorption band relative to any background absorption of the sample, or relative to the absorption bands due to other defects simultaneously present, seems desirable for intense MCDA signals. However, if the absorption is too high, then not enough light will pass through the sample and arrive at the detector. One, therefore, looses sensitivity in spite of the enhanced MCDA effect.

The optimum range for the optical density (O.D.) of the sample is approximately O.D. = 0.9, which is seen as follows [9.5]. From (4.3.3,5) we have for the MCDA

$$\varepsilon = \frac{d}{4}(\alpha_r - \alpha_l), \tag{9.1.1}$$

$$\alpha_{r,l} = -\frac{1}{d} \ln\left(\frac{I_{r,l}}{I_0}\right) = -\frac{1}{d} \ln\left(\frac{N_{r,l}}{N_0}\right), \tag{9.1.2}$$

with N representing the number of detected photons. The shot noise is

$$|\Delta N| = \sqrt{N}, \tag{9.1.3}$$

$$|\Delta\alpha| = \frac{1}{d}\left|\frac{\Delta N}{N}\right| = \frac{1}{d\sqrt{N}}, \tag{9.1.4}$$

$$|\Delta\varepsilon| = \frac{d}{4}2|\Delta\alpha|, \tag{9.1.5}$$

assuming $\alpha_r \approx \alpha_l$, as is usually observed. It follows:

$$|\Delta\varepsilon| = \frac{1}{2}\frac{\sqrt{N}}{N} = \frac{1}{2\sqrt{N_0}}\,e^{\left(\frac{\alpha d}{2}\right)}. \tag{9.1.6}$$

In order to optimize with respect to concentration, we introduce the concentration related quantities α' and ε' by setting $\alpha = c\alpha'$ and $\varepsilon = c\varepsilon'$. One obtains for the SNR

$$\text{SNR} = \left|\frac{\varepsilon}{\Delta\varepsilon}\right| = 2\sqrt{N_0}\,\varepsilon'\,c\,\exp\left(-\frac{\alpha'cd}{2}\right). \tag{9.1.7}$$

The optimum SNR, as a function of concentration, is obtained by calculating $\delta(\text{SNR})/\delta c = 0$:

$$\frac{\delta(\text{SNR})}{\delta c} = 2\sqrt{N_0}\,\varepsilon'\left(1 - \frac{\alpha'dc}{2}\right)\exp\left(-\frac{\alpha'cd}{2}\right). \tag{9.1.8}$$

Equation (9.1.8) vanishes for $\alpha'dc = \alpha d = 2$, which corresponds to an O.D. of $|\log(e^{-2})| = 0.87$.

It is, therefore, not advantageous to work with either too low or too high optical densities. If the oscillator strength of the defect under study is low, then one should try to use a thick enough sample in order to have $\alpha d = 2$. With the cavity design of Sect. 9.4.9 this is usually possible.

When using the modulation technique to measure the MCDA or the MCPE, modulation frequencies of 30 ... 50 kHz give good noise separations from the $1/f$ noise from microphonics. One can measure changes of ε of about 10^{-5} from the ultraviollett (250 nm) to the near infrared (1600 nm) with a band width of 1 Hz. Even an MCDA as small as 10^{-4} still gives good sensitivity for ODMR experiments.

9.1.2 Optically Detected EPR

From the discussion in Chap. 4 it is clear that one should choose a low temperature and a high magnetic field. Experiments at 4.2 K are not feasible because of the boiling of He. One should, therefore, work below the lambda point. However, such a low temperature may cause the spin–lattice relaxation time T_1 to become too long (for example, of the order of seconds, if not minutes). This prevents one from modulating the microwaves or rf, and from detecting the ODEPR signal with the appropriate lock–in technique. It may, therefore, be advantageous to work at higher temperature in an exchange gas sample chamber, provided the MCDA is sufficiently large, and to be able to use the advantages of modulation spectroscopy. A modulation technique is particularly necessary for the optical detection of ENDOR. Since the MCDA is proportional to B/T at low temperature, small temperature drifts can be misinterpreted as weak ENDOR signals. If no effect modulation of either microwaves or rf is possible, then a slow "jump" modulation technique may be a help. In this method, one changes the static magnetic field (as fast as possible) from a value on the EPR line to a value off the EPR line, forwards and backwards, and records the difference (however, see Sect. 9.4.8).

For high magnetic fields one needs high microwave frequencies. It is common to use 24 GHz (K–band, 18 ... 26.5 GHz) or 35 GHz (Q–band). For still higher frequencies, quasi–optical resonators must be used. However, since the sample size is required to be correspondingly small, then there may not be enough light absorption. In addition, the smaller the microwave components become, the more difficult it is to avoid microphonics [9.6].

The population differences between the magnetic sublevels can also be influenced by optical pumping effects, as described in Sect. 4.8. Since these effects depend on the spin–mixing parameters and optical selection rules, no general prediction on the sensitivity can be made. However, it is certainly advantageous to have the experimental capability for modulated optical pumping with polarized and unpolarized light, or varying modulation frequencies. The optimum modulation frequency will depend on the spin–lattice relaxation times, and, thus, for a given system, on the temperature of the experiment. On the other hand, optical pumping can also effectively shorten the spin-lattice relaxation time (4.8.1,3). It can therefore be helpful, in case of a long T_1, to effectively shorten it in this way in order to be able to apply modulation techniques.

To detect relaxed excited states by the MCPE method, the population difference of the magnetic sublevels and, therefore, the signal amplitudes, are mostly determined by the optical pumping. Low temperature is essential to make the radiative lifetime from the RES as long as possible in order to be able to change the populations by microwave transitions. Radiative lifetimes below 100 ns are too short for the microwave powers below 1 W and loaded cavity quality factors in the range 3000 ... 5000 typically available at present.

9.2 ODMR Spectrometers Monitoring Light Emission

This kind of spectrometer is usually based on a conventional EPR spectrometer with an electromagnet to provide the static magnetic field. Therefore, there is rarely a problem with field homogeneity. In most cases, one measures the microwave–induced change in emitted light intensity using a suitable cavity with slots or a small opening (Sect. 9.4.9). A typical set up is shown in Fig. 9.1. If the gap between the pole pieces permits, the exciting light is focused onto the sample with mirrors. Light pipes are also used, especially if there are space problems between cryostat and pole pieces. Sometimes light

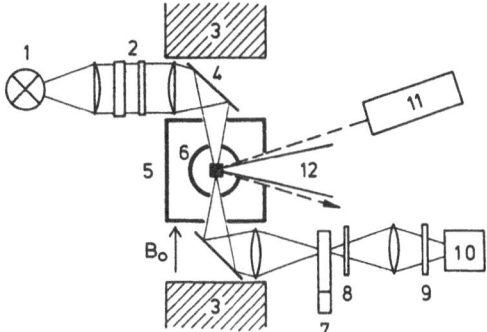

Fig. 9.1. Block diagram of an emission type ODMR spetrometer. The components are: (1) light source; (2) filters, polarizers; (3) pole pieces of magnet; (4) mirrors; (5) cryostat; (6) microwave cavity; (7) photo–elastic modulator; (8) linear polarizer under 45°; (9) filters; (10) photo-detector; (11) optional laser excitation; (12) alternative emission aperture

pipes are used to vertically illuminate the sample through a bottom window in the cryostat. However, no polarization effects can be studied in this kind of arrangement. Because of space problems, filters are widely used instead of monochromators. Which kind of optical cavity is used depends on the way the emission is excited. One suitable design is described in Sect. 9.4.9. Another type is a rectangular cavity with an opening in the upper half. This is conveniently used when exciting the emission with a laser from the same side as the emission is collected (Fig. 9.1) [9.2]. Of course, in this way measurements of polarization changes are inconvenient. A measurement of only the change of emission intensity has the disadvantage that, for example, in triplet systems the sign of the fine–structure constant cannot be determined (Sect. 4.10).

Circularly polarized emitted light is analyzed with a circular light analyzer. One can either monitor a fixed polarization, or the MCPE as the difference of right and left circular polarization, using a photo–acoustic modulator (for details see Sect. 9.4.5). In order to selectively feed the magnetic sublevels in the excited states, one should be able to pump with polarized light collinearly with the magnetic field. It is quite common to use laser light to excite the emission. Laser light has the advantage of high intensity and is easy to focus onto the sample. A dye laser system has the further advantage, particularly for semiconductors, of being able to excite emissions with sub–band–gap light (Sect. 9.4.9). The measurement of the MCPE in an arrangement like that of Fig. 9.1 has the disadvantage that mirrors will cause a loss of polarization information to a degree which is a function of the spectral range. These difficulties can be avoided by using pierced pole pieces with a hole parallel to the magnetic field in order to realize a *Faraday* configuration. This, however, has the restriction of a rather small optical aperture. Other special optical arrangements have been suggested to overcome these difficulties [9.7,8].

9.3 ODMR Spectrometers Monitoring Magnetic Circular Properties of Absorption and Emission

9.3.1 General Description of the Spectrometer

The apparatus is basically an optical spectrometer in which the light propagates parallel to the magnetic field (*Faraday* configuration). It is composed of a light source, monochromators, image forming optical components such as lenses or mirrors, light detectors, polarizers and modulators for the light polarization. The following basic quantities have to be measured:

(i) optical absorption

(ii) optical emission

(iii) magnetic circular dichroism of the absorption (MCDA) (or the *Faraday* rotation, which gives the equivalent information)

(iv) magnetic circular polarization of the emission (MCPE)

These measurements can be performed with a combined MCDA/MCPE spectrometer (Fig. 9.2). The spectrometer is built such that various interchangeable light sources (1) and a monochromator (3) provide excitation light over a wide spectral range. The light is amplitude modulated for absorption measurements. The circular polarization is modulated between right and left circular polarization by a combination of a linear polarizer and a birefringent acousto–optical modulator [9.9] (for details see Sect. 9.4.5). There are different interchangeable light detectors. Emission can be measured either collinearly with the magnetic field or at right angles through a second monochromator (3'). In order to measure the MCPE, one places the combination of birefringent modulator and linear polarizer into the emission light path immediately behind the sample (Fig. 9.2). The heart of the spectrometer is the cryostat with the superconducting split-coil magnet. For the ODMR measurements described in Chap. 4, additional equipment is necessary. The following experiments are performed:

(i) ODEPR detected by MCDA or MCPE

(ii) ODENDOR detected by MCDA or MCPE

(iii) MCDA and MCPE measurements tagged by EPR or ENDOR

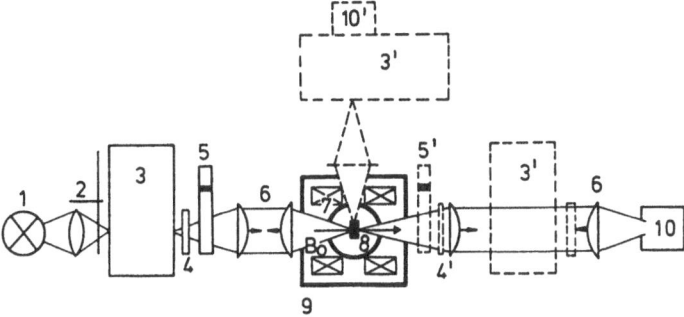

Fig. 9.2. Block diagram of an MCDA- (and MCPE)–type ODMR spectrometer. The components are: (1) light source; (2) chopper; (3) monochromator; (3') monochromator–position when measuring MCPE; (4) linear polarizer under 45°; (4') position of linear polarizer when measuring MCPE; (5) photo–elastic modulator; (5') position of photo–elastic modulator when measuring MCPE; (6) lenses, computer controlled in position; (7) split–coil supraconducting magnet; (8) microwave cavity; (9) cryostat; (10) photo–detector; (10') photo-detector when measuring MCPE

For these experiments one needs to equip the above MCDA/MCPE spectrometer with a microwave source, which can be modulated, and a special microwave cavity into which the sample is placed. For ENDOR an rf field must be fed into the cavity. Also, the sample must be rotated to change the crystal orientation with respect to the magnetic field. Although none of these additional components are novel to EPR or ENDOR, their combination with the optical MCDA/MCPE spectrometer causes difficulties which will be described below.

For most experiments, computer control of most parameters of the apparatus is advisable, for some (for example, the tagging experiments) it is necessary. These parameters are optical wavelength, sample temperature, magnetic field, optical focus across a wide spectral range (typically from 210 nm ... 1700 nm), crystal orientation, modulation of circular polarization as a function of wavelength, microwave power and microwave modulation, rf and its modulation, and finally, the recording of the signals. It proved advantageous to assign these tasks to several specialized computers. For example, the control of the stepping motors driving the monochromators or positioning the lenses, and similar tasks, can be done by processors. The central computer then has more freedom for an economical control of the whole procedure. Specific programs to measure absorption, MCDA, tagged MCDA, ODENDOR and ODEPR were developed. A master control program was realized which allows the performance of specific non–standardized experiments.

A particular problem arises for the optical part of the spectrometer used for the spatially resolved MCDA and ODEPR experiments. One possibility is to narrow down the image of the lamp on the sample, or else to use a focused laser beam. The focused light spot must then be moved across the sample. For ODEPR experiments the sample cannot be moved in the cavity since such a motion would detune the cavity resonance. This can only be done for an MCDA experiment. When moving the focused light spot across the sample, there is the possibility that unwanted signals appear as a function of position because of windows (under stress), lenses and the light path. A limitation to the SNR is caused by a limit to the maximum light intensity per area in the narrow light spot on the sample, under which the defects are not destroyed. There can also be local heating effects, which would drive the spin system out of thermal equilibrium, and there may be optical pumping effects, which have to be taken into account before conclusions on the spatial defect distribution can be drawn. Up to now, there is little experience with such kinds of spatially resolved MCDA and ODEPR experiments.

9.3.2 Measurement of Magnetic Circular Dichroism of Absorption

From (4.3.6) one obtains for small MCDA

$$\varepsilon \approx \frac{(I_l - I_r)}{2(I_l + I_r)}. \tag{9.3.1}$$

Right circular polarized light, I_r, means, that the electrical vector rotates clockwise when looking toward the light source [9.10]. Various definitions of the circular dichroism and phase conventions for light waves are used in the literature. If one deals with quantum mechanical transition matrix elements, one has to stick to one convention carefully. Circular polarized light is produced by a combination of a linear polarizer under 45° and a $\lambda/4$ plate (Fig. 9.3a). By changing the orientation of the linear polarizer by 90° one could produce I_r and I_l and measure the dichroism by forming the difference between two subsequent measurements $I_r/[2(I_l + I_r)] - I_l/[2(I_l + I_r)]$ according to (9.3.1). This is not sensitive enough. A fast modulation between I_l and I_r, and subsequent demodulation of the signal is preferred. Such a fast modulation is achieved by the use of a birefringent acousto–optical modulator with frequencies of 30 ... 50 kHz [9.9,11–14] (Fig. 9.3b). The resulting time dependent signal can be calculated by means of the *Mueller* or *Jones* formalism [9.10,15] since, apart from the photo–acoustic modulator, the polarizing effects of the optical components in the light path to the detector also have to be taken into account. Each polarizing component (such as cold windows, lenses, etc.) is represented by characteristic matrices that act one after another on the light, which is represented by a vector. Apart from magneto–optical

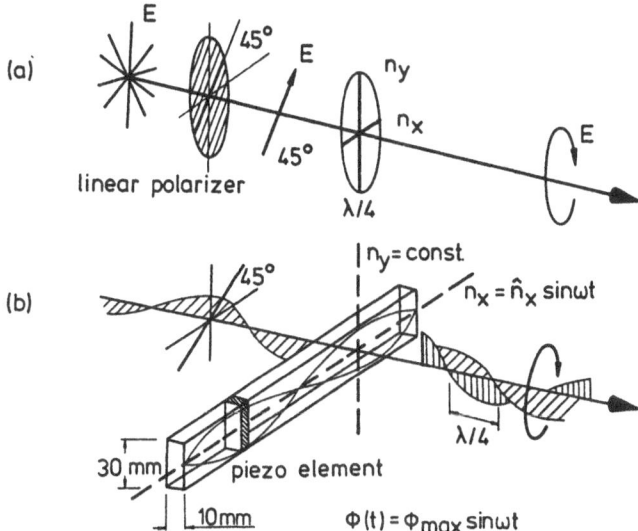

Fig. 9.3a,b. Generation of circularly polarized light **a** arrangement of linear polarizer and $\lambda/4$–plate. **b** arrangement of linear polarizer and photo–elastic modulator

properties, the sample itself may have birefringent effects if it does not have an isotropic crystal structure, or if it develops birefringence under stress.

For an idealized spectrometer with completely circular polarized light of only small divergence one obtains for the signal $I(t)$ at the detector, assuming isotropic absorption and a magnetic circular dichroism [9.5]

$$I(t) = I_0 \exp\left(-\frac{\alpha_r + \alpha_l}{2}d\right) [\cosh(2\varepsilon) - \sinh(2\varepsilon)\sin\phi(t)]. \quad (9.3.2)$$

$\phi(t)$ is the sinusoidal phase difference from the phase modulator. Since $\phi(t) = \phi_{max}\sin\omega t$ (Fig. 9.3b), the last term in (9.3.2) is of the form $\sin(\phi_{max}\sin(\omega t))$. It can be expanded into a *Fourier* series with the coefficients $J_1(\phi_{max})$, which are *Bessel* functions of first order

$$\sin\phi(t) = \sin(\phi_{max}\sin(\omega t))$$
$$= 2J_1(\phi_{max})\sin(\omega t) + 2J_3(\phi_{max})\sin(3\,\omega t) + \dots. \quad (9.3.3)$$

Equation (9.3.2) contains the time independent part

$$I_{dc} = I_0 \exp\left[-\frac{(\alpha_r + \alpha_l)}{2}d\right]\cosh(2\varepsilon), \quad (9.3.4)$$

which can be measured as an average photo–current at the detector. A lock–in detector selects the first component of (9.3.3) and suppresses the odd harmonic terms:

$$I_{ac} = -I_0 \exp\left[-\frac{(\alpha_r + \alpha_l)}{2}d\right][\sinh(2\varepsilon)][2J_1(\phi_{max})]. \qquad (9.3.5)$$

The ratio of (9.3.4) and (9.3.5) is proportional to $\tanh(2\varepsilon)$ and contains a calibration factor K

$$\frac{I_{ac}}{I_{dc}} = 2KJ_1(\phi_{max})\tanh(2\varepsilon), \qquad (9.3.6)$$

$$\tanh(2\varepsilon) = \frac{1}{2KJ_1(\phi_{max})} \cdot \frac{I_{ac}}{I_{dc}}. \qquad (9.3.7)$$

For $\varepsilon < 0.1$ it is sufficient to consider

$$\varepsilon \approx \frac{1}{4KJ_1(\phi_{max})} \cdot \frac{I_{ac}}{I_{dc}}. \qquad (9.3.8)$$

The largest signal is obtained for $\phi_{max} = 105°$, where J_1 has a first maximum $J_1(105°) = 0.5818$ (not for $90°$). ϕ_{max} depends on the wavelength. It must either be kept experimentally at the maximum value when changing the wavelength, or corrections must be introduced mathematically after the experiment (Sect. 9.4.5).

A calibration of the apparatus is possible by use of a known circular dichroism [9.16–18] using (9.3.7). However, if one calibrates using a circular polarizer corresponding to an MCDA of infinity, the calibration signal is orders of magnitude larger than the magneto–optical MCDA, thus, one must be careful with the linearity of the measurement instruments. Above, it is assumed that no depolarizing elements are in the light path and that the light is parallel. It is also assumed that the sample has no linear birefringence or linear dichroism in addition to circular dichroism and rotation. All these conditions are normally not fulfilled in the ideal way assumed to derive (9.3.2). In particular, one has to deal with stress birefringence in non–perfect cryostat windows. An additional birefringence before the sample influences the signal, but not one behind the sample, if the detector is completely isotropic. A phase shift of the light behind the sample can be registered by the detector as a false signal, if the detector is also dichroic, for example, if it has a dichroic window (which can happen). In those cases, one must recalculate the signal (9.3.2) with the help of the characteristic matrices. One must take the *Mueller* formalism with 4×4 matrices if one has to deal with depolarizations. In the *Jones* formalism one chooses a right/left circular basis for the light vector. For an ideal isotropic configuration all matrices following the modulator are then diagonal 2×2 matrices. Non–isotropic elements, such as linear dichroism and birefringence, have non–diagonal elements and result in a different time dependent signal. In particular, there can be additional terms $\cos[\phi_{max}\sin(\omega t)]$ which, in the series expansion, give *Fourier* components with even multiples of the modulation frequency. Then I_{ac}/I_{dc} is no longer proportional to the MCDA, but a function of all polarizations. In such cases it is not possible to correct a measurement by the simple subtraction of a *background*. A practical check as to whether such a case is met, would be to control the signal demodulating it at twice the modulation frequency. Such a signal should be zero if there are no off–diagonal terms. It is not zero, if linear birefringence and linear dichroisms are there, especially at the surface of the sample. This is the case, for instance, if the light incidence is not perpendicular to the sample surface.

A particularly bad problem is the linear birefringence and dichroism of the sample itself. The problem can be overcome for optical uniaxial crystals if the c–axis of the crystal is oriented parallel to the propagation of light.

Otherwise the signal (9.3.2) cannot be simply obtained by considering the matrices of the MCDA and of the linear dichroism in series independently; their operators do not commute. One has to compute a new matrix which considers both effects simultaneously. It is obtained from the differential propagation matrices [9.15,19] which are differential with respect to the coordinate along the light path (also [9.20–22]), and have to be added for both contributions. One then has to solve the differential equation for the light propagation, which can be a difficult task. Fortunately, all this is relevant only if one is interested in the correctness of the magnitude of the MCDA. For example, if one uses the MCDA of a defect as its characteristic quantity, and performs a mapping investigation over a given sample (such as a semiconductor wafer), these difficulties have to be seriously considered. A large wafer held at low temperature will undoubtedly be susceptible to stress dichroism or birefringence. However, for ODMR measurements one is usually not interested in the correct value of the MCDA. This is also true when rotating the crystal.

9.3.3 Measurement of
Magnetic Circular Polarization of Emission

The circular polarization of emission P_e is defined as the difference of intensities of right and left circular polarized light relative to the sum of the intensities I_{tot} of all polarizations, including unpolarized light.

$$P_e = \frac{(I_l - I_r)}{I_{tot}}. \tag{9.3.9}$$

The degree of circular polarization of the emission can be measured with a modulation technique analogous to that of the MCDA measurements (Fig. 9.4). The combination of photo–acoustic modulator and linear polarizer is placed behind the sample. One obtains a signal analogous to that of (9.3.2), again

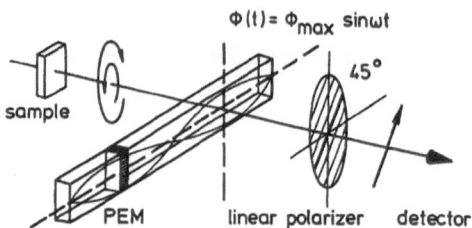

Fig. 9.4. Generation of alternating right and left circulating polarized light

assuming ideal conditions (see above)

$$P_e = \frac{1}{K' J_1(\phi_{max})} \frac{I_{ac}}{I_{dc}}. \qquad (9.3.10)$$

K' is a calibration factor. For elliptical emitted light there are also even components in the *Fourier* expansion of the $\sin \phi(t)$ term. One measurement is not sufficient to determine the polarization [9.7,23,24]. For ordinary ODEPR measurements this is not important, since one is only interested in the field position of the lines. However, if one wants to do tagging experiments for anisotropic centers, and determine the optical anisotropic transition probabilities by tagging the EPR lines of centers at various orientations, then one is forced to fully analyze the polarization of the emission.

9.4 Experimental Details of the Components of an MCDA/MCPE ODMR Spectrometer

The ODMR spectrometer is composed of the following components:

(i) The MCDA/MCPE spectrometer

(ii) A cryo–system with superconducting magnet and sample cooling

(iii) Microwave source, microwave components and cavity

(iv) Radio–frequency system for ENDOR

(v) Measurement control and signal processing system

Experimental aspects and special features of the single components needed will be discussed below in some detail, based upon recent experiences.

9.4.1 Light Sources

For MCDA measurements one needs continuous light sources without intensive lines and little stray light. A high illumination intensity per unit area (irradiance) is wanted. Halogen lamps which have only thermal statistical noise have proved to be stable and luminous. Their irradiances' area is determined by the typical halogen process, while the filament size determines the wattage. The highest light irradiances are achieved by high pressure arc lamps (xenon, mercury). New lamps can have a particularly nasty low–frequency noise due to arc jumps. Xenon lamps have stray light in the infrared which must be suppressed by water filters. The high light intensity of mercury high pressure lamps in several lines is advantageous in the UV, but not so suitable for scanning experiments like tagging in the UV. Deuterium lamps (200 W electrical power) in the UV limit the spectral range for the spectrometer down to about 210 nm. This UV source is weak. Combined with the poor transmission of most optical components for the UV, they are a limiting factor of the sensitivity in the UV. As a result, measurements are very difficult below 210 ... 220 nm. The use of tunable lasers has the advantage of a simple optical arrangement and high light intensities. However, the high intensity cannot normally be used for MCDA experiments, due to a frequency dependent high amplitude jitter. This noise destroys the small difference in light intensities for right and left polarized light in each half phase of the modulated light. This difference is usually much smaller than 1%. The use of noise reducing systems is, therefore, advisable in connection with the use of lasers. Noise suppressions of 40 db up to 1 kHz, and 26 db up to about 1 MHz, are commercially available.

The requirements in amplitude stability for MCPE experiments are met by most lasers because the effect is observed directly via variations of emission intensities. Also, the high intensity of lasers is required to compensate for spatial and quantum efficiency losses, and the limited position of the emitted light reaching the detector.

9.4.2 Monochromators

The spectral resolution depends on the problems studied. For many solid state defects a 1/4 m monochromator is sufficient. For semiconductors with many ZPL transitions, a higher resolution may be required. Since MCPE experiments must be performed mostly in the *Faraday* configuration, a double monochromator is necessary to suppress stray light in the emission light path. The usable aperture is limited, mainly by the effective size of the photo–acoustic modulator(see below), and the cryostat windows and their geometry.

9.4.3 Imaging Systems

With quartz lenses there are no problems with polarization in the range of
200 ... 3000 nm. Mirrors cannot be used between sample and polarizers
where the polarization information is critical. Light guides are valuable to feed
exciting light to the sample under difficult geometrical conditions. However,
the polarization information is lost. Under certain conditions, it is preserved
in single–mode light pipes [9.14]. For continuous measurements in a wide
spectral range, for example, for tagging between 220 nm and 1700 nm, one
needs to install a lens with controllable positioning because of the chromatic
aberrations of the quartz lenses. (The focal length varies by 20% between 220
and 600 nm). Also, the optical path in optically dense materials (e.g., the
polarizer) has to be taken into account. The aperture of the system for both
sample illumination and the collection of emitted light should be made as large
as possible, and controllable by aperture irises. For MCDA measurements it
must be made absolutely sure that no light passes the sample.

9.4.4 Linear Polarizers

For the UV and the visible spectral range, prism polarizers of varying construction can be used. Often it suffices to use foil polarizers in the visible and
near infrared range, since for ODMR the extinction ratio is not critical. Difficulties arise near 1000 nm. For longer wavelengths grating polarizers must
be used (e.g., MgF_2).

9.4.5 Photo–Elastic Modulator

Circularly polarized light is generated with a combination of a linear polarizer,
inclined by 45°, and a $\lambda/4$ plate (Fig. 9.3a). A modulation of right and left circularly polarized light can be generated either by rotating both components
relative to each other, or by changing the phase difference in the $\lambda/4$ plate by
±90°. Precise control of the phase difference is necessary for each wavelength.
It is mechanically difficult to produce the fast rotating polarizers necessary for
a good signal. Inhomogeneities and beam offsets lead to difficulties. Achroma-
tic $\lambda/4$ plates are also difficult to produce [9.25]. Acousto–optical modulators
(birefringent acousto–optical modulator, phase modulator) are advantageous.
In a synthetic fused quartz slab, one generates an acoustical longitudinal pressure wave with a coupled piezo element. This makes the isotropic quartz slab
birefringent, with the principal axes parallel and perpendicular to the acoustic wave. The frequency range is 30 ... 100 kHz. Other materials having an
acousto–optical effect, such as CaF_2, can also be used. Quartz modulators
working at about 50 kHz are commercially available [9.9,11,12]. They can be
used well beyond 1000 nm as $\lambda/4$ modulators. Fig. 9.3b shows a design with

a larger usable aperture. The quartz slab oscillates in a resonance at about 30 kHz with a usable aperture of 30 × 30 mm². The piezo element is glued to the slab where there is a pressure maximum. This requires a good glue such as epoxy with high tensile strength and low acoustical losses. Such modulators were used up to 1000 nm. Beyond that wavelength the modulation amplitude is held constant, and a numerical correction is introduced for the signal [9.5]. The modulation amplitude corresponding to a phase angle of 105° (maximum of the J_1 *Bessel* function) is dependent on the wavelength and is controlled by a computer. It is necessary for MCDA measurements to maintain the correct modulation amplitude when the spectral dependence of the MCDA is to be measured. For ODEPR or ODENDOR measurements this is less critical. One can also use *Pockels* cells for the same purpose, especially when working with lasers with well collimated light. The advantage is that one can work with different frequencies. Voltages of several thousand volts are necessary for phase differences of 90°.

9.4.6 Detectors

Detectors should have a high quantum efficiency rather than a high noise equivalent power (NEP), since one usually works at high light levels. Suitable photomultipliers are excellent in the UV and the visible spectral range. When working with a superconducting magnet, the photomultipliers must be protected sufficiently from the magnetic field. In a distance of 1.5 m from a 5 T coil, mu–metal protection is sufficient. Semiconductor detectors should also be placed at least this far away. No photomultipliers should be used which have polarization effects at the photocathode. Between 700 nm and 1000 nm, Si diodes are preferable because they withstand higher light intensity levels. Suitable low noise amplifiers for the diodes are available. In the near infrared up to about 1800 nm, the best results are obtained with cooled Ge detectors. The pre–amplifier should also be liquid nitrogen cooled. For MCDA measurements, fast pre–amplifiers with 100 kHz band width are necessary. It is important that detector and detector–window have no circular polarization effects. Beyond 1600 nm one can use InAs or InSb detectors. However, the thermal background radiation becomes more limiting the longer the wavelength.

9.4.7 Cryostat

There should be separate cryo–systems for the cavity with the sample and the superconducting magnet, in order to be able to change samples when the magnet is cold. As discussed before, the sample should be held at a temperature below the lambda point of liquid helium (e.g. 1.5 ... 1.7 K). The sample and cavity are immersed in liquid helium. Measurements in cold

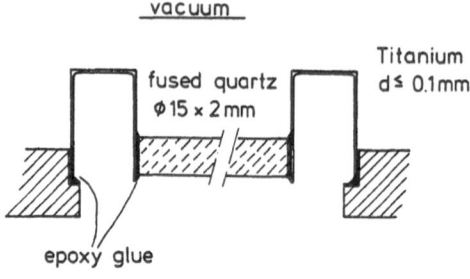

Fig. 9.5. Construction of a stress free cold optical window. (After [9.26])

He gas should also be possible. The space available for the cavity should have at least a 30 mm^2 square cross section for a cylindrical K–band cavity. This is large enough for a rectangular X–band cavity if a change of microwave frequency is needed. For ODEPR, the thermal load from light and microwaves necessitates pumping on the He bath at a rate of about 30 m^3/h. When measuring ODENDOR the load is up to 1 W which requires strong roots pumps. Control of the sample temperature is critical, since the MCDA effect is proportional to $1/T$. Temperature drifts of the signal, due to the thermal load, can be of the order of the signal itself, and are often hard to distinguish from real signals. The vapor pressure of the He bath is, therefore, computer controlled by suitable valves. The temperature can be measured via the He vapor pressure or with glass–carbon resistors (1.5 ... 300 K).

It is very important that the cold windows at the sample compartment are isotropic and stress–free to avoid birefringence. A cryostat of suitable construction is shown in Fig. 9.5 following *Mollenauer et al.* [9.26]. Titanium is used because it is nonmagnetic and has a small thermal expansion coefficient. Synthetic quartz windows can be used from the UV to 3 μm. Beyond that, compressed amorphous ZnS can be employed. The epoxies used to fix the window should not be exposed to X–irradiation which renders them brittle. This must be borne in mind when irradiating the sample in situ to produce radiation damage defects.

9.4.8 Magnet

Split coil superconducting magnets in a *Helmholtz* arrangement have proved to be advantageous. Apart from the longitudinal opening (light path) they allow two perpendicular bores, a vertical one for the sample and cavity, and a horizontal one for an optical path perpendicular to the magnetic field. This allows the construction of a very versatile apparatus. However, the size of the openings perpendicular to the axis of the magnet critically influences the homogeneity of the magnetic field. The homogeneity of the magnet is less critical than in conventional EPR since the sample volume measured is much smaller (it is determined by the diameter of the light beam and the sample thickness). The filling factor is not the determining factor for the signal intensity as it is in EPR (Chaps. 2,8). However, it is difficult to calibrate the magnetic field. The reproducibility of the field is limited. Calibration with a *Hall* probe is usually not good enough for the determination of precise g values. A proton resonance gauss–meter does not operate because of the poor homogeneity. It seems best to calibrate the field relative to defects with known g factors. A disadvantage of the superconducting magnet is that rapid field changes are not possible because of the high inductance of the coils [9.27].

9.4.9 Microwave System and Cavity

No microwave bridge is necessary for ODEPR measurements. The microwave system contains a microwave source, uniline, frequency meter, an attenuator, microwave modulator and a directional coupler for tuning the cavity. More than 50 db attenuation is often needed. At present, it seems that klystrons are preferable to other sources. A power level of 600 ... 1000 mW is desirable, which is not yet obtainable from semiconductor sources in K–band. *Gunn*, *Impatt* and YIG oscillators show strong temperature drifts which must be compensated by AFC circuits. Amplification of a weaker microwave source by means of a travelling wave tube (TWT) is possible, but has several disadvantages. One is the low noise figure (35 db max). The broad band noise background causes particular problems when ODENDOR measurements are intended. One cannot burn a narrow hole into an inhomogeneous EPR line (Chap. 5). Therefore, TWT tubes should only be used if the high microwave power is really necessary. High B_1 amplitudes at the sample are achieved, even for moderate microwave powers of 200 ... 300 mW, in a cavity with a good quality factor (about 5000). In contrast to conventional EPR, there is a need for a good optical access to the sample. An rf field amplitude must also be provided at the sample for ENDOR. Mechanically reliable cavities can be produced for K–band, while in Q–band there are problems with both the cavity and the sample size. The sample is so small that it causes problems with the whole optical imaging system. Figure 9.6 shows a successful design for a cylindrical cavity working in the TE_{011} mode, which has a wide optical access. It was derived from an earlier design of a cavity consisting of a series of coaxial discs held at a specific separation from each other [9.28]. The cavity can be described as consisting of two top hats held opposite to each other. The separation of the hats, as well as the width of their brims, must be chosen such that the cut off wavelength condition is fulfilled, i.e., no microwaves can escape. At 24 GHz the distance is 4 mm, the flanges (brims) are 5 mm. It is advantageous if the bottom of the resonator can be adjusted to match

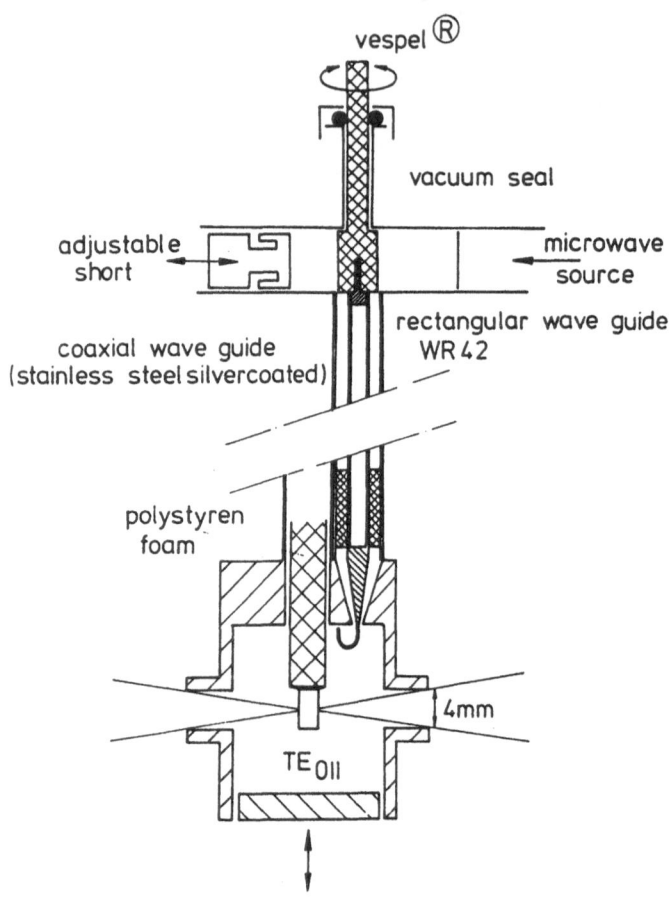

Fig. 9.6. Cylindrical cavity with wide optical access for ODEPR experiments and top part of the microwave coaxial guide (for details see text)

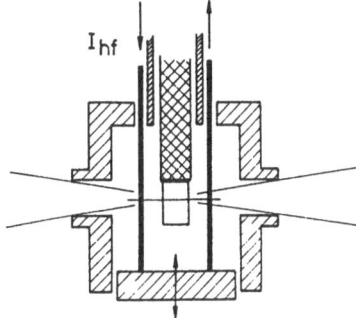

Fig. 9.7. Cylindrical cavity to measure ODENDOR with two *Helmholtz* loops for the introduction of an rf field

frequency changes resulting from the dielectric constant of the sample. The microwave coupling is achieved with a rotatable hook attached to a coaxial microwave guide. This is not very critical (Fig. 9.6). Optimum microwave coupling is achieved by turning the hook into a suitable position. The upper part of the coaxial wave guide is terminated by a microwave "chimney" (for details see Fig. 9.6). The sample is inserted through a tube into the cavity, and attached to a piece of polystyrene or a similar material with small electrical losses. With such an arrangement one can obtain a loaded quality factor at low temperature of about 5000. The sample size in K–band is $3 \times 3 \times 4$ mm^3 maximum. The sample can be rotated by means of stepping motors, and also shifted up and down with a resolution of 0.05 mm for spatially resolved measurements of the MCD and ODEPR.

9.4.10 Radio–Frequency System for ODENDOR

In principle, the considerations of Chap. 8 as to how to bring a sufficiently large rf field amplitude B_2 to the sample, also apply here. In both cases, cylindrical cavities are used. One important difference arises here in that the temperature must be kept considerably lower than in conventional ENDOR. As a consequence of the dimension of the ODMR cryostat with the superconducting magnet, the transfer lines for the rf are much longer (approximately 100 cm). One must avoid losses along those transfer lines as much as possible since they add to the thermal load of the superfluid helium bath. The rf transfer lines should have a low impedance. In one successful solution, they consisted of a Teflon strip of about 1 cm width which had a thin copper coating on both sides. Alternatively, a 50 Ω transfer line with a terminating resistor outside the cryostat, was also used successfully. The four ENDOR rods are fixed to the adjustable bottom of the cavity, but can be moved within insulators (Teflon, Vespel) at the top of the cavity (Fig. 9.7). The transformer for the impedance matching of the rf source to the leads and ENDOR rods is located outside the cryostat. The rf lead system can also be used to introduce a 90° phase shifted oscillating magnetic field. This produces a longitudinal field modulation which, in case of an inhomogeneous EPR line, enhances the ODEPR signal by sweeping over many spin packets (Sect. 4.11).

9.4.11 Control and Registration Electronics

The MCDA signal is lock–in demodulated at 30 ... 50 kHz. The modulation of ODEPR and ODENDOR signals requires a suitable double lock–in technique (with a large band width for the first lock–in detector). The MCD signal I_{ac}/I_{dc} is calculated by the computer as the ratio of the first (I_{ac}) and zeroth (I_{dc}) *Fourier* component of the modulated light signal. Calibration factors and stress modulator corrections are immediately performed by the computer. If I_{dc} contains a large dark current, which is the case when working with biased IR detectors, one must introduce an additional amplitude modulation of the light and the corresponding demodulation. When using a photomultiplier, one can regulate I_{dc} to be constant by simply adjusting the voltage of the photomultiplier. When only measuring ODEPR or ODENDOR, it is sufficent to register only I_{ac}. In this way one gains measurement time especially if I_{dc} is noisy. It is often observed that a non–resonant background signal appears when applying high microwave power (see also [9.2]). This background can only be suppressed by separate measurements on and off resonance, and finding the difference with the computer. Unfortunately, a "jump" modulation cannot be applied with the superconducting magnet because it does not allow rapid changes of the magnetic field. The transfer of the measured signals to the computer can be done advantageously with a two–channel A/D converter through an IEEE 488 Bus interface. The computer controllable gate time of the A/D converter determines the integration time of the lock–in amplifiers, which are used in the broad band mode. This results in a more efficient noise reduction than one would achieve with an RC filter at the output of the amplifier. As discussed in Chap. 6, digital filters can also be applied with the computer. For further experimental details see also [9.29,30].

Appendices

A. Nuclear g–Factors, Quadrupole Constants, NMR Frequencies

z	Isotope[a]	Natural abundance [%]	Spin	NMR frequency for 0.35 T field [MHz]	g_n	Electric quadrupole moment Q in multiples of $|e| \times 10^{-24}$ cm^2
0	^{1}n*	...	1/2	10.20769	-3.82608422	....
1	^{1}H	99.985	1/2	14.90218	5.5856912	....
	^{2}H	0.0148	1	2.287575	0.8574376	0.002875
	^{3}H*	...	1/2	15.89525	5.957920	....
2	^{3}He	0.000138	1/2	11.35266	-4.255248	....
3	^{6}Li	7.5	1	2.193167	0.8220514	-0.000644
	^{7}Li	92.5	3/2	5.791950	2.170961	-0.040
4	^{9}Be	100.	3/2	2.094	-0.7850	0.053
5	^{10}B	19.8	3	1.60133	0.600216	0.08608
	^{11}B	80.2	3/2	4.782043	1.792424	0.040
6	^{13}C	1.11	1/2	3.74795	1.40482	....
7	^{14}N	99.63	1	1.077201	0.4037607	0.0193^v
	^{15}N	0.366	1/2	1.511052	-0.5663784	....
8	^{17}O	0.038	5/2	2.02099	-0.757516	-0.026
9	^{19}F	100.	1/2	14.02721	5.257732	....
10	^{21}Ne	0.27	3/2	1.17708	-0.441197	0.1029
11	^{22}Na*	...	3	1.553	0.5820	.?
	^{23}Na	100.	3/2	3.944228	1.478391	0.108
12	^{25}Mg	10.00	5/2	0.91291	-0.34218	0.22^v
13	^{27}Al	100.	5/2	3.886094	1.456601	0.150
14	^{29}Si	4.67	1/2	2.9630	-1.1106	....
15	^{31}P	100.	1/2	6.03804	2.26320	....
16	^{33}S	0.75	3/2	1.1448	0.42911	-0.064
17	^{35}Cl	75.77	3/2	1.461795	0.5479157	-0.08249
	^{36}Cl*	...	2	1.71477	0.642735	-0.0180
	^{37}Cl	24.23	3/2	1.216790	0.4560820	-0.06493
18	^{39}Ar*	...	7/2	0.99	-0.37	.?
19	^{39}K	93.26	3/2	0.6963030	0.2609909	0.054
	^{40}K*	0.0117	4	0.86582	-0.32453	-0.067
	^{41}K	6.73	3/2	0.3821910	0.1432542	0.060

z	Isotope[a]	Natural abundance [%]	Spin	NMR frequency for 0.35 T field [MHz]	g_n	Electric quadrupole moment Q in multiples of $|e| \times 10^{-24}$ cm^2
20	^{41}Ca*	...	7/2	1.215641	-0.4556514	.?
	^{43}Ca	0.135	7/2	1.00424	-0.376414	<0.23[v]
21	^{45}Sc	100.	7/2	3.62586	1.35906	-0.22[u]
22	^{47}Ti	7.4	5/2	0.84144	-0.31539	0.29[u]
	^{49}Ti	5.4	7/2	0.841667	-0.315477	0.24[u]
23	^{50}V	0.250	6	1.48495	0.556593	0.209[u]
	^{51}V	99.750	7/2	3.91747	1.46836	-0.0515[u]
24	^{53}Cr	9.50	3/2	0.8396	-0.3147	-0.0285/+0.022
25	^{53}Mn*	...	7/2	3.828	1.435	.?
	^{55}Mn	100.	5/2	3.6868	1.3819	0.33[u]
26	^{57}Fe	2.15	1/2	0.4818	0.1806	....
27	^{59}Co	100.	7/2	3.516	1.318	0.42
	^{60}Co*	...	5	2.025	0.7589	0.44[u]
28	^{61}Ni	1.13	3/2	1.3340	-0.50001	0.162
29	^{63}Cu	69.2	3/2	3.959	1.484	-0.222
	^{65}Cu	30.8	3/2	4.237	1.588	-0.195
30	^{67}Zn	4.10	5/2	0.934604	0.350312	0.150[u]
31	^{69}Ga	60.1	3/2	3.58673	1.34439	0.168
	^{71}Ga	39.9	3/2	4.55729	1.70818	0.106
32	^{73}Ge	7.8	9/2	0.5214100	-0.1954371	-0.19[u]
33	^{75}As	100.	3/2	2.56026	0.959647	0.29[u]
34	^{77}Se	7.6	1/2	2.8528	1.0693	....
	^{79}Se*	...	7/2	0.7758	-0.2908	0.8[u]
35	^{79}Br	50.69	3/2	3.746469	1.404266	0.293[u]
	^{81}Br	49.31	3/2	4.038446	1.513706	0.27
36	^{83}Kr	11.5	9/2	0.575481	-0.215704	0.260[u]
	^{85}Kr*	...	9/2	0.5957	0.2233	0.45[u]
37	^{85}Rb	72.17	5/2	1.44402	0.541253	0.273
	^{87}Rb	27.83	3/2	4.89369	1.83427	0.130
38	^{87}Sr	7.0	9/2	0.64806	-0.24291	0.15
39	^{89}Y	100.	1/2	0.7332410	-0.2748361	....
40	^{91}Zr	11.2	5/2	1.39118	-0.521448	.?
41	^{93}Nb	100.	9/2	3.6583	1.3712	-0.28[u]
42	^{95}Mo	15.9	5/2	0.9754	-0.3656	-0.019[u]
	^{97}Mo	9.6	5/2	0.9962	-0.3734	0.2[v]
43	^{99}Tc*	...	9/2	3.3701	1.2632	0.34[v]
44	^{99}Ru	12.7	5/2	0.664	-0.249	0.076[u]
	^{101}Ru	17.0	5/2	0.744	-0.279	0.44[u]
45	^{102}Rh*	...	(6)	1.83	0.685	.?
	^{103}Rh	100.	1/2	0.4717	-0.1768	....
46	^{105}Pd	22.2	5/2	0.683	-0.256	0.66[u]
47	^{107}Ag	51.83	1/2	0.606282	-0.227249	....
	^{109}Ag	48.17	1/2	0.698309	-0.261743	....
48	^{111}Cd	12.8	1/2	3.17597	-1.19043	....
	^{113}Cd*	12.2	1/2	3.3226	-1.2454	....
49	^{113}In	4.3	9/2	3.27791	1.22864	0.846
	^{115}In*	95.7	9/2	3.28498	1.23129	0.861

z	Isotope[a]	Natural abundance [%]	Spin	NMR frequency for 0.35 T field [MHz]	g_n	Electric quadrupole moment Q in multiples of $|e| \times 10^{-24}$ cm^2
50	^{115}Sn	0.38	1/2	4.9028	-1.8377	....
	^{117}Sn	7.75	1/2	5.34139	-2.00208	....
	^{119}Sn	8.6	1/2	5.58812	-2.09456	....
51	^{121}Sb	57.3	5/2	3.5897	1.3455	-0.33[u]
	^{123}Sb	42.7	7/2	1.9443	0.72876	-0.68
	^{125}Sb*	...	7/2	2.005	0.7514	.?
52	^{123}Te	0.89	1/2	3.9314	-1.4736	....
	^{125}Te	7.0	1/2	4.7398	-1.7766	....
53	^{127}I	100.	5/2	3.00221	1.12530	-0.789[v]
	^{129}I*	...	7/2	1.9979	0.74886	-0.553[v]
54	^{129}Xe	26.4	1/2	4.15115	-1.55595	....
	^{131}Xe	21.2	3/2	1.23055	0.461240	-0.120[u]
55	^{133}Cs	100.	7/2	1.968518	0.7378477	-0.003
	^{134}Cs*	...	4	1.9967	0.74842	0.389
	^{135}Cs*	...	7/2	2.0828	0.78069	0.051
	^{137}Cs*	...	7/2	2.1658	0.81180	0.052
56	^{133}Ba*	...	1/2	4.11	-1.54	....
	^{135}Ba	6.59	3/2	1.4909	0.55884	0.20[u]
	^{137}Ba	11.2	3/2	1.6679	0.62515	0.34[u]
57	^{137}La*	...	7/2	2.054	0.7700	0.26
	^{138}La*	0.089	5	1.9817	0.74278	0.51
	^{139}La	99.911	7/2	2.1215	0.79520	0.20
59	^{141}Pr	100.	5/2	4.3	1.6	-0.041[u]
60	^{143}Nd	12.2	7/2	0.8207	-0.3076	-0.56
	^{145}Nd	8.3	7/2	0.507	-0.190	-0.29
61	^{147}Pm	...	7/2	2.01	0.752	0.66[u]
62	^{147}Sm*	15.1	7/2	0.6195	-0.2322	-0.18[u]
	^{149}Sm	13.9	7/2	0.5109	0.1915	0.056[u]
	^{151}Sm*	...	5/2	0.379	0.142	0.52[v]
63	^{151}Eu	47.9	5/2	3.706	1.389	1.53[u]
	^{152}Eu*	...	3	1.7265	0.64713	3.16[u]
	^{153}Eu	52.1	5/2	1.637	0.6134	3.92[u]
	^{154}Eu*	...	3	1.783	0.6683	3.9[u]
	^{155}Eu*	...	5/2	2.06	0.772	.?
64	^{155}Gd	14.8	3/2	0.4597	-0.1723	1.30[u]
	^{157}Gd	15.7	3/2	0.6011	-0.2253	1.34
65	^{157}Tb*	...	3/2	3.5	1.3	.?
	^{158}Tb*	...	3	1.563	0.5860	2.7
	^{159}Tb	100.	3/2	3.580	1.342	1.34
66	^{161}Dy	19.0	5/2	0.504	-0.189	2.47
	^{163}Dy	2.49	5/2	0.710	0.266	2.51
67	^{165}Ho	100.	7/2	3.180	1.192	2.73[u]
68	^{167}Er	22.9	7/2	0.4317	-0.1618	2.827[u]
69	^{169}Tm	100.	1/2	1.24	-0.466	....
	^{171}Tm*	...	1/2	1.229	-0.4606	....
70	^{171}Yb	14.4	1/2	2.637	0.9885	....
	^{173}Yb	16.2	5/2	0.72554	-0.27195	2.8

z	Isotope[a]	Natural abundance [%]	Spin	NMR frequency for 0.35 T field [MHz]	g_n	Electric quadrupole moment Q in multiples of $|e| \times 10^{-24}$ cm^2
71	^{173}Lu*	...	7/2	1.78	0.669	.?
	^{174}Lu*	...	(1)	5.18	1.94	.?
	^{175}Lu	97.39	7/2	1.7059	0.63943	5.68^u
	^{176}Lu*	2.61	7	1.21	0.452	8.0^u
72	^{177}Hf	18.6	7/2	0.6048	0.2267	4.5
	^{179}Hf	13.7	9/2	0.3799	-0.1424	5.1
73	^{181}Ta	99.9877	7/2	1.8070	0.67729	3.44^u
74	^{183}W	14.3	1/2	0.6284800	0.2355694	....
75	^{185}Re	37.40	5/2	3.4011	1.2748	2.33^u
	^{187}Re*	62.60	5/2	3.4357	1.2878	2.22^u
76	^{187}Os	1.6	1/2	0.3498	0.1311	....
	^{189}Os	16.1	3/2	1.30	0.488	0.8
77	^{191}Ir	37.3	3/2	0.259	0.097	0.78^u
	^{193}Ir	62.7	3/2	0.285	0.107	0.70^u
78	^{195}Pt	33.8	1/2	3.2522	1.2190	....
79	^{197}Au	100.	3/2	0.261371	0.097968	0.594^u
80	^{199}Hg	16.8	1/2	2.699321	1.011770	....
	^{201}Hg	13.2	3/2	0.996423	-0.373483	0.42
81	^{203}Tl	29.5	1/2	8.656103	3.244514	....
	^{204}Tl*	...	2	0.1211	0.0454	.?
	^{205}Tl	70.5	1/2	8.7385	3.2754	....
82	^{207}Pb	22.1	1/2	3.1343	1.1748	....
83	^{207}Bi*	...	9/2	2.59	0.970	-0.50^v
	^{209}Bi	100.	9/2	2.50	0.938	-0.46^u
84	^{209}Po*	...	1/2	4.0	1.5	....
89	^{227}Ac*	...	3/2	1.9	0.73	1.7^v
90	^{229}Th*	...	5/2	0.43	0.16	4.4^v
91	^{231}Pa*	...	3/2	3.58	1.34	.?
92	^{233}U*	...	5/2	0.69	0.26	3.5/7.9^v
	^{235}U*	0.720	7/2	0.27	-0.10	4.3^v
93	^{237}Np*	...	5/2	3.2	1.2	4.5^v
94	^{239}Pu*	...	1/2	1.08	0.406	....
	^{241}Pu*	...	5/2	0.755	-0.283	5.6^u
95	^{241}Am*	...	5/2	1.72	0.644	4.9^u
	^{243}Am*	...	5/2	1.72	0.644	4.9^u
96	^{243}Cm*	...	5/2	0.43	0.16	.?
	^{245}Cm*	...	7/2	0.3	0.1	.?
	^{247}Cm*	...	9/2	0.21	0.08	.?

[a] Stable isotopes and those with half–lives > 1 yr
* Means radioactive
u Means polarization correction not made
v Means unclear whether the correction was made

Table courtesy of
Prof. J. A. Weil and D. S. Rao,
Dept. of Chemistry, University of Saskatchewan
Saskatoon, Canada, S7N OWO

Planck's constant $h = 6.626093 \times 10^{-34}$ Joules $\times$ sec
free electron g value $g_e = 2.0023193044$
Bohr Magneton $\mu_B = 9.274026 \times 10^{-24}$ Joules/Tesla
Nuclear Magneton $\mu_n = 5.050824 \times 10^{-27}$ Joules/Tesla

B. The Cayley Transformation Formula

As pointed out in Chap. 6, all necessary symmetry transformations can be achieved by the application of rotation matrices $R(\mathbf{v}, \theta)$ which provide rotation around the direction of an unity vector $\mathbf{v}$ by an angle θ (*Cayley* formula [B.1]). If the absolute value of θ follows the condition:

$$|\theta| \# n\pi, \tag{B.1}$$

where $n = 1,2,3,..$
then $R(\mathbf{v}, \theta)$ can be derived in the following way:
Let

$$\mathbf{v} = (v_x, v_y, v_z), \tag{B.2}$$

where

$$v_x^2 + v_y^2 + v_z^2 = 1, \tag{B.3}$$

and let

$$a = v_x \tan(\theta/2), \tag{B.4}$$

$$b = v_y \tan(\theta/2), \tag{B.5}$$

$$c = v_z \tan(\theta/2), \tag{B.6}$$

and

$$X = 1 + a^2 + b^2 + c^2. \tag{B.7}$$

It follows:

$$R(\mathbf{v}, \theta) = \frac{1}{X} \begin{bmatrix} 1 + a^2 - b^2 - c^2 & 2(ab - c) & 2(ac + b) \\ 2(ba + c) & 1 + b^2 - c^2 - a^2 & 2(bc - a) \\ 2(ca - b) & 2(cb + a) & 1 + c^2 - a^2 - b^2 \end{bmatrix} \quad (B.8)$$

In numerical calculations (B.1) causes no problems. For example, $\theta = \pi$ can be replaced by $\theta = \pi - 10^{-6}$ with no noticeable error.

If $\mathbf{v}$ is perpendicular to a mirror plane, a mirror operation can be achieved by a 180° rotation ($\theta = \pi - 10^{-6}$) around $\mathbf{v}$ followed by an inversion operation.

C. Algorithm for the Subtraction of an Unknown Background

The ENDOR spectrum consisting of the superposition of ENDOR peaks and a background signal is given by a set of numbers $S_i^{(0)}, i = 1, .., N$. $S^{(0)}$ is the signal and the index i corresponds to equidistant frequency positions. It is assumed that all ENDOR lines are positive and that the background signal varies "slowly" with frequency. No further knowledge about the frequency dependence of the background signal is necessary.

In a first step, the ENDOR spectrum is filtered by assigning to each index i the average value of k ($k < N, k \geq 3$, odd integer) values $S_i^{(0)}$ around i to the index i. One obtains a new spectrum $S_i^{*(0)}$.
Let

$$l = (k - 1)/2. \quad (C.1)$$

The filtered spectrum is simply obtained as:

$$S_i^{*(0)} = \frac{1}{k} \sum_{j=i-l}^{i+l} S_j^{(0)}, \quad (C.2)$$

where $i > l$ and $i \leq (N - l)$. $S^{*(0)}$ can be very efficiently calculated using a recursive computer algorithm. To also extend (C.2) to the signal values at both ends of the ENDOR spectrum, the spectrum is extrapolated linearly for $i < 1$ and $i > N$. For a suitable selection of the value k (the filter width) the ENDOR peaks in the spectrum $S^{(0)}$ are slightly broadened by the application of (C.2), whereas the shape of the background signal is not noticeably affected. The broadened ENDOR peaks in the spectrum $S^{*(0)}$ are smaller compared to those in $S^{(0)}$.

In a second step, the two spectra $S^{(0)}$ and $S^{*(0)}$ are compared with each other and $S^{(0)}$ is modified in the following way:
For each index i where $S_i^{*(0)}$ is smaller than $S_i^{(0)}$ the value of $S_i^{(0)}$ is replaced by that of $S_i^{*(0)}$. One obtains a new spectrum $S_i^{(1)}$ as:

$$S_i^{(1)} = S_i^{*(0)} \text{ if } S_i^{*(0)} < S_i^{(0)}, \text{ else } S_i^{(1)} = S_i^{(0)}. \quad (C.3)$$

The spectrum $S^{(1)}$ is essentially equal to $S^{(0)}$; however, the top of each ENDOR peak in $S^{(0)}$ has been cut down to the corresponding smaller value in $S^{*(0)}$. The new spectrum $S^{(1)}$ is slightly more similar to the background signal than the original spectrum $S^{(0)}$.

The process defined by (C.1) to (C.3) is now repeated for the spectrum $S^{(1)}$ by replacing $S^{(0)}$ by $S^{(1)}$. After up to 100 iterations, the spectrum $S^{(100)}$ is remarkably equal to the background signal in $S^{(0)}$ and can be easily subtracted from $S^{(0)}$. The resulting spectrum contains only the ENDOR lines without the background signal. The speed of convergence and the performance of the algorithm can be improved by augmenting the filter width (given by k) gradually along with the iterations. A further slight improvement of the performance can be achieved by the application of more complex filters as described in Appendix D, but then much more computer time is needed. If the original ENDOR spectrum is severely obscured by noise, prefiltering of the spectrum before subtracting the background is useful.

There are no good algorithms for subtracting background signals in cases where the lines can be positive or negative in the same spectrum, as for Double ENDOR or ODMR. For alternative algorithms to subtract background signals see [C.1].

D. Digital Filters for Application in ENDOR Spectra

In principle, all the polynomial filters described in [D.1-3], and other filter algorithms (for example [D.4]) can be used to diminish high frequency components of noise in ENDOR spectra. The situation is, however, different if digital filters are used to prepare the data for the application of a subsequent deconvolution algorithm as described in Appendix E. Any digital filter distorts the shape of the lines in the spectra somewhat, and these distortions affect the performance of deconvolution algorithms. For the algorithm in Appendix E a digital filter turned out to be the best choice, which simply consists of a convolution with a *Gaussian* line of suitable half width. Such a filter conserves only the first moment of the lines in the spectrum, but higher moments are of no interest for deconvolution.

A problem is that a convolution of a spectrum with a *Gaussian* function needs very much computer time, since no recursive algorithm is possible. It is, however, possible to achieve nearly the same effect by subsequent convolutions of the spectrum with rectangular functions of different suitable lengths, and unity height. Convoluting a spectrum with a rectangular function of unity height simply means the filter algorithm of ([C.2], Appendix C) which is very efficiently calculated recursively. The lengths of these different filters [parameter k in (C.1)] are determined by the shape of the *Gaussian* function. They can be visualized by the lengths of stripes which are obtained by "cutting" the area of the *Gaussian* function parallel to the x-axis into stripes of constant width. In a similar way, any other function used for a convolution can be approximated in this way for a very efficient calculation of the convolution integral.

E. Deconvolution of ENDOR Spectra

Best results for decreasing the width of ENDOR lines without excessive generation of noise were obtained by the application of an iterating deconvolution algorithm [E.1–3]. The amount of deconvolution, and at the same time the amount of noise generation, can be adjusted by the number of iterations used. The algorithm works as follows: The spectrum is given by a set of numbers S_i, $i = 1, .., N$, where S is the signal at the frequency position numbered by i. The spectrum should contain no background, all lines must be positive and high frequency components of noise are suppressed by a suitable digital filter (see Appendix D). New spectra $S_i^{(n+1)}$ are obtained by:

$$S_i^{(n+1)} = S_i + (S_i^{(n)} - AS_i^{(n)}), \tag{E.1}$$

where $n = 0, 1, 2, ..$, $S_i^{(0)} = S_i$ and A is an operator which convolutes the spectrum $S_i^{(n)}$ with a line shape function. This function must have the shape and the half width of a single line in the original spectrum S_i (the magnitude of the lines plays no role). The convolution can be calculated as described in Appendix D. After some iterations, the spectrum $S^{(n)}$ nearly converges to the deconvoluted one. About 20 iterations are already sufficient in many practical cases.

In experimental spectra all lines never have the same shape. This causes distortions and pseudoeffects in the spectra $S^{(n)}$. These effects can, however, be greatly suppressed by not allowing the expression $(S_i^{(n)} - AS_i^{(n)})$ in (E.1) to become negative for any i and n. This can be described by a "positivity operator" P in the following way:

$$S_i^{(n+1)} = S_i + P(S_i^{(n)} - AS_i^{(n)}), \tag{E.2}$$

where $PX = X$ if $X \geq 0$ and $PX = 0$ if $X < 0$.

This operator P does not distort line positions, it just modifies the relative height of lines slightly. An additional effect is that the generation of noise is significantly suppressed by the application of P. An exact deconvolution is only possible for *Gaussian* line shapes in S_i. In practical calculations it turned out, however, that best results were always obtained if the line shape function used for the convolution (operator A) was, as much as possible, similar to the shape of the experimental lines. The shape of many ENDOR lines can be described by a mixture between a *Gaussian* and a *Lorentzian* function. This mixture can be obtained for the calculation of A by convoluting a *Lorentzian* function with a *Gaussian* function. The mixing can then be described by one parameter determining the relative half widths of these two functions.

There are a total number of three parameters which must be optimized experimentally for deconvoluting a spectrum:

(i) the number of iterations,

(ii) the half width of the convolution function (used for A), and

(iii) the *Gaussian – Lorentzian* mixing parameter.

Also, EPR spectra can be deconvoluted to some extent after integration of the measured spectra.

F. Peak Search Algorithm

The spectrum is given by a set of numbers S_i, $i = 1, .., N$, where i numbers the frequency position of the value S_i. It is assumed that all lines in the spectrum are positive and that high frequency components of noise have been suppressed by the application of a suitable digital filter (see Appendix D). The spectrum may contain a background.

Peak positions or lines in the spectrum are then easily found by searching for local minima of the second derivative of the spectrum. This can be very efficiently achieved with a computer by the following procedure (also [F.1]): One draws a straight line of a length corresponding to k frequency steps ($k \geq 3$, odd integer) between two points, S_i and $S_{i+(k-1)}$, in the spectrum. This line is moved over the entire spectrum and the vertical distance of each dot S_l, $i - (k-1)/2 \leq l \leq i + (k-1)/2$ to the line is monitored. If the maximum of these k distance values occurs at a frequency position in the middle of the straight line (at the index i) then a peak of the data is present at the frequency number i. At the same time, the distance value at the peak position can be taken as some indication for the peak height, independent from background signals. This algorithm works even if lines are superimposed to some extent in the spectrum. However, in this case the line positions found are no longer exact. The performance of the algorithm can be optimized by a suitable selection of the parameter k (roughly the half width of a peak).

A great improvement of the accuracy in determining the position of superimposed peaks can be achieved by deconvoluting the spectra (see Appendix E).

G. Simulation of EPR Spectra

In a first order treatment of the electron *Zeeman* interaction, the fine–structure interaction, and all hf and shf interactions, the EPR spectrum of a defect can successively be obtained by the following procedure (also [G.1]):

One starts without any nuclear interactions and calculates the EPR spectrum just due to the electron *Zeeman* interaction and the fine–structure interaction. If there is no fine–structure interaction present, the spectrum consists only of a single *Lorentzian* line. Theoretically, the half width of this line is of the order of the half width of an ENDOR line.

Then nuclei are added one after the other, and for each additional nucleus the existing EPR spectrum is modified in the calculation according to the new interaction. In a first order treatment the interaction with each neighbor nucleus can be considered independently from all the other interactions. The symmetry positions of the neighbor nuclei play no role since only the actual interaction energies W_{shf} with each neighbor nucleus are taken for the calculation. For the same reason, it makes no difference whether an additional nucleus belongs to the same shell as the preceding one or not.

This procedure is relatively straightforward as long as all neighbor nuclei have only one isotope. If there are different isotopes at least for one nucleus of the defect the algorithm is more complex:
Consider the complete EPR spectrum EPR^N for the interaction with N neighbor nuclei (here it makes no difference how many of these N nuclei have different isotopes). Now, if the nucleus $N + 1$ is added, which has k isotopes $i_1, .., i_k$ with abundances $a_1, .., a_k$, one obtains k new EPR spectra $EPR_1^{(N+1)}, .., EPR_k^{(N+1)}$, where each new spectrum $EPR_l^{(N+1)}, l = 1, .., k$ is obtained by modifying the existing EPR spectrum EPR^N due to the interaction with the isotope i_l. For each isotope with no magnetic moment the corresponding new spectrum is identical to the existing one. The intensity of each new EPR spectrum is proportional to the abundance of the corresponding isotope (this is also true for isotopes with no magnetic moment). The complete new EPR spectrum $EPR^{(N+1)}$ is now obtained by adding all new EPR spectra $EPR_l^{(N+1)}$, which are multiplied by the abundance of their corresponding isotope:

$$EPR^{(N+1)} = \sum_{l=1}^{k} a_l EPR_l^{(N+1)} \qquad (G.1)$$

This spectrum, $EPR^{(N+1)}$, is now ready for the modification according to the interaction with the nucleus $N + 2$.

The modification of an existing EPR spectrum EPR^N by the interaction with an isotope i_l of a new nucleus $N + 1$ can be very efficiently calculated in the following way:
Let the spectrum EPR^N be stored in the center of a long array S_i^N, $i = 1, .., n$, where the index i numbers all the magnetic field values which will

be significant for the complete final spectrum. An additional array $S_i^{(N+1)}$ is empty at this moment. The isotope i_l of the nucleus $N+1$ may have the hf or shf interaction W_{shf_l} and the nuclear spin I_l. The product $m_{I_l} W_{shf_l}$, where $m_{I_l} = -I_l, .., I_l$ may be expressed by an integer number $j_{m_{I_l}}$ which means that the interaction $m_{I_l} W_{shf_l}$ is expressed by $j_{m_{I_l}}$ magnetic field steps in the array S_i^N. Depending on m_{I_l}, the number $j_{m_{I_l}}$ may be positive or negative. The array S_i^N is now shifted by $j_{m_{I_l}}$ steps on the magnetic field scale and for each value of $j_{m_{I_l}}$ the corresponding shifted spectrum $S_{(i+j_{m_{I_l}})}^N$ is added to the array $S_i^{(N+1)}$:

$$S_i^{(N+1)} = \sum_{m_{I_l}=-I_l}^{+I_l} S_{(i+j_{m_{I_l}})}^N \tag{G.2}$$

This array $S_i^{(N+1)}$ contains now the modified EPR spectrum $EPR_i^{(N+1)}$, according to the interaction with the isotope i_l of the nucleus $N+1$. The complete new EPR spectrum $EPR^{(N+1)}$ follows from (G.1).

If a defect exists in different orientations, then the complete EPR spectra are calculated for each orientation, and these individual spectra are added to give the full EPR spectrum of the defect.

With the above algorithm one automatically obtains the integrated or the original EPR spectrum, depending on the type of the first spectrum stored in the array. If, in the beginning of the process, the spectrum for the electron *Zeeman* interaction and the fine–structure interaction is stored in the integrated form, then one finally obtains the integrated full EPR spectrum, and vice versa.

With the above algorithm it is also very easy to calculate ENDOR–induced EPR spectra. Formally, the only difference between the EPR and the EI–EPR spectrum is that the EI–EPR spectrum contains for one isotope of one nucleus and one defect orientation only two nuclear quantum states instead of $2I+1$ states. This can be easily accounted for in (G.2). If a fine–structure splitting is present in the EPR spectrum for $S > 1/2$, then the selection of special m_S states by EI–EPR instead of all $2S+1$ states must be taken into account [G.2] when calculating the first EPR spectrum (electron *Zeeman* and fine–structure interactions) in the above algorithm.

When calculating an EPR spectrum, one usually does not know the interactions with all neighbor nuclei down to the lowest interaction energies for distant neighbors. The effect of these unknown small interactions can be simulated in the calculation by starting with *Gaussian* lines of given half width, instead of extremely narrow *Lorentzian* lines, when calculating the first EPR spectrum in the above process.

Since only a relatively small number of simple operations is necessary for the calculations, very little computer time is necessary for the calculation of an EPR spectrum using the above algorithm.

It should, however, be stressed that this simple and efficient algorithm is only applicable if the first order solution of the spin *Hamiltonian* is sufficient to describe the spectra.

H. Values of $\psi_{ns}(0)$ and $< r^{-3} >_{np}$ for Selected Ions

This appendix contains amplitudes of $\psi_{ns}(0)$ and values of $<r^{-3}>_{np}$ for several common constituents of ionic crystal. These are presented as typical examples of the quantities needed in the calculation of shf interactions.

Table H.1. Amplitudes of $\psi_{ns}(0)$ (in atomic units)

ion \ $\psi_{ns}(0)$	$1s$	$2s$	$3s$	$4s$
Li$^+$	2.62			
Na$^+$	19.8	4.84		
K$^+$	45.6	13.1	4.32	
Rb$^+$	443.9	39.5	16.0	5.40
F$^-$	14.6	3.29		
Cl$^-$	38.5	10.8	3.19	
Br$^-$	115.1	36.1	14.4	4.29

Table H.2. $<r^{-3}>_{np}= \int_0^\infty \frac{1}{r^3}|P_{np}|^2 dr$ (in atomic units) where $P_{np}(r) = r R_{np}(r)$, with $R(r)$ being the radial wave function of the np–orbital.

ion \ np	$2p$	$3p$	$4p$
Na$^+$	17.1		
K$^+$	153.7	12.92	
Rb$^+$	1516	228.1	20.23
F$^-$	6.40		
Cl$^-$	102.8	5.523	
Br$^-$	1261	180.9	10.24

Table H.3. $\int_0^\infty \frac{1}{r^3} P_{np} P_{n'p} dr$ (in atomic units)

ion \ $np.n'p$	$4p.3p$	$3p.2p$
K$^+$		41.0
Rb$^+$	69.2	
Cl$^-$		22.3
Br$^-$	50.5	

The quantities in Tables H.1–3 were calculated with the non–relativistic *Slater* orbitals given by *Clementi* and *Roetti* [H.1].

References

Chapter 1

1.1 C.P. Slichter: *Principles of Magnetic Resonance*, Springer Ser. in Solid State Sci., Vol. 1 3rd ed. (Springer, Berlin, Heidelberg, New York 1990)

1.2 A. Abragam, B. Bleaney: *Electron Paramagnetic Resonance of Transition Ions* (Clarendon, Oxford 1970)

1.3 J.E. Wertz and J.R. Bolton: *Electron Spin Resonance: Elementary Theory and Practical Applications* (McGraw–Hill, New York 1972)

1.4 R.T. Schumacher: *Introduction to Magnetic Resonance* (Benjamin, New York 1970)

1.5 G.E. Pake, T.L. Estle: *The Physical Principles of Electron Paramagnetic Resonance*, 2nd ed. (Benjamin, New York 1973)

1.6 W.B. Fowler (ed): *Physics of Color Centers* (Academic, New York 1968)

1.7 K. Seeger: *Semiconductor Physics* Springer Ser. in Solid State Sci. Vol. 40 5th ed. (Springer, Berlin, Heidelberg, New York 1991)

1.8 Y. Yang, F. Lüty: Phys. Rev. Lett. **51**, 419 (1983)

1.9 B.K. Meyer, D.M. Hofmann, J.R. Niklas, J.-M. Spaeth: Phys. Rev. B **36**, 1332 (1987)

1.10 M. Godlewski, W.M. Shan, B. Monemar, H.P. Gislason: in *Semi-insulating III–V-Materials*, ed. by G. Grossmann and L. Ledebo (Adam Hilger, Bristol, Philadelphia 1988) p. 325

1.11 J. Michel, J.R. Niklas, J.-M. Spaeth, C. Weinert: Phys. Rev. Lett. **57**, 611 (1986)

1.12 J. Bourgoin, M. Lannoo: *Point Defects in Semiconductors II: Experimental Aspects*, Springer Ser. in Solid State Sci., Vol. 35 (Springer, Berlin, Heidelberg, New York 1983)

1.13 G. Feher: Phys. Rev. **103**, 834 (1956)

1.14 G. Feher: Phys. Rev. **114**, 1219 (1959)

1.15 N.V. Karlov, J. Margerie, Y. Merle d'Aubigné: J. Physique **24**, 717 (1963)

1.16 L.F. Mollenauer, S. Pan: Phys. Rev. B **6**, 772 (1972)

1.17 B.C. Cavenett: Adv. in Phys. **30**, 475 (1981)

1.18 W.B. Lynch, O.W. Pratt: Magn. Res. Rev. (GB) **10**, 111 (1985)

1.19 S. Geschwind: "Optical Techniques in EPR in Solids", in *Electron Paramagnetic Resonance*, ed. by S. Geschwind (Plenum, New York 1972)

1.20 K.P. Dinse, C.J. Winscon: "Optically Detected ENDOR Spectroscopy", in *Triplet State ODMR Spectroscopy*, ed. by R.H. Clarke (Wiley, New York 1984)

1.21 D.M. Hofmann, B.K. Meyer, F. Lohse, J.-M. Spaeth: Phys. Rev. Lett. **53**, 1187 (1984)

1.22 D.M. Hofmann: "Strukturaufklärung des EL2 Defektes in Galliumarsenid mit optisch nachgewiesener Elektronen–Kern–Doppelresonanz"; Doctoral Dissertation, Universität–GH Paderborn(1987)

1.23 H.G. Grimmeiss, E. Janzen, H. Ennen, O. Schirmer, J. Schneider, R. Wörner, E. Holm, E. Sirtl, P. Wagner: Phys. Rev. B **24**, 4571 (1981)

1.24 J.R. Niklas, J.-M. Spaeth: Solid State Commun. **46**, 121 (1983)

1.25 J.-M. Spaeth: "Atomic Hydrogen as a Model Defect in Alkali Halides", in *Defects in Insulating Crystals*, ed. by V.M. Tuchkevich and K.K. Shvarts (Springer, Berlin, Heidelberg, New York 1981) p. 232

1.26 A.M. Stoneham: *Theory of Defects in Insulators and Semiconductors* (Clarendon, Oxford 1975)

Chapter 2

2.1 C.P. Slichter: *Principles of Magnetic Resonance*, Springer Ser. in Solid State Sci., Vol. 1 3rd ed. (Springer, Berlin, Heidelberg, New York 1990)

2.2 A. Abragam, B. Bleaney: *Electron Paramagnetic Resonance of Transition Ions* (Clarendon, Oxford 1970)

2.3 J.E. Wertz and J.R. Bolton: *Electron Spin Resonance: Elementary Theory and Practical Applications* (McGraw–Hill, New York 1972)

2.4 R.T. Schumacher: *Introduction to Magnetic Resonance* (Benjamin, New York 1970)

2.5 G.E. Pake, T.L. Estle: *The Physical Principles of Electron Paramagnetic Resonance*, 2nd ed. (Benjamin, New York 1973)

2.6 N. Bloembergen: *Nuclear Magnetic Relaxation* (Benjamin, New York 1961)

2.7 R. Orbach, H.J. Stapleton: "Electron Spin Lattice Relaxation", in *Electron Paramagnetic Resonance*, ed. by S. Geschwind (Plenum, New York 1972)

2.8 C.P. Poole, H.H. Farach: *Relaxation in Magnetic Resonance* (Academic, New York 1971)

2.9 C. Kittel: *Introduction to Solid State Physics*, 3rd ed. (Wiley, New York 1967)

2.10 K.-H. Hellwege: *Einführung in die Festkörperphysik* (Springer, Berlin, Heidelberg, New York 1981)

Chapter 3

3.1 A. Abragam, B. Bleaney: *Electron Paramagnetic Resonance of Transition Ions* (Clarendon, Oxford 1970)

3.2 G.E. Pake, T.L. Estle: *The Physical Principles of Electron Paramagnetic Resonance*, 2nd ed. (Benjamin, New York 1973)

3.3 W. Känzig, M.H. Cohen: Phys. Rev. Lett. **3**, 509 (1959)

3.4 H. Seidel, H.C. Wolf: "ESR and ENDOR Spectroscopy of Color Centers in Alkali Halide Crystals", in *Physics of Color Centers*, ed. by W.B. Fowler (Academic, New York 1968)

3.5 H.R. Zeller, W. Känzig: Helv. Phys. Acta. **40**, 845 (1967)

3.6 S.H. Muller, M. Springer, E.G. Sievers, C.A.J. Ammerlaan: Solid State Commun. **25**, 987 (1978)

3.7 J. Michel, N. Meilwes, J.R. Niklas, J.-M. Spaeth: in *Shallow Impurities in Semiconductors*, Proc. of the 3rd Int. Conf., Lingköping, 1988

3.8 J. Michel, J.R. Niklas, J.-M. Spaeth: Phys. Rev. B **40**, 1732 (1989)

3.9 C. Rudowicz: Magn. Res. Rev. **13**, 1 (1987)

3.10 H.H. Towner, J.M. Kim, H.S. Story: J. Chem. Phys. **56**, 3676 (1971)

3.11 H. Söthe: private communication

3.12 H. Söthe, L.G. Rowan, J.-M. Spaeth: J. Phys.: Condens. Matter **23**, 3591 (1989)

3.13 U. Kaufmann, J. Schneider: Solid State Commun. **25**, 1113 (1978)

3.14 Y. Ueda, J.R. Niklas, J.-M. Spaeth, U. Kaufmann and J. Schneider: Solid State Commun. **46**, 127 (1983)

3.15 J.J. Rousseau: "Etude de la Dynamique du Réseau et des Changements de Phase Structuraux dans les Fluopérovskites"; Doctoral Dissertation, Université du Maine et Université Pierre et Marie Curie, Paris (1977)

3.16 P. Studzinski: "Paramagnetische Ionen zur Untersuchung struktureller Phasenübergänge – eine ENDOR–Untersuchung"; Doctoral Dissertation, Universität–GH Paderborn (1985)

3.17 F. Agullo–Lopéz, C.R.A. Catlow, P.D. Townsend: "Point Defects in Materials" (Academic, London 1988) Chap. 6

3.18 C.P. Slichter: *Principles of Magnetic Resonance*, Springer Ser. in Solid State Sci., Vol. 1 3rd ed. (Springer, Berlin, Heidelberg, New York 1990)

3.19 H.G. Grimmeiss, E. Janzen, H. Ennen, O. Schirmer, J. Schneider, R. Wörner, E. Holm, E. Sirtl, P. Wagner: Phys. Rev. B **24**, 4571 (1981)

3.20 R. Wörner, O.F. Schirmer: Solid State Commun. **51**, 665 (1984)

3.21 S. Greulich–Weber, J.R. Niklas, J.-M. Spaeth: J. Phys.: Condens. Matter **35**, 1 (1989)

3.22 E. Goovaerts, J. Andriessen, S.V. Nistor, D. Schoemaker: Phys. Rev. B **24**, 29 (1981)

3.23 F.J. Ahlers, J.-M. Spaeth, J. Phys. C: Solid State Phys. **19**, 4693 (1986)

3.24 J. Casas–Gonzáles, H.W. den Hartog, R. Alcalá: Phys. Rev. B **21**, 3826 (1980)

3.25 P. Studzinski, J. Casas–Gonzáles, J.–M. Spaeth: J. Phys. C: Solid State Phys. **17**, 5411 (1984)

3.26 E. Zoritá: "Espectroscopia de resonancia magnetica de impurezas de niquel en crystales de $KMgF_3$, $RbCaF_3$ y K_2MgF_4"; Doctoral Dissertation, Universidad de Zaragoza (1988)

3.27 E. Zoritá, P.J. Alonso, R. Alcalá: Phys. Rev. B **35**, 3116 (1987)

3.28 J.–M. Spaeth: Z. Physik **192**, 107 (1966)

3.29 G. Heder, J.R. Niklas, J.–M. Spaeth: Phys. Stat. Sol. (b) **100**, 567 (1980)

3.30 J.–M. Spaeth: "Atomic Hydrogen as a Model Defect in Alkali Halides", in *Defects in Insulating Crystals*, ed. by V.M. Tuchkevich and K.K. Shvarts (Springer, Berlin, Heidelberg, New York 1981) p. 232

3.31 G. Heder, J.–M. Spaeth: Phys. Stat. Sol. (b) **125**, 523 (1984)

3.32 L.O. Schwan, W. Geigle, H. Paus: Z. Phys. B **35**, 43 (1979)

Chapter 4

4.1 S. Geschwind: "Optical Techniques in EPR in Solids", in *Electron Paramagnetic Resonance*, ed. by S. Geschwind (Plenum, New York 1972)

4.2 I.Y. Chan: "Zero Field ODMR Techniques – Phosphorescence Detection", in *Triplet State ODMR Spectroscopy: Techniques and Applications to Biophysical Systems*, ed. by R.H. Clarke (Wiley, New York 1982)

4.3 D.W. Pratt: "ODMR Studies of Excited Triplet States in High Fields", in *Triplet state ODMR Spectroscopy*, ed. by R.H. Clarke (Wiley, New York 1982)

4.4 D.B. Fichten: "Zero Phonon Transitions", in *Physics of Color Centers*, ed. by W.B. Fowler (Academic, New York 1968) Chap. 5

4.5 H. Henry, C.P. Slichter: "Moments in the Degeneracy of Optical Spectra", in *Physics of Color Centers*, ed. by W.B. Fowler (Academic, New York 1968) Chap. 6

4.6 J. Bourgoin, M. Lannoo: *Point Defects in Semiconductors II: Experimental Aspects*, Springer Ser. in Solid State Sci., Vol. 35 (Springer, Berlin, Heidelberg, New York 1983)

4.7 M. Lannoo, J.C. Bourgoin: "Point Defects in Semconductors I, Theoretical Aspects", Springer Ser. Solid State Sci., Vol. 22 (Springer, Berlin, Heidelberg, New York 1981)

4.8 B.K. Meyer, J.–M. Spaeth: Phys. Rev. B **32**, 1409 (1985)

4.9 A.M. Stoneham: *Theory of Defects in Solids: Electronic Structure of Defects in Insulators and Semiconductors* (Clarendon, Oxford 1975)

4.10 T.H. Keil: Phys. Rev. A **140**, 601 (1965)

4.11 B. Henderson, F. Imbush: Contemp. Phys. **29**, 235 (1988)

4.12 D.B. Fitchen, R.H. Silsbey, T.A. Fulton, E.L. Wolf: Phys. Rev. Lett. **11**, 2775 (1963)

4.13 A.E. Hughes: "Optical Techniques and an Introduction to the Symmetry Properties of Point Defects", in *Defects and their Structure in Nonmetallic Solids*, ed. by B. Henderson and A.E. Hughes, ASI series B19 (Plenum, New York 1976)

4.14 A.A. Maradudin: Solid State Physics **18**, 274 (Academic, New York 1966)

4.15 P.J. Dean, D.C. Herbert: "Bound Excitons in Semiconductors", in *Excitons*, ed. by Y.K. Cho, Topics in Current Phys., Vol. 14 (Springer, Berlin, Heidelberg, New York 1979) Chap. 3

4.16 G.L. Bir, E.E. Pikus: *Symmetry and Strain Induced Effects in Semiconductors* (Keter, Jerusalem 1974)

4.17 A.A. Kaplyanskii: Optics Spectroscopy **16**, 329 (1964)

4.18 D.L. Dexter, C.C. Klick, G.A. Russel: Phys. Rev. **100**, 603 (1955)

4.19 R.H. Bartram, A.M. Stoneham: Solid State Commun. **17**, 1593 (1975)

4.20 A.A. Jahn, E. Teller: Proc. Roy. Soc. A **161**, 220 (1937)

4.21 A.A. Jahn: Proc. Roy. Soc. A **164**, 117 (1938)

4.22 F.S. Ham: "Jahn–Teller Effects in Electron Paramagnetic Resonance Spectra", in *Electron Paramagnetic Resonance*, ed. by S. Geschwind (Plenum, New York 1972)

4.23 R. Engelman: *The Jahn–Teller Effects in Molecules and Crystals* (Wiley, New York 1972)

4.24 N.V. Karlov, J. Margerie, V. Merle d'Aubigné: J. Phys. Rad. **24**, 77 (1963)

4.25 L.F. Mollenauer, S. Pan: Phys. Rev. B **6**, 772 (1972)

4.26 L.I. Schiff: *Quantum Mechanics* (McGraw–Hill, New York 1949)

4.27 H. Paus: "Farbzentren und zweiwertige Fremdkationen in Alkalihalogenid Kristallen"; Habilitationsschrift, Univ. Stuttgart (1980)

4.28 F.J. Ahlers, F. Lohse, J.-M. Spaeth, L.F. Mollenauer: Phys. Rev. B **28**, 1249 (1983)

4.29 W. Gellermann, F. Lüty, C.R. Pollack: Opt. Commun. **39**, 391 (1981)

4.30 B.K. Meyer, J.-M. Spaeth, M. Scheffler: Phys. Rev. Lett. **52**, 851 (1984)

4.31 B.K. Meyer, D.M. Hofmann, J.R. Niklas and J.-M. Spaeth: Phys. Rev. B **36**, 1332 (1987)

4.32 M. Fockele, F. Lohse, J.-M. Spaeth, R.H. Bartram: J. Phys.: Condens. Matter **1**, 13 (1989)

4.33 M. Fockele, F. Lohse, J.-M. Spaeth: Israel Journ. of Chem. **29**, 13 (1989)

4.34 F.J. Ahlers, F. Lohse, Th. Hangleiter, J.-M. Spaeth, R.H. Bartram: J. Phys. C: Solid State Phys. **17**, 4877 (1984)

4.35 B. Clerjaud, C. Nand, B. Deveaud, B. Lambert, B. Plot, G. Bremond, C. Benjeddon, G. Guillot, A. Nouailhat: J. Appl. Phys. **59**, 4207 (1985)

4.36 A. Görger, B.K. Meyer, J.-M. Spaeth, A. Hennel: "Transition Elements in III–V Semiconductors – A Study With Optically Detected Magnetic Resonance", in *Semi-Insulating III-V Materials*, ed. by G. Grossmann and L. Ledebo (Adam Hilger, Bristol 1988) p. 331

4.37 M. Bauemler, B.K. Meyer, J. Schneider: Mat. Sci. Forum **38–41**, 797 (1989)

4.38 G. Hörsch, H.J. Paus: Opt. Commun. **60**, 69 (1987)

4.39 M. Heinemann, B.K. Meyer, J.-M. Spaeth, K. Löhnert: "The occupation of the two charge states of EL2 in LEC–grown GaAs–Wafers – a mapping investigation", in *Defect Recognition and Image Processing in III–V Compounds II*, ed. by E.R. Weber (Elsevier, New York 1987)

4.40 J.-M. Spaeth, D.M. Hofmann, M. Heinemann, B.K. Meyer: in J. Int. Phys. Conf. Ser. **91**, 391 (1988) Chap. 4

4.41 H. Winnacker, Th. Vetter, F.X. Zach: "Mapping of the Deep Donor EL2 in Semi–Insulating GaAs in both its Charge States: $EL2^0$ and $EL2^+$", in *Semi–Insulating III–V Materials*, ed. by G. Grossmann and L. Ledebo (Adam Hilger, Bristol 1988) p. 583

4.42 H. Pannepucci, L.F. Mollenauer: Phys. Rev. B **178**, 589 (1969)

4.43 E.S. Sabitzky, C.H. Anderson: Phys. Rev. B **1**, 1028 (1970)

4.44 M. Heinemann: "Homogenität und thermische Stabilität von EL2–Defekten in Galliumarsenid"; Diplomarbeit, Universität–GH Paderborn (1987)

4.45 B.K. Meyer, J.-M. Spaeth: Defect and Diffusion Forum (Trans. Tech. Publ.) **62 / 63**, 39 (1989)

4.46 J. Hage, J.R. Niklas, J.-M. Spaeth: J. Electron. Mater. A **14**, 1051 (1984)

4.47 J. Hage, J.R. Niklas, J.-M. Spaeth: J. Phys. C: Semcond. Sci. and Technol. **4**, 773 (1989)

4.48 H. Katayama–Yoshida, A. Zunger: Phys. Rev. B **33**, 2961 (1986)

4.49 A. Abragam, B. Bleaney: *Electron Paramagnetic Resonance of Transition Ions* (Clarendon, Oxford 1970) p. 797

4.50 A. Juhl, A. Hoffman, D. Bimberg, H.J. Schulz: Appl. Phys. Lett. **50**, 1292 (1987)

4.51 A. Görger, J.-M. Spaeth: Semicond. Sci. Technol. **6**, 800 (1991)

4.52 J.-M. Spaeth, F. Lohse: J. Phys. Chem. Sol. **51**, 861 (1990)

4.53 P.D. Devor, I.J. D'Haenens, C.K. Asawa: Phys. Rev. Lett. **8**, 432 (1962)

4.54 C. Baldacchini, U.M. Grassano, A. Tanga: Phys. Rev. B **16**, 5570 (1977)

4.55 A. Winnacker, K.E. Mauser, B. Niesert: Z. Physik B **26**, 97 (1977)

4.56 K.E. Mauser, B. Niesert, A. Winnacker: Z. Physik B **26**, 107 (1977)

4.57 H. Ohkura: Crystal Lattice Defects and Amorphous Materials **12**, 401 (1985)

4.58 H.J. Reyer, K. Hahn, Th. Vetter, A. Winnacker: Z. Phys. B **33**, 357 (1979)

4.59 F.J. Ahlers, J.-M. Spaeth: J. Phys. C: Solid State Phys. **19**, 4693 (1986)

4.60 B.K. Meyer, G. Heder, F. Lohse, J.-M. Spaeth: Solid State Commun. **43**, 325 (1982)

4.61 B.K. Meyer, J.-M. Spaeth: J. Phys. C: Solid State Phys. **17**, 2213 (1984)

4.62 C. Hermann, G. Lampel: Phys. Rev. Lett. **27**, 373 (1971)

4.63 C. Weisbuch, C. Hermann: Phys. Rev. B **15**, 816 (1977)

4.64 B.C. Cavenett: Adv. in Phys. **30**, 475 (1981)

4.65 J.E. Nicholls, J.J. Davies, B.C. Cavenett, J.R. James, D.J. Dunstan: J. Phys. C: Solid State Phys. **12**, 361 (1979)

4.66 B.K. Meyer, Th. Hangleiter, J.-M. Spaeth, G. Strauch, Th. Zell, A. Winnacker, R.H. Bartram: J. Phys. C: Solid State Phys. **18**, 1503 (1985)

4.67 D.G. Thomas, M. Gershenson, F.A. Trumbore: Phys. Rev. A **133**, 269 (1964)

4.68 P.J. Dean: in *Progress in Solid State Chemistry*, Vol. 8, ed. by J.O. McCalden and Y. Somorjai (Pergamon, Oxford 1973) p. 1

4.69 D. Block, A. Hervé, R.T. Cox: Phys. Rev. B **25**, 6049 (1982)

4.70 J.J. Davies: J. Phys. C: Solid State Phys. **16**, 867 (1983)

4.71 J. Kluge, J. Donecker: Phys. Stat. Sol. (a) **81**, 675 (1984)

4.72 P. Edel, C. Hennies, Y. Merle d'Aubigné, R. Romestain, Y. Twarowski: Phys. Rev. Lett. **28**, 1268 (1972)

4.73 P. Dawson, C.M. McDonagh, B. Henderson, L.S. Welch: J. Phys. C: Solid State Phys. **11**, 983 (1978)

4.74 W.B. Lynch, O.W. Pratt: Magn. Res. Rev. (GB) **10**, 111 (1985)

4.75 D. Henderson: Semicond. Insulators **3**, 299 (1978)

4.76 F.J. Ahlers, F. Lohse, J.-M. Spaeth: Solid State Commun. **43**, 321 (1982)

4.77 P.J. Dean, W. Schairer, M. Lorentz, T.N. Morgan: J. Lum. **9**, 343 (1974)

4.78 J.J. Lappe, "Magnetische Resonanzuntersuchungen an extrinsischen und intrinsischen Donatoren in Galliumphosphid"; Doctoral Dissertation, Universität–GH Paderborn (1990)

4.79 P.J. Dean, D.C. Herbert: "Bound Excitons in Semiconductors", in *Excitons*, ed. by Y.K. Cho, Topics in Current Phys., Vol. 14 (Springer, Berlin, Heidelberg, New York 1979) p. 105

4.80 J.H. Schulman, W.D. Compton: *Color Centers in Solids* (Pergamon, New York 1962) p. 56

4.81 K.S. Song, R.T. Williams: *Self Trapped Excitons* (Springer, Berlin, Heidelberg, New York 1992) p. 1

Chapter 5

5.1 G. Feher: Phys. Rev. **103**, 500 (1956)

5.2 G. Feher: Phys. Rev. **103**, 834 (1956)

5.3 H. Seidel: Z. Physik **165**, 218 (1961)

5.4 H. Seidel: Z. Physik **165**, 239 (1961)

5.5 H. Seidel: Z. angew. Physik **14**, 21 (1962)

5.6 C. Kevan, W. Kispert: *Electron Spin Double Resonance Spectroscopy* (Wiley, New York 1976)

5.7 B.K. Meyer, J.-M. Spaeth, M. Scheffler: Phys. Rev. Lett. **52**, 851 (1984)

5.8 D.M. Hofmann: "Strukturaufklärung des EL2 Defektes in Galliumarsenid mit optisch nachgewiesener Elektronen–Kern–Doppelresonanz";
Doctoral Dissertation, Universität–GH Paderborn(1987)

5.9 J.J. Lappe, "Magnetische Resonanzuntersuchungen an extrinsischen
und intrinsischen Donatoren in Galliumphosphid"; Doctoral Dissertation, Universität–GH Paderborn (1990)

5.10 D.Y. Jeon, J.F. Donegan, G.D. Watkins: Phys. Rev. B **39**, 3207 (1988)

5.11 S. Greulich–Weber, J.R. Niklas, E. Weber, J.–M. Spaeth: Phys. Rev. B
30, 6292 (1984)

5.12 J.–M. Spaeth: Z. Physik **192**, 107 (1966)

5.13 G. Heder, J.R. Niklas, J.–M. Spaeth: Phys. Stat. Sol. (b) **100**, 567
(1980)

5.14 J.R. Niklas, J.–M. Spaeth: Phys. Stat. Sol. (b) **101**, 221 (1980)

5.15 J.–M. Spaeth, J.R. Niklas: Rec. Development in Condensed Matter Physics **1**, 393 (1981)

5.16 M. Yuste, C. Taurel, M. Rahmani, D. Lemoyne: J. Phys. Chem. Solids
37, 961 (1976)

5.17 J.R. Niklas, G. Heder, M. Yuste, J.–M. Spaeth: Solid State Commun.
26, 169 (1978)

5.18 R.U. Bauer, J.R. Niklas, J.–M. Spaeth: Phys. Stat. Sol. (b) **118**, 557
(1983)

5.19 R.C. DuVarney, J.R. Niklas, J.–M. Spaeth: Phys. Stat. Sol. (b) **103**,
329 (1980)

5.20 R.U. Bauer, J.R. Niklas, J.–M. Spaeth: Phys. Stat. Sol. (b) **119**, 171
(1983)

5.21 R. Biehl, M. Plato, K. Möbius: J. Chem. Phys. **63**, 3515 (1975)

5.22 K.P. Dinse, R. Biehl, K. Möbius: J. Chem. Phys. **61**, 4335 (1974)

5.23 P. Studzinski, J.R. Niklas, J.–M. Spaeth: Phys. Stat. Sol. (b) **101**, 673
(1980).

5.24 W.T. Doyle, A.B. Wolbarst: J. Phys. Chem. Sol. **36**, 549 (1975)

5.25 H. Söthe, P. Studzinski, J.–M. Spaeth: Phys. Stat. Sol. (b) **130**, 339
(1985).

5.26 H. Söthe, "ENDOR–Untersuchungen am $F_H(F^-)$–Zentrum in Kalium-
chlorid"; Diplomarbeit, Universität–GH Paderborn (1984)

5.27 D. Studzinski, J.–M. Spaeth: J. Phys. C: Solid State Phys. **19**, 6441
(1986)

5.28 S. Greulich–Weber, J.R. Niklas, J.–M. Spaeth: J. Phys.: Condens. Matter **35**, 1 (1989)

Chapter 6

6.1 J.–M. Spaeth: Z. Physik **192**, 107 (1966)

6.2 J.–M. Spaeth, M. Sturm: Phys. Stat. Sol. (b) **42**, 739 (1970)

6.3 W.T. Doyle, A.B. Wolbarst: J. Phys. Chem. Sol. **36**, 549 (1975)

6.4 G. Heder, J.R. Niklas, J.-M. Spaeth: Phys. Stat. Sol. (b) **100**, 567 (1980)

6.5 J.R. Niklas, J.-M. Spaeth: Solid State Commun. **46**, 121 (1983)

6.6 S. Greulich–Weber, J.R. Niklas, E. Weber, J.-M. Spaeth: Phys. Rev. B **30**, 6292 (1984)

6.7 S. Greulich–Weber, J.R. Niklas, J.-M. Spaeth: J. Phys. C: Solid State Phys. **17**, 911 (1984)

6.8 A. Cayley: Journal f. d. reine u. angew. Math. **32**, 119 (1846)

6.9 C. Rudowicz: Magn. Res. Rev. **13**, 1 (1987)

6.10 H.A. Buckmaster: Can. J. Phys. **42**, 386 (1964)

6.11 J.R. Niklas, J.-M. Spaeth, G.D. Watkins: Proc. of the MRS Conf., San Francisco 1985, **46**, 237 (1985)

6.12 F. Beeler, M. Scheffler, O. Jepsen, O. Gunnarsson: Phys. Rev. Lett. **54**, 2525 (1985)

6.13 H.A. Jahn, E. Teller: Proc. Roy. Soc. A **161**, 220 (1937)

6.14 J. Sierro: J. Phys. Chem. Sol. **28**, 417 (1967)

6.15 R.C. DuVarney, J.R. Niklas, J.-M. Spaeth: Phys. Stat. Sol. (b) **128**, 673 (1985)

6.16 Th. Müssig, J.R. Niklas, F. Granzer, J.-M. Spaeth: Crystal Lattice Defects and Amorphous Materials **16**, 169 (1987)

6.17 J. Michel, J.R. Niklas, J.-M. Spaeth, C. Weinert: Phys. Rev. Lett. **57**, 611 (1986)

6.18 J. Michel, J.R. Niklas, J.-M. Spaeth: Phys. Rev. B **40**, 1732 (1989)

6.19 R. Wörner, O.F. Schirmer: Solid State Commun. **51**, 665 (1984)

6.20 S. Greulich–Weber, J.R. Niklas, J.-M. Spaeth: J. Phys.: Condens. Matter **35**, 1 (1989)

6.21 H. Seidel: "Superhyperfeinstruktur–Analyse paramagnetischer Störstellen in Kristallen mit Elektronen–Kern–Doppelresonanz (ENDOR)"; Habilitationsschrift, Univ. Stuttgart (1966)

6.22 J. Hage, J.R. Niklas, J.-M. Spaeth: J. Electron. Mater. A **14**, 1051 (1984)

6.23 J. Hage, J.R. Niklas, J.-M. Spaeth: J. Phys. C: Semcond. Sci. and Technol. **4**, 773 (1989)

6.24 J.E. Feuchtwang: Phys. Rev. **126**, 1628 (1962)

6.25 D. Schoemaker: Phys. Rev. **174**, 1060 (1968)

6.26 J. Hage, J.R. Niklas, J.-M. Spaeth: Mater. Sci. Forum **10–12**, 259 (1986)

6.27 J. Hage: "Magnetische Resonanzuntersuchungen an intrinsischen Defekten und Übergangsmetallzentren in GaAs und GaP"; Doctoral Dissertation, Universität–GH Paderborn (1987)

6.28 G.G. Belford, R.L. Belford, J.F. Burkhalter: J. Magn. Res. **11**. 251 (1973)

6.29 M.I. Scullane, L.K. White, N.D. Chasteen: J. Magn. Res. **47**, 383 (1982)

6.30 G. Heder, J.–M. Spaeth, A.H. Harker: J. Phys. C: Solid State Phys. **13**, 4965 (1980)
6.31 J.–M. Spaeth, F. Koschnick: J. Phys. Chem. Sol. **52**, 1 (1991)
6.32 H. Overhof, M. Scheffler, C.M. Weinert: Materials Science Forum **38– 41**, 293 (1989)
6.33 H.H. Assenault, P. Marmet: Rev. Sci. Inst. **48**, 512 (1977)
6.34 M.U.A. Bromba, H. Ziegler: Anal. Chem. **51**, 1760 (1979)
6.35 H. Ziegler: Appl. Spectrosc. **35**, 88 (1981)
6.36 M.U.A. Bromba, H. Ziegler: Anal. Chem. **53**, 1583 (1981)
6.37 R.W. Schafer, R.M. Merserean, M.A. Richards: Proc. of the IEEE **69**, 432 (1981)
6.38 B.R. Frieden: "Image Enhancement and Restoration",in *Topics in Applied Physics*, Vol. 6, ed. by T.S. Huang (Springer, Berlin, Heidelberg, New York 1979)
6.39 J.R. Niklas: "Elektronen–Kern–Doppelresonanz–Spektroskopie zur Strukturuntersuchung von Festkörperstörstellen"; Habilitationsschrift, Universität–GH Paderborn (1983)

Chapter 7

7.1 F.S. Ham: in *Electron Paramagnetic Resonance*, ed. by S. Geschwind (Plenum, New York 1972) p. 1–119
7.2 L.L. Foldy, S.A. Wouthuysen: Phys. Rev. **78**, 29 (1950)
7.3 J.D. Bjorken, S.D. Drell: *Relativistic Quantum Mechanics* (McGraw–Hill, New York 1964) pp. 45–62
7.4 M.E. Rose: *Elementary Theory of Angular Momentum* (Wiley, New York 1957)
7.5 C. Schwartz: Phys. Rev. **97**, 380 (1955)
7.6 M. Weissbluth: *Atoms and Molecules* (Academic, New York 1978)
7.7 G. Breit, I.I. Rabi: Phys. Rev. **38**, 2082 (1931)
7.8 S. Sugano, Y. Tanabe, H. Kamimura: *Multiplets of Transition Metal Ions in Crystals* (Academic, New York, London 1970)
7.9 A. Abragam, B. Bleaney: *Electron Paramagnetic Resonance of Transition Ions* (Oxford University Press, 1970; Dover, New York 1986)
7.10 G.F. Koster, H. Statz: Phys. Rev. **113**, 445 (1959)
7.11 T. Ray: Proc. R. Soc. A **277**, 76 (1964)
7.12 J. Owen, J.H.M. Thornley: Rep. Prog. Phys. **29**, 675 (1966)
7.13 E. Siánek, Z. Sroubek: in *Electron Paramagnetic Resonance*, ed. by S. Geschwind (Plenum, New York 1972) pp. 535–575
7.14 A.J. Freeman, R.E. Watson: in *Treatise on Magnetism*, ed. by G. Rado and H. Suhl (Academic, New York 1965)
7.15 E. Clementi, C. Roetti: "Roothaan–Hartree–Fock Atomic Wavefunctions, Basic Functions and their Coefficients for Ground and Excited

States of Neutral and Ionized Atoms, Z≤54", in *Atomic Data and Nuclear Data Tables* **14**, 177 (1974)

7.16 H.B.G. Casimir, *On the Interaction Between Atomic Nuclei and Electrons* (Freeman, San Francisco 1963) p. 54

7.17 J.H. Mackey, D.E. Wood: J. Chem. Phys. **52**, 4914 (1970)

7.18 W. Marshall: in *Symposium on Paramagnetic Resonance*, Vol. 1 (Academic, New York 1963) p. 347

7.19 M. Stapelbroek, O.R. Gilliam, R.H. Bartram, Phys. Rev. B **16**, 37 (1977)

7.20 F.J. Adrian, A.N. Jette: J. Chem. Phys. **81**, 2415 (1984)

7.21 F.J. Adrian, A.N. Jette, J.-M. Spaeth: Phys. Rev. B **31**, 3923 (1985)

7.22 A. Szabo, N.S. Ostlund: *Modern Quantum Chemistry* (McGraw-Hill, New York 1989)

7.23 C.C.J. Roothaan: Rev. Mod. Phys. **23**, 69 (1951)

7.24 D.M. Chipman: J. Chem. Phys. **71**, 761 (1976)

7.25 D. Feller, E.R. Davidson: J. Chem. Phys. **80**, 1006 (1984)

7.26 D.M. Chipman: J. Chem. Phys. **91**, 5455 (1989)

7.27 A. Abragam: *The Principles of Nuclear Magnetism* (Oxford University Press, Oxford 1961) p. 166

7.28 J. Owen, J.H.M. Thornley: Rep. Prog. Phys. **29**, 675 (1966)

7.29 J.C. Phillips, L. Kleinman: Phys. Rev. **116**, 287 (1959)

7.30 C.F. Melius, W.A. Goddard III: Phys. Rev. A **10**, 1528 (1974)

7.31 W.R. Wadt, P.J. Hay: J. Chem. Phys. **82**, 284 (1985)

7.32 K.N. Shrivistava: Phys. Rep. **20**, 137 (1975)

7.33 J.C. Slater: Phys. Rev. **76**, 1592 (1949)

7.34 W. Kohn: in *Solid State Physics*, eds. F. Seitz and D. Turnbull, Vol. 5 (Academic, New York 1957)

7.35 B.S. Gourary, F.J. Adrian: Phys. Rev. **105**, 1180 (1957)

7.36 B.S. Gourary, A.E. Fein: J. Appl. Phys. Suppl. **33**, 331 (1962)

7.37 R.H. Bartram, A.M. Stoneham, P. Gash: Phys. Rev. **176**, 1014 (1968)

7.38 G.F. Koster, J.C. Slater: Phys. Rev. **95**, 1167 (1954)

7.39 G.A. Baraff, M. Schlüter: Phys. Rev. B **19**, 4965 (1979)

7.40 J.C. Slater: *Quantum Theory of Molecules and Solids* Vol. IV (McGraw-Hill, New York 1974)

7.41 J. Callaway, N.H. March: *Solid State Physics* **38**, 136 (1984)

7.42 B.S. Gourary, F.J. Adrian: *Solid State Physics*, eds. F. Seitz and D. Turnbull, Vol. 10 (Academic, New York 1960)

7.43 H. Eschrig: Phys. Stat. Sol. (b) **96**, 329 (1979)

7.44 P.-O. Löwdin: J. Chem. Phys. **18**, 365 (1950)

7.45 R.F. Wood, U. Öpik: Phys. Rev. **162**, 736 (1967)

7.46 R.F. Wood: Phys. Stat. Sol. (b) **42**, 849 (1970)

7.47 H. Söthe, L.G. Rowan, J.-M. Spaeth: J. Phys. Condensed Matter **23**, 3591 (1989)

7.48 H. Söthe, J.-M. Spaeth, F. Lüty: Reviews of Solid State Science **4**, 499 (1990)

7.49 P. Studzinski, J. Casas–Gonzales, J.-M. Spaeth: J. Phys. C: Solid State Phys. **17**, 5411 (1984)

7.50 J.-M. Spaeth: Z. Physik **192**, 107 (1966)

7.51 J.-M. Spaeth, H. Seidel: Phys. Stat. Sol. (b) **46**, 323 (1971)

7.52 R.U. Bauer, J.R. Niklas, J.-M. Spaeth: Phys. Stat. Sol. (b) **118**, 557 (1983)

7.53 F. Koschnick, Th. Hangleiter, J.-M. Spaeth, R.S. Eachus: J. Radiation Effects and Defects in Solids **119–121**, 837 (1991)

7.54 J.-M. Spaeth: Phys. Stat. Sol. (b) **34**, 171 (1969)

7.55 Ch. Hoentzsch, J.-M. Spaeth: Phys. Stat. Sol. (b) **88**, 581 (1978)

7.56 P. Studzinski, J.R. Niklas, J.-M. Spaeth: Phys. Stat. Sol. (b) **101**, 673 (1980)

7.57 G. Heder, J.R. Niklas, J.-M. Spaeth: Phys. Stat. Sol. (b) **100**, 567 (1980)

7.58 J.-M. Spaeth: Crystal Lattice Defects and Amorphous Materials **12**, 381 (1985)

7.59 R.M. Sternheimer, D. Tycko: Phys. Rev. **93**, 734 (1954)

7.60 R.M. Sternheimer, H.M. Foley: Phys. Rev. **102**, 731 (1956)

7.61 Y. Ueda, J.R. Niklas, J.-M. Spaeth, U. Kaufmann, J. Schneider: Solid State Commun. **46**, 127 (1983)

7.62 B.K. Meyer, D.M. Hofmann, J.R. Niklas, J.-M. Spaeth: Phys. Rev. B **36**, 1332 (1987)

Chapter 8

8.1 D.H. Whiffen: Molec. Phys. **10**, 595 (1966)

8.2 Ch. Hoentzsch, J.R. Niklas, J.-M. Spaeth: Rev. Sci. Instr. **49**, 1100 (1978)

8.3 H. Seidel: Z. angew. Physik **14**, 21 (1962)

8.4 J.R. Niklas: "Elektronen–Kern–Doppelresonanz–Spektroskopie zur Strukturuntersuchung von Festkörperstörstellen"; Habilitationsschrift, Universität–GH Paderborn (1983)

8.5 K. Möbius, R. Biehl: in *Multiple Electron Resonance Spectroscopy*, ed. by M.M. Dorio and J.H. Freed (Plenum, New York 1979)

8.6 H. Seidel: Z. Physik **165**, 218 (1961)

8.7 C.P. Poole, Jr.: *Electron Spin Resonance* (Interscience, New York 1967)

Chapter 9

9.1 B.C. Cavenett: Adv. in Phys. **30**, 475 (1981)

9.2 S. Geschwind: "Optical Techniques in EPR in Solids", in *Electron Paramagnetic Resonance*, ed. by S. Geschwind (Plenum, New York 1972)

9.3 K.P. Dinse, C.J. Winscon: "Optically Detected ENDOR Spectroscopy", in *Triplet State ODMR Spectroscopy*, ed. by R.H. Clarke (Wiley, New York 1984)

9.4 L.F. Mollenauer, S. Pan: Phys. Rev. B **6**, 772 (1972)

9.5 F. Lohse: private communication

9.6 H.W. van Kesteren, W.T. Wenckebach, J.A.J.M. Disselhorst: J. Phys. E: Sci. Instrum. **20**, 648 (1987)

9.7 J. Donecker, J. Kluge: J. Phys. D: Appl. Phys. **19**, L 199 (1986)

9.8 J.J. Davies: Contemp. Phys. **17**, 275 (1976)

9.9 S.N. Jasperson, S.E. Schnattely: Rev. Sci. Instrum. **40**, 761 (1969)

9.10 R.M.A. Azzam, N.M. Bashara: *Ellipsometry and Polarized Light* (North-Holland, Amsterdam, New York, Oxford 1977)

9.11 J.C. Kemp: J. Opt. Soc. Am. **59**, 950 (1970)

9.12 J.C. Kemp: in *Polarized Light and Its Interaction With Modulating Devices*, ed. Hinds International Inc. Hillsboro, USA, 1987

9.13 M. Billardon, J. Badoz: C. R. Acad. Sci. Paris **262**, 1672 (1966)

9.14 L.F. Mollenauer, D. Downie, H. Engstrom, W B. Grant, Appl. Optics **8**, 661 (1969)

9.15 R.M.A. Azzam: J. Opt. Soc. Am. **68**, 1756 (1978)

9.16 Y. Shindo, M. Nakagawa: Rev. Sci. Instrum. **56**, 32 (1985)

9.17 R. Takakuwa: Jasco Application Notes **1/4**, 1 (Japan Spectroscopic Co. Ltd., Tokyo)

9.18 K. Tuzimura, T. Konno, H. Meguro, M. Hatano, T. Murakami, K. Kashiwabara, K. Saito, Y. Kondo, T.M. Suzuki: Anal. Biochem. **81**, 167 (1977)

9.19 R.C. Jones: J. Opt. Soc. Am. **38**, 671 (1948)

9.20 H. Kubo, R. Nagata: J. Opt. Soc. Am. **73**, 1719 (1983)

9.21 H. Kubo, R. Nagata: J. Opt. Soc. Am. A **2**, 30 (1985)

9.22 H.G. Jerrard: *Optics and Laser Technology*, (Butterworth & Co. 1982)

9.23 B. Drevillon, J. Perrin, R. Marbot, A. Violet, J.L. Dalby: Rev. Sci. Instrum. **53**, 969 (1982)

9.24 S.C. Rashleigh, R.H. Stolen: Laser Focus, 1983

9.25 J.M. Beckers: Applied Optics **10**, 973 (1971)

9.26 L.F. Mollenauer, C.D. Grandt, H. Panepucci: Rev. Sci. Instrum. **39**, 1958 (1968)

9.27 O. Burghaus, E. Haindl, M. Plato, K. Möbius: J. Phys. E: Sci. Instrum. **18**, 294 (1985)

9.28 M. Chamel, R. Chicault, Y. Merle d'Aubigné: J. Phys. E: Sci. Instrum.
9, 87 (1967)
9.29 E.H. Izen, F.A. Modine: Rev. Sci. Instrum. **43**, 1563 (1972)
9.30 M. Gehrtz, C. Bräuchle, J. Voitländer: J. Phys. E: Sci. Instrum. **17**,
1046 (1984)

Appendix B

B.1 A. Cayley: Journal f. d. reine u. angew. Math. **32**, 119 (1846)

Appendix C

C.1 H.H. Assenault, P. Marmet: Rev. Sci. Inst. **48**, 512 (1977)

Appendix D

D.1 M.U.A. Bromba, H. Ziegler: Anal. Chem. **51**, 1760 (1979)
D.2 H. Ziegler: Appl. Spectrosc. **35**, 88 (1981)
D.3 M.U.A. Bromba, H. Ziegler: Anal. Chem. **53**, 1583 (1981)
D.4 P. Marmet: Rev. Sci. Instrum. **50**, 79 (1979)

Appendix E

E.1 R.W. Schafer, R.M. Marsereau, M.A. Richards: Proc. of the IEEE **69**,
432 (1981)
E.2 B.R. Frieden: "Image Enhancement and Restoration",in *Topics in Applied Physics*, Vol. 6, ed. by T.S. Huang (Springer, Berlin, Heidelberg, New York 1979)
E.3 J.R. Niklas: "Elektronen–Kern–Doppelresonanz–Spektroskopie zur Strukturuntersuchung von Festkörperstörstellen"; Habilitationsschrift, Universität–GH Paderborn (1983)

Appendix F

F.1 J.R. Niklas: "Elektronen–Kern–Doppelresonanz–Spektroskopie zur Strukturuntersuchung von Festkörperstörstellen"; Habilitationsschrift, Universität–GH Paderborn (1983)

Appendix G

G.1 J.R. Niklas: "Elektronen–Kern–Doppelresonanz–Spektroskopie zur Strukturuntersuchung von Festkörperstörstellen"; Habilitationsschrift, Universität–GH Paderborn (1983)

G.2 J.R. Niklas, J.–M. Spaeth: Phys. Stat. Sol. (b) **101**, 221 (1980)

Appendix H

H.1 E. Clementi, C. Roetti: "Roothaan–Hartree–Fock Atomic Wavefunctions, Basic Functions and their Coefficients for Ground and Excited States of Neutral and Ionized Atoms, $Z \leq 54$", in *Atomic Data and Nuclear Data Tables* **14**, 177 (1974)

Subject Index

absorption, intracenter, 80
absorptive signal, 289
abundance, 173
acceptor
– donor–acceptor pair recombination, 122
– pairs of donors and acceptors, 126
acousto–optical birefringent modulator, 314, 316
AFC (automatic frequency control), 286
– circuits, 325
– cut off frequency, 299
– stability, 294
alkali halides, hydrogen centers in, 273
amplification factor, 263
amplifier, broad band power, 306, 307
amplifier, microwave, 295
amplitude noise, 284
angular dependence, ENDOR, 169, 173
angular momentum, 12
– operator, 14, 15
anion vacancy sites, 68
anisotropic shf constant, 273
antibonding molecular orbitals, 248
antisite
– PP_3Y_p antisite defect in GaP, 130
aperture, optical, 313
approximation
– *Franck–Condon*, 82
– *Born–Oppenheimer*, 258
atomic hydrogen on interstitial sites, 68
automatic coupling control, 303
automatic frequency control (AFC), 286
– circuits, 325
– cut off frequency, 299

– stability, 294
axial triplet state defect, 130
axis system, principal, 150

background signal, 228
background, subtraction, 334
Bartram, Stoneham and *Gash* (BSG), 261
Belford, 219
birefringence, stress, 318
birefringent acousto–optical modulator, 314, 316
Bloch equation, 26, 28
Bohr magneton, 11, 235, 333
bonding molecular orbitals, 248
Born–Oppenheimer approximation, 258
Breit–Rabi formula, 243
Brillouin function, 24, 107
broad band power amplifier, 306, 307
by–pass, relaxation by–passes, 161

CaF_2, Ni^+ centers in, 63
calculated EPR spectrum, 206
CaO, F center in, 130
cation vacancy sites, 68
cavity, 281
– coupling parameter, 290
– cylindrical, 304, 327
– ENDOR microwave, 301
– microwave, 325
Cayley transformation formula, 178, 333
CC–diagram (configuration coordinate diagram), 81
center
– F center in CaO, 130
– F_A and F_AA in CaO, 130

Springer Series in Solid-State Sciences

Editors: M. Cardona   P. Fulde   K. von Klitzing   H.-J. Queisser

Springer Series in Solid-State Sciences

Editors: M. Cardona   P. Fulde   K. von Klitzing   H.-J. Queisser